All You Need in Maths!

Jan van de Craats and Rob Bosch

# ALL YOU NEED IN MATHS!

PEARSON

ISBN 978-90-430-3285-8
NUR 173

A publication of Pearson Benelux BV,
P.O. Box 75598, 1070 AN Amsterdam
Website: www.pearson.nl – e-mail: amsterdam@pearson.com

Original title: *Basisboek wiskunde – Tweede editie*
Pearson Benelux BV, Amsterdam, 2009

Translated from the Dutch by Jan van de Craats

Illustrations and LaTeX-layout: Jan van de Craats
Cover: Inkahootz, Amsterdam

Jan van de Craats is professor emeritus at the University of Amsterdam
Rob Bosch is assistant professor at the Nederlandse Defensie Academie at Breda and Den Helder

## How to use this book

This is an exercise book. Each chapter starts with exercises, printed on the left-hand page. You can immediately start making them, since the first ones are always easy. The exercises gradually become more difficult. Once you have finished an exercise, you can check your answer at the end of the book.

On the right-hand pages, the theory behind the exercises is explained in a clear and concise manner. Use this information if and when required. If you come across unknown terms or concepts, consult the index to find the page where they are explained.

Important formulas, definitions and theorems are printed in blue. Most of these are also included in the *Formula overview*, which can be found on page 309.

## The Greek alphabet

| | | | | | | | | |
|---|---|---|---|---|---|---|---|---|
| $\alpha$ | A | alpha | $\iota$ | I | iota | $\rho$ | P | rho |
| $\beta$ | B | beta | $\kappa$ | K | kappa | $\sigma$ | $\Sigma$ | sigma |
| $\gamma$ | $\Gamma$ | gamma | $\lambda$ | $\Lambda$ | lambda | $\tau$ | T | tau |
| $\delta$ | $\Delta$ | delta | $\mu$ | M | mu | $\upsilon$ | Y | upsilon |
| $\epsilon$ | E | epsilon | $\nu$ | N | nu | $\varphi$ | $\Phi$ | phi |
| $\zeta$ | Z | zeta | $\xi$ | $\Xi$ | xi | $\chi$ | X | chi |
| $\eta$ | H | eta | o | O | omicron | $\psi$ | $\Psi$ | psi |
| $\vartheta$ | $\Theta$ | theta | $\pi$ | $\Pi$ | pi | $\omega$ | $\Omega$ | omega |

# Contents

# Preface

This book covers the basic mathematics you need to successfully embark on a university or college career in technology, natural sciences, computer and information science, economics, business and management studies, and related disciplines. All subjects treated in this book are important for the disciplines mentioned, although for business and management science some aspects may be left aside, for instance most of the chapters 17 (Trigonometry), 22 (Integration techniques) and 23 (Applications).

By basic mathematics we mean elementary algebra, number sequences, equations, geometry, functions and differential and integral calculus. Probability and statistics – applied mathematical disciplines with their own specific approach – are not included.

### Practicing is essential

In our didactic view, practicing is essential. Indeed, to acquire any skill – be it soccer, playing the piano or mastering a new language – there is only one way to success: intensive practicing. For soccer, it is essential to train regularly, to master the piano, you must practice your scales and do finger exercises, and before you can speak a new language, you must learn grammar and lots of words. Without basic technical skills, nothing can be achieved. The same applies to maths.

### Why learn mathematics?

Of course, most users will primarily be interested in applying mathematics in their own field of study. But maths is an indispensable general tool. Browsing through any science book, you will see an avalanche of formulas expressing the laws of that domain in mathematical language. Using mathematical techniques, these laws are manipulated and combined to derive new results. In particular, you see simple algebraic transformations, but also the use of logarithms, exponential and trigonometric functions, differentiation, integration and more advanced mathematical techniques. These are the tools that the user must learn to handle. Substituting numerical values into a formula to get a numerical answer is only of minor importance. What really matters, is how the underlying ideas can lead to new insights.

The main goal of learning mathematics for higher education, therefore, should be the acquisition of universal mathematical skills. Universal, since the same techniques can be applied in the most diverse situations and disciplines. Developing competence in manipulating formulas is of paramount importance. Familiarity with mathematical functions, their graphs and their properties is essential. Manipulating numbers is only a very small portion of this vast repertory. This is why the (graphing) calculator plays a very modest role in this book. When such a tool (or the use of a computer) is necessary, it will be explicitly stated. But, most exercises can – and should – be solved using pen and paper only.

### Whom this book is for

We primarily wrote this book for students who lack confidence in maths because they feel their knowledge is defective. This book can help them refresh their skills. But it may also be used as a course book. The balanced structure and step-by-step progression, with lots of exercises and concise explanations of the theory, also make it suitable for self-study or distance learning. Nevertheless, it will always be difficult to learn a discipline like mathematics completely on your own: the value of a good teacher guiding you through difficult matters should not be underestimated.

### How this book is organized

All chapters (except the last three) are organized in the same way: exercises on the left-hand pages and the accompanying explanations on the right-hand pages. The user is explicitly invited to begin with the exercises first. If you stumble upon difficult or unknown concepts or notations, or you don't understand or know certain details, you should consult the text on the right-hand page and, if necessary, the index at the end of the book. The exercises have been chosen with great care. They start out easy, with many similar problems to establish basic skills, and gradually become more challenging. If you have solved all the exercises in a chapter, you can be sure that you fully master the subject.

In our explanations we do not treat the theory with full mathematical rigour. If you want to learn more about the mathematical background of certain definitions and results, we refer you to the last three chapters. These chapters do not have exercises, but contain deeper explanations and proofs. They have deliberately been placed at the end of the book, and will be appreciated only by those who have already reached a certain mathematical level. However, if you are only interested in applications, you can skip these last chapters: all you need, is covered in the preceding chapters. For completeness sake, the book contains an appendix with all the important formulas and a list of answers to the exercises.

We hope that the English edition of our book will serve many students as a reliable guide in mathematics.

## Acknowledgements

Many readers and users commented on a preliminary Dutch internet version of *Basisboek wiskunde*, pointing out obscurities and errors. Special thanks should go to Frank Heierman, who scrutinized the whole text and put forward suggestions for improvements. But we are also indebted to Henk Pfaltzgraff, Hans De Prez, Erica Mulder, Rinse Poortinga, Jaap de Jonge, Jantine Bloemhof and Wouter Berkelmans for their comments. Chris Zaal, André Heck and Wybo Dekker provided valuable advice on technical and layout matters.

Since the appearance of the *first edition* of *Basisboek wiskunde* in 2005, we again received useful comments and suggestions for improvements. In this respect we are particularly grateful to Lia van Asselt, Henri Ruizenaar, Frank Arnouts, Doortje Goldbach, Abdelhak El Jazouli, Robbert van Aalst and René van Hassel. We also appreciated comments by Jan Essers, Jan Los, Wim Caspers, Adri van den Boom, C.E. van Wijk, G.J.J. Baas, Erik Beijeman, Hermien Beverdam, A. Dolfing and Marjan van der Vegt. In 2009, we implemented most of these comments and suggestions into the *second edition* of our book, which was the foundation of this English translation.

During the years, many users have pointed out errors in the list of answers: Nabi Abudaldah, A.S. Tigelaar, N.J. Schoonderbeek, Mathijs Schuts, Evert van de Vrie, Niël Dogger, Paul Bles, J. Bon, Max van den Aker, Evelien de Greef, Bas Bemelmans, Kevin de Berk, Veditam Bishoen, Loek Spitz, Robert van Eekhout, Tim de Graaff, Rik Kaasschieter, Jerry van Ulden, Theo de Jong, Vincent Temmerman, Eva van Herel, Stephan den Bleker, Michiel van Lieshout and Dick van de Loo. A list of answers without errors is our ultimate goal; all contributions towards this ideal are very welcome!

Oosterhout and Breda, The Netherlands, May 2014,

Jan van de Craats and Rob Bosch

# I Numbers

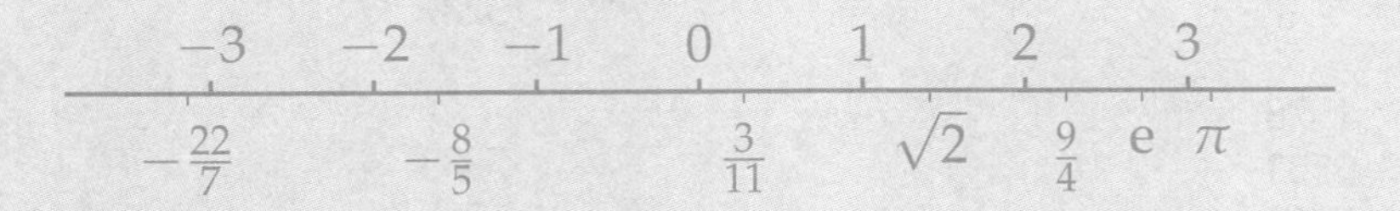


Part I is about the arithmetic of numbers. There are various types of numbers: positive numbers, negative numbers, integers, fractions, rational and irrational numbers. Examples of irrational numbers are $\sqrt{2}$, $\pi$ and $\mathrm{e}$. In higher mathematics, one also uses imaginary and complex numbers, but in this book we restrict ourselves to *real numbers*, i.e. numbers that can be represented as points on a number line.

The first two chapters are a recap of your skills in primary school arithmetic: addition, subtraction, multiplication, and division of integers and fractions. Chapter 3 treats the properties of powers and roots.

# 1 Calculating with integers

Perform the following calculations:

1.1

a.
$$\begin{array}{r} 873 \\ 112 \\ 1718 \\ 157 \\ 3461 \\ \hline \ldots \end{array} +$$

b.
$$\begin{array}{r} 1578 \\ 9553 \\ 7218 \\ 212 \\ 4139 \\ \hline \ldots \end{array} +$$

1.2

a.
$$\begin{array}{r} 9134 \\ 4319 \\ \hline \ldots \end{array} -$$

b.
$$\begin{array}{r} 4585 \\ 3287 \\ \hline \ldots \end{array} -$$

c.
$$\begin{array}{r} 7033 \\ 1398 \\ \hline \ldots \end{array} -$$

1.3 Calculate:

a. $34 \times 89$

b. $67 \times 46$

c. $61 \times 93$

d. $55 \times 11$

e. $78 \times 38$

1.4 Calculate:

a. $354 \times 83$

b. $67 \times 546$

c. $461 \times 79$

d. $655 \times 102$

e. $178 \times 398$

Find the quotient and the remainder by using a long division:

1.5

a. $154 : 13$

b. $435 : 27$

c. $631 : 23$

d. $467 : 17$

e. $780 : 37$

1.6

a. $2334 : 53$

b. $6463 : 101$

c. $7682 : 59$

d. $6178 : 451$

e. $5811 : 67$

1.7

a. $15457 : 11$

b. $4534 : 97$

c. $63321 : 23$

d. $56467 : 179$

e. $78620 : 307$

1.8

a. $42334 : 41$

b. $13467 : 101$

c. $35641 : 99$

d. $16155 : 215$

e. $92183 : 83$

## Addition, subtraction and multiplication

The sequence

$$1, 2, 3, 4, 5, 6, 7, 8, 9, 10, 11, 12, \ldots$$

enumerates the *positive integers*. Every child learns to count in this way. Addition, subtraction and multiplication with such numbers by hand are learned in primary school. Examples are given to the right.

$$\begin{array}{rl} 341 & \\ 295 & \\ 718 & \\ 12 & \\ 1431 & \\ \hline 2797 & + \end{array} \qquad \begin{array}{rl} 8135 & \\ 3297 & \\ \hline 4838 & - \end{array} \qquad \begin{array}{rl} 431 & \\ 728 & \\ \hline 3448 & \times \\ 862\phantom{0} & \\ 3017\phantom{00} & \\ \hline 313768 & \end{array}$$

## Long division

Division by hand is done by *long division*. To the right, the long division for 83218 : 37, i.e. 83218 divided by 37, is shown. The *quotient* 2249 is constructed digit by digit, from left to right, above the dividend. Each digit of the quotient is positioned exactly above the corresponding product in the 'tail' of the calculation. The *remainder* 5 is found below, at the end of the calculation. The long division shows that

$$\begin{array}{r|l} & 2249 \quad \leftarrow \text{quotient} \\ \hline 37 & 83218 \\ & 74\phantom{218} \\ \hline & 92\phantom{18} \\ & 74\phantom{18} \\ \hline & 181\phantom{8} \\ & 148\phantom{8} \\ \hline & 338 \\ & 333 \\ \hline & 5 \quad \leftarrow \text{remainder} \end{array}$$

$$83218 = 2249 \times 37 + 5$$

This can also be written as

$$\frac{83218}{37} = 2249 + \frac{5}{37}$$

The right-hand side is usually simplified to $2249\frac{5}{37}$, which yields

$$\frac{83218}{37} = 2249\frac{5}{37}$$

Decompose the following numbers into prime factors:

1.9

a. 24
b. 72
c. 250
d. 96
e. 98

1.10

a. 288
b. 1024
c. 315
d. 396
e. 1875

1.11

a. 972
b. 676
c. 2025
d. 1122
e. 860

1.12

a. 255
b. 441
c. 722
d. 432
e. 985

1.13

a. 2000
b. 2001
c. 2002
d. 2003
e. 2004

1.14

a. your age in months
b. your year of birth
c. your PIN code

Find all divisors of the following numbers. Proceed accurately and systematically, since otherwise you risk missing a few. It is a good idea to write down the prime factorization of each number first.

1.15

a. 12
b. 20
c. 32
d. 108
e. 144

1.16

a. 72
b. 100
c. 1001
d. 561
e. 196

## Divisors and prime numbers

Sometimes, a division yields a quotient without a remainder, i.e. with remainder 0. For instance, $238 : 17 = 14$. This means that $238 = 14 \times 17$. The numbers 14 and 17 are called *divisors* of 238 and the notation $238 = 14 \times 17$ is called a *decomposition into factors* or *factorization* of 238. The words 'divisor' and 'factor' are synonymous in this respect.

The divisor 14 itself can also be decomposed, namely as $14 = 2 \times 7$, but a further decomposition of 238 is not possible, since 2, 7 and 17 are *prime numbers*, i.e. numbers that cannot be decomposed into smaller factors. In this way, we have found the *prime factorization* $238 = 2 \times 7 \times 17$.

Since $238 = 1 \times 238$ is also a decomposition of 238, the numbers 1 en 238 are divisors of 238. Every number has 1 and itself as divisor. The interesting, *proper* divisors, however, are the divisors that are greater than 1 and smaller than the number itself. The prime numbers are the numbers greater than 1 that have no proper divisors. The sequence of prime numbers starts as follows:

$$2, 3, 5, 7, 11, 13, 17, 19, 23, 29, 31, 37, 41, 43, 47, 53, 59, 61, 67, 71, 73, 79, \ldots$$

Every integer greater than 1 can be decomposed into prime factors. To the right, examples are given of how to find such a *prime factorization* by systematically looking for bigger and bigger prime factors. Each time you find one, you divide by it and proceed with the quotient.

$$\begin{array}{r|l} 180 & 2 \\ 90 & 2 \\ 45 & 3 \\ 15 & 3 \\ 5 & 5 \\ 1 & \end{array} \qquad \begin{array}{r|l} 585 & 3 \\ 195 & 3 \\ 65 & 5 \\ 13 & 13 \\ 1 & \end{array} \qquad \begin{array}{r|l} 3003 & 3 \\ 1001 & 7 \\ 143 & 11 \\ 13 & 13 \\ 1 & \end{array}$$

You are done as soon as you end up with a quotient of 1. The prime factors are then collected on the right-hand side of the ladder. The three ladder diagrams yield the prime factorization:

$$\begin{aligned} 180 &= 2 \times 2 \times 3 \times 3 \times 5 = 2^2 \times 3^2 \times 5 \\ 585 &= 3 \times 3 \times 5 \times 13 = 3^2 \times 5 \times 13 \\ 3003 &= 3 \times 7 \times 11 \times 13 \end{aligned}$$

As shown, it is convenient to write prime factors that occur more than once in power notation: $2^2 = 2 \times 2$ and $3^2 = 3 \times 3$. Some more examples are given below (you may construct the ladder diagrams for yourself):

$$\begin{aligned} 120 &= 2 \times 2 \times 2 \times 3 \times 5 = 2^3 \times 3 \times 5 \\ 81 &= 3 \times 3 \times 3 \times 3 = 3^4 \\ 48 &= 2 \times 2 \times 2 \times 2 \times 3 = 2^4 \times 3 \end{aligned}$$

Find the greatest common divisor (gcd) of:

1.17

a. 12 and 30
b. 24 and 84
c. 27 and 45
d. 32 and 56
e. 34 and 85

1.18

a. 45 and 225
b. 144 and 216
c. 90 and 196
d. 243 and 135
e. 288 and 168

1.19

a. 1024 and 864
b. 1122 and 1815
c. 875 and 1125
d. 1960 and 6370
e. 1024 and 1152

1.20

a. 1243 and 1244
b. 1721 and 1726
c. 875 and 900
d. 1960 and 5880
e. 1024 and 2024

Find the least common multiple (lcm) of:

1.21

a. 12 and 30
b. 27 and 45
c. 18 and 63
d. 16 and 40
e. 33 and 121

1.22

a. 52 and 39
b. 64 and 80
c. 144 and 240
d. 169 and 130
e. 68 and 51

1.23

a. 250 and 125
b. 144 and 216
c. 520 and 390
d. 888 and 185
e. 124 and 341

1.24

a. 240 and 180
b. 276 and 414
c. 588 and 504
d. 315 and 189
e. 403 and 221

Find the gcd and the lcm of:

1.25

a. 9, 12 and 30
b. 24, 30 and 36
c. 10, 15 and 35
d. 18, 27 and 63
e. 21, 24 and 27

1.26

a. 28, 35 and 49
b. 64, 80 and 112
c. 39, 52 and 130
d. 144, 168 and 252
e. 189, 252 and 315

## The gcd and the lcm

Two numbers can have common divisors. The *greatest common divisor* (gcd) is of particular interest. If the decomposition into prime factors of both numbers is known, their gcd can be found immediately. For instance, on page 9 we calculated the following prime factorizations:

$$\begin{aligned}180 &= 2^2 \times 3^2 \times 5\\ 585 &= 3^2 \times 5 \times 13\\ 3003 &= 3 \times 7 \times 11 \times 13\end{aligned}$$

From this we see that

$$\begin{aligned}\gcd(180, 585) &= \gcd(2^2 \times 3^2 \times 5\,,\, 3^2 \times 5 \times 13) = 3^2 \times 5 = 45\\ \gcd(180, 3003) &= \gcd(2^2 \times 3^2 \times 5\,,\, 3 \times 7 \times 11 \times 13) = 3\\ \gcd(585, 3003) &= \gcd(3^2 \times 5 \times 13\,,\, 3 \times 7 \times 11 \times 13) = 3 \times 13 = 39\end{aligned}$$

The *least common multiple* (lcm) of two numbers is the smallest number that is a multiple of both numbers. In other words, it is the smallest number that may be divided without a remainder by both numbers. The lcm can also be found immediately, based on the prime factorizations of the numbers. For instance:

$$\text{lcm}(180, 585) = \text{lcm}(2^2 \times 3^2 \times 5\,,\, 3^2 \times 5 \times 13) = 2^2 \times 3^2 \times 5 \times 13 = 2340$$

A convenient property of the gcd and the lcm of two numbers is the fact that, when they are multiplied, their product equals the product of the two numbers. For instance:

$$\gcd(180, 585) \times \text{lcm}(180, 585) = 45 \times 2340 = 105300 = 180 \times 585$$

The gcd and the lcm can also be calculated for more than two numbers, based on their prime factorizations. For instance:

$$\begin{aligned}\gcd(180, 585, 3003) &= 3\\ \text{lcm}(180, 585, 3003) &= 2^2 \times 3^2 \times 5 \times 7 \times 11 \times 13 = 180180\end{aligned}$$

**A smart idea**
There is a method to calculate the gcd of two numbers without prime factorization. In many cases, this method is much faster. The basic idea is that the gcd of two numbers is also a divisor of their *difference*. (Do you know why?)
For instance, $\gcd(4352, 4342)$ must also be a divisor of $4352 - 4342 = 10$. The number 10 only has the prime factors 2 and 5. Obviously, 5 is not a common divisor of both numbers, but 2 is, so we have $\gcd(4352, 4342) = 2$. If you are smart, you may save yourself lots of time using this idea!

# 2 Calculating with fractions

2.1 Simplify:

a. $\frac{15}{20}$

b. $\frac{18}{45}$

c. $\frac{21}{49}$

d. $\frac{27}{81}$

e. $\frac{24}{96}$

2.2 Simplify:

a. $\frac{60}{144}$

b. $\frac{144}{216}$

c. $\frac{135}{243}$

d. $\frac{864}{1024}$

e. $\frac{168}{288}$

2.3 Write with common denominator:

a. $\frac{1}{3}$ and $\frac{1}{4}$

b. $\frac{2}{5}$ and $\frac{3}{7}$

c. $\frac{4}{9}$ and $\frac{2}{5}$

d. $\frac{7}{11}$ and $\frac{3}{4}$

e. $\frac{2}{13}$ and $\frac{5}{12}$

2.4 Write with common denominator:

a. $\frac{1}{6}$ and $\frac{1}{9}$

b. $\frac{3}{10}$ and $\frac{2}{15}$

c. $\frac{3}{8}$ and $\frac{5}{6}$

d. $\frac{5}{9}$ and $\frac{7}{12}$

e. $\frac{3}{20}$ and $\frac{1}{8}$

2.5 Write with common denominator:

a. $\frac{1}{3}$, $\frac{1}{4}$ and $\frac{1}{5}$

b. $\frac{2}{3}$, $\frac{3}{5}$ and $\frac{2}{7}$

c. $\frac{1}{4}$, $\frac{1}{6}$ and $\frac{1}{9}$

d. $\frac{2}{10}$, $\frac{1}{15}$ and $\frac{5}{6}$

e. $\frac{5}{12}$, $\frac{7}{18}$ and $\frac{3}{8}$

2.6 Write with common denominator:

a. $\frac{2}{27}$, $\frac{5}{36}$ and $\frac{5}{24}$

b. $\frac{7}{15}$, $\frac{3}{20}$ and $\frac{5}{6}$

c. $\frac{4}{21}$, $\frac{3}{14}$ and $\frac{7}{30}$

d. $\frac{4}{63}$, $\frac{5}{42}$ and $\frac{1}{56}$

e. $\frac{5}{78}$, $\frac{5}{39}$ and $\frac{3}{65}$

Determine which of the given fractions is the greatest by writing them with a common denominator:

2.7

a. $\frac{5}{18}$ and $\frac{6}{19}$

b. $\frac{7}{15}$ and $\frac{5}{12}$

c. $\frac{9}{20}$ and $\frac{11}{18}$

d. $\frac{11}{36}$ and $\frac{9}{32}$

e. $\frac{20}{63}$ and $\frac{25}{72}$

2.8

a. $\frac{4}{7}$ and $\frac{2}{3}$

b. $\frac{14}{85}$ and $\frac{7}{51}$

c. $\frac{26}{63}$ and $\frac{39}{84}$

d. $\frac{31}{90}$ and $\frac{23}{72}$

e. $\frac{37}{80}$ and $\frac{29}{60}$

## Rational numbers

The sequence ..., −3, −2, −1, 0, 1, 2, 3, ... is the sequence of all integer numbers. The *number line* drawn below is a geometric representation of this sequence.

$$-3 \quad -2 \quad -1 \quad 0 \quad 1 \quad 2 \quad 3$$

The *rational numbers*, i.e. the numbers that can be written as a fraction, also have their place on the number line. The number line below shows some rational numbers.

$$-3 \quad -2 \quad -1 \quad 0 \quad 1 \quad 2 \quad 3$$

$$-\tfrac{22}{7} \quad -\tfrac{8}{5} \quad \tfrac{3}{11} \quad \tfrac{5}{3} \quad \tfrac{28}{9}$$

A fraction contains two integers, the *numerator* and the *denominator*, separated by a horizontal line or a slanting line. For example, in the fraction $\frac{28}{6}$ the numerator is 28 and the denominator is 6. The denominator of a fraction cannot be zero. A rational number is a number that can be written as a fraction, but this notation is not unique: if you multiply the numerator and denominator by the same integer (not equal to 0) or divide both by a common factor greater than 1, this does not change its place on the number line. For instance:

$$\frac{28}{6} = \frac{14}{3} = \frac{-14}{-3} = \frac{70}{15}$$

Fractions like $\frac{-5}{3}$ and $\frac{22}{-7}$ are usually written as $-\frac{5}{3}$ and $-\frac{22}{7}$, respectively. Integers can also be written as fractions, e.g. $7 = \frac{7}{1}$, $-3 = -\frac{3}{1}$ and $0 = \frac{0}{1}$. This shows that integers are rational numbers as well.

Dividing the numerator and the denominator of a fraction by the same factor is called *simplifying* the fraction. For instance, $\frac{28}{6}$ may be simplified to $\frac{14}{3}$ by dividing both the numerator and denominator by 2. A fraction is called *irreducible* or *in lowest terms* if the greatest common divisor (gcd) of the numerator and denominator equals 1. Thus, $\frac{14}{3}$ is in lowest terms, while $\frac{28}{6}$ is not. Every fraction can be written in lowest terms by dividing the numerator and denominator by their gcd.

Two fractions always can be rewritten with a common denominator. For instance, $\frac{4}{15}$ and $\frac{5}{21}$ can be written with a common denominator $15 \times 21 = 315$ since $\frac{4}{15} = \frac{84}{315}$ and $\frac{5}{21} = \frac{75}{315}$. But taking the least common multiple (lcm) of the denominators, in this case $\mathrm{lcm}(15, 21) = 105$, yields the simplest fractions with a common denominator, namely $\frac{28}{105}$ and $\frac{25}{105}$.

Calculate:

2.9

a. $\frac{1}{3}+\frac{1}{4}$

b. $\frac{1}{5}-\frac{1}{6}$

c. $\frac{1}{7}+\frac{1}{9}$

d. $\frac{1}{9}-\frac{1}{11}$

e. $\frac{1}{2}+\frac{1}{15}$

2.10

a. $\frac{2}{3}+\frac{3}{4}$

b. $\frac{3}{5}-\frac{4}{7}$

c. $\frac{2}{7}+\frac{3}{4}$

d. $\frac{4}{9}-\frac{3}{8}$

e. $\frac{5}{11}+\frac{4}{15}$

2.11

a. $\frac{1}{6}+\frac{1}{4}$

b. $\frac{1}{9}-\frac{2}{15}$

c. $\frac{3}{8}+\frac{1}{12}$

d. $\frac{1}{3}+\frac{5}{6}$

e. $\frac{4}{15}-\frac{3}{10}$

2.12

a. $\frac{2}{45}+\frac{1}{21}$

b. $\frac{5}{27}-\frac{1}{36}$

c. $\frac{5}{72}+\frac{7}{60}$

d. $\frac{3}{34}+\frac{1}{85}$

e. $\frac{7}{30}+\frac{8}{105}$

2.13

a. $\frac{1}{3}+\frac{1}{4}+\frac{1}{5}$

b. $\frac{1}{2}-\frac{1}{3}+\frac{1}{7}$

c. $\frac{1}{4}-\frac{1}{5}+\frac{1}{9}$

d. $\frac{1}{2}-\frac{1}{7}-\frac{1}{3}$

e. $\frac{1}{8}+\frac{1}{3}-\frac{1}{5}$

2.14

a. $\frac{1}{2}+\frac{1}{4}+\frac{1}{8}$

b. $\frac{1}{3}+\frac{1}{6}+\frac{1}{4}$

c. $\frac{1}{12}+\frac{1}{8}-\frac{1}{2}$

d. $\frac{1}{9}-\frac{1}{12}+\frac{1}{18}$

e. $\frac{1}{10}-\frac{1}{15}+\frac{1}{6}$

2.15

a. $\frac{1}{2}-\frac{1}{4}+\frac{1}{8}$

b. $\frac{1}{3}+\frac{1}{6}-\frac{1}{4}$

c. $\frac{1}{12}-\frac{1}{8}-\frac{1}{2}$

d. $\frac{1}{9}-\frac{1}{12}-\frac{1}{18}$

e. $\frac{1}{10}+\frac{1}{15}+\frac{1}{6}$

2.16

a. $\frac{1}{3}-\frac{1}{9}+\frac{1}{27}$

b. $\frac{1}{2}+\frac{1}{10}-\frac{2}{15}$

c. $\frac{1}{18}-\frac{7}{30}-\frac{3}{20}$

d. $\frac{3}{14}-\frac{1}{21}+\frac{5}{6}$

e. $\frac{2}{5}-\frac{3}{10}+\frac{4}{15}$

2.17

a. $\frac{2}{5}-\frac{1}{7}-\frac{1}{10}$

b. $\frac{3}{2}+\frac{2}{3}-\frac{5}{6}$

c. $\frac{8}{21}-\frac{2}{7}+\frac{3}{4}$

d. $\frac{2}{11}-\frac{5}{13}+\frac{1}{2}$

e. $\frac{4}{17}-\frac{3}{10}+\frac{2}{5}$

## Addition and subtraction of fractions

Adding two fractions with the same denominator is simple: the numerators are added up while the denominator is left unchanged. The same applies to subtracting fractions with the same denominator. Examples:

$$\frac{5}{13}+\frac{12}{13}=\frac{17}{13} \qquad \text{and} \qquad \frac{5}{13}-\frac{12}{13}=\frac{-7}{13}=-\frac{7}{13}$$

If the denominators are different, the fractions first have to be written with a common denominator. Again, it is best to take the least common multiple (lcm) of the denominators. Examples:

$$\begin{aligned}
\frac{2}{5}+\frac{8}{3} &= \frac{6}{15}+\frac{40}{15} &= \frac{46}{15}\\
-\frac{7}{12}+\frac{4}{15} &= -\frac{35}{60}+\frac{16}{60} &= -\frac{19}{60}\\
-\frac{13}{7}-\frac{18}{5} &= -\frac{65}{35}-\frac{126}{35} &= -\frac{191}{35}
\end{aligned}$$

Adding or subtracting more than two fractions works in the same way: it is best to write them with a common denominator. And again it is most economical to take the lcm of the denominators as their common denominator. For example:

$$\frac{2}{3}+\frac{3}{10}-\frac{2}{15}=\frac{20}{30}+\frac{9}{30}-\frac{4}{30}=\frac{25}{30}=\frac{5}{6}$$

As shown in this example, it may be possible to simplify the final answer.

### Fractions and rational numbers

A fraction is a *notation* of a rational number. Multiplying the numerator and denominator by the same factor changes the notation of the fraction, but not the rational number that it represents. One might also say that the *value* of the fraction is not changed when you multiply the numerator and denominator by the same factor. The fractions $\frac{5}{2}$, $\frac{15}{6}$ and $\frac{50}{20}$ all have the same value, and on the number line they all occupy the same place, namely halfway between 2 en 3.

In practice, we often use the word 'fraction' where, strictly speaking, we should say the 'value of the fraction'. Indeed, we do the same when writing $\frac{5}{2}=\frac{15}{6}$ or when saying that $\frac{5}{2}$ equals $\frac{15}{6}$.

Calculate:

2.18

a. $\frac{2}{3} \times \frac{5}{7}$

b. $\frac{4}{9} \times \frac{2}{5}$

c. $\frac{2}{13} \times \frac{5}{7}$

d. $\frac{9}{13} \times \frac{7}{2}$

e. $\frac{1}{30} \times \frac{13}{10}$

2.19

a. $\frac{2}{3} \times \frac{9}{2}$

b. $\frac{8}{9} \times \frac{3}{4}$

c. $\frac{14}{15} \times \frac{10}{7}$

d. $\frac{25}{12} \times \frac{18}{35}$

e. $\frac{36}{21} \times \frac{28}{27}$

2.20

a. $\frac{63}{40} \times \frac{16}{27}$

b. $\frac{49}{25} \times \frac{30}{21}$

c. $\frac{99}{26} \times \frac{39}{44}$

d. $\frac{51}{36} \times \frac{45}{34}$

e. $\frac{46}{57} \times \frac{38}{69}$

2.21

a. $\frac{2}{3} \times \frac{6}{5} \times \frac{15}{4}$

b. $\frac{6}{35} \times \frac{15}{4} \times \frac{14}{9}$

c. $\frac{26}{33} \times \frac{22}{9} \times \frac{15}{39}$

d. $\frac{18}{49} \times \frac{35}{12} \times \frac{4}{21}$

e. $\frac{24}{15} \times \frac{4}{27} \times \frac{45}{16}$

2.22

a. $\frac{2}{3} : \frac{5}{7}$

b. $\frac{1}{3} : \frac{1}{2}$

c. $6 : \frac{1}{5}$

d. $\frac{6}{5} : \frac{10}{9}$

e. $\frac{4}{5} : \frac{5}{7}$

2.23

a. $\frac{2}{3} : \frac{4}{9}$

b. $\frac{7}{10} : \frac{21}{15}$

c. $10 : \frac{5}{3}$

d. $\frac{12}{25} : \frac{18}{35}$

e. $\frac{24}{49} : \frac{36}{49}$

2.24

a. $\dfrac{\frac{2}{3}}{\frac{3}{4}}$

b. $\dfrac{\frac{6}{5}}{\frac{9}{10}}$

c. $\dfrac{\frac{12}{7}}{\frac{9}{14}}$

2.25

a. $\dfrac{\frac{1}{2} + \frac{1}{3}}{\frac{1}{4} + \frac{1}{6}}$

b. $\dfrac{\frac{5}{9} + \frac{3}{10}}{\frac{3}{4} - \frac{8}{9}}$

c. $\dfrac{\frac{4}{3} - \frac{3}{4}}{\frac{2}{3} + \frac{3}{2}}$

2.26

a. $\dfrac{\frac{2}{7} + \frac{5}{6}}{\frac{1}{5} + \frac{3}{4}}$

b. $\dfrac{\frac{1}{6} - \frac{5}{3}}{\frac{2}{7} - \frac{2}{5}}$

c. $\dfrac{\frac{3}{5} - \frac{11}{12}}{\frac{6}{7} + \frac{3}{11}}$

## Multiplication and division of fractions

The *product* of two fractions is a fraction where the numerator is the product of the two numerators and the denominator is the product of the two denominators. Examples:

$$\frac{5}{13} \times \frac{12}{7} = \frac{5 \times 12}{13 \times 7} = \frac{60}{91} \qquad \text{and} \qquad \frac{8}{7} \times \frac{-5}{11} = \frac{8 \times (-5)}{7 \times 11} = -\frac{40}{77}$$

The rule for dividing fractions is: *dividing by a fraction is the same as multiplying by the inverted fraction.* The inverted fraction is obtained by interchanging the numerator and the denominator. Examples:

$$\frac{5}{13} : \frac{12}{7} = \frac{5}{13} \times \frac{7}{12} = \frac{35}{156} \qquad \text{and} \qquad \frac{8}{7} : \frac{-5}{11} = \frac{8}{7} \times \frac{11}{-5} = -\frac{88}{35}$$

You can also use a different notation for division of fractions, namely with a horizontal fraction line. For example:

$$\frac{\frac{5}{13}}{\frac{12}{7}} \qquad \text{instead of} \qquad \frac{5}{13} : \frac{12}{7}$$

This gives us a 'fraction' with a fraction in both the numerator and the denominator.

### Other notations for fractions

Instead of a horizontal line between numerator and denominator, a slanting line can also be used: $1/2$ instead of $\frac{1}{2}$. Sometimes, it is better to use the slanting line for typographical reasons. These notation are also used together, mostly for typographical reasons. Examples:

$$\frac{5/13}{12/7} \qquad \text{or} \qquad \frac{5}{13} \Big/ \frac{12}{7}$$

In some cases, it may be convenient to write fractions in a *mixed form*, i.e. writing the integer part separately, as $2\frac{1}{2}$ instead of $\frac{5}{2}$. However, this notation is very awkward when you are multiplying or dividing fractions. This is the reason why we hardly ever use mixed forms of fractions in this book.

# 3 Powers and roots

Write each of the following expressions as an integer or as a fraction in lowest terms:

3.1

a. $2^3$
b. $3^2$
c. $4^5$
d. $5^4$
e. $2^8$

3.2

a. $(-2)^3$
b. $(-3)^2$
c. $(-4)^5$
d. $(-5)^4$
e. $(-2)^6$

3.3

a. $2^{-3}$
b. $4^{-2}$
c. $3^{-4}$
d. $7^{-1}$
e. $2^{-7}$

3.4

a. $2^0$
b. $9^{-1}$
c. $11^{-2}$
d. $9^{-3}$
e. $10^{-4}$

3.5

a. $(-4)^3$
b. $3^{-5}$
c. $(-3)^{-3}$
d. $2^4$
e. $(-2)^{-4}$

3.6

a. $(-2)^0$
b. $0^2$
c. $12^{-1}$
d. $(-7)^2$
e. $(-2)^{-7}$

3.7

a. $(\frac{2}{3})^2$
b. $(\frac{1}{2})^4$
c. $(\frac{4}{5})^3$
d. $(\frac{2}{7})^2$

3.8

a. $(\frac{2}{3})^{-2}$
b. $(\frac{1}{2})^{-3}$
c. $(\frac{7}{9})^{-1}$
d. $(\frac{3}{2})^{-4}$

3.9

a. $(\frac{4}{3})^{-2}$
b. $(\frac{1}{2})^{-4}$
c. $(\frac{4}{5})^{-1}$
d. $(\frac{2}{3})^{-5}$

3.10

a. $(\frac{1}{4})^{-1}$
b. $(\frac{6}{5})^0$
c. $(\frac{4}{3})^3$
d. $(\frac{5}{2})^{-4}$

3.11

a. $(\frac{6}{7})^2$
b. $(\frac{8}{7})^0$
c. $(\frac{6}{7})^{-2}$
d. $(\frac{2}{7})^3$

3.12

a. $(\frac{4}{9})^3$
b. $(\frac{5}{3})^{-3}$
c. $(\frac{5}{11})^2$
d. $(\frac{3}{6})^{-5}$

## Exponentiation

For any number $a \neq 0$ and any positive integer $k$ one defines

$$\begin{aligned} a^k &= \overbrace{a \times a \times \cdots \times a}^{k \text{ times}} \\ a^0 &= 1 \\ a^{-k} &= \frac{1}{a^k} \end{aligned}$$

This way, $a^n$ is defined for each integer $n$. The number $a$ is called the *base* and $n$ is called the *exponent*. Examples:

$$\begin{aligned} 7^4 &= 7 \times 7 \times 7 \times 7 = 2401 \\ \left(-\frac{1}{3}\right)^0 &= 1 \\ \left(\frac{3}{8}\right)^{-1} &= \frac{1}{\frac{3}{8}} = \frac{8}{3} \\ 10^{-3} &= \frac{1}{10^3} = \frac{1}{1000} \end{aligned}$$

Properties:

$$\begin{aligned} a^n \times a^m &= a^{n+m} \\ a^n : a^m &= a^{n-m} \\ (a^n)^m &= a^{n \times m} \\ (a \times b)^n &= a^n \times b^n \\ \left(\frac{a}{b}\right)^n &= \frac{a^n}{b^n} \end{aligned}$$

A special case is base 0. In the example above, we assumed $a \neq 0$ to exclude the possibility of negative integer exponents resulting in a 'fraction' with denominator 0. For positive integers $n$, however, we take $0^n = 0$. In mathematics, it is also usual to define $0^0 = 1$. This last definition is simply a matter of convenience, ensuring that certain formulas keep their validity for 0 as well. An example is the formula $a^0 = 1$, which now holds for all $a$, including $a = 0$. Don't ask for 'deeper' justifications; it is only a matter of convenience!

Write the following expressions in the standard form, i.e. in the form $a\sqrt{b}$ where $a$ is an integer and $\sqrt{b}$ is an irreducible root:

3.13
a. $\sqrt{36}$
b. $\sqrt{81}$
c. $\sqrt{121}$
d. $\sqrt{64}$
e. $\sqrt{169}$

3.14
a. $\sqrt{225}$
b. $\sqrt{16}$
c. $\sqrt{196}$
d. $\sqrt{256}$
e. $\sqrt{441}$

3.15
a. $\sqrt{8}$
b. $\sqrt{12}$
c. $\sqrt{18}$
d. $\sqrt{24}$
e. $\sqrt{50}$

3.16
a. $\sqrt{72}$
b. $\sqrt{32}$
c. $\sqrt{20}$
d. $\sqrt{98}$
e. $\sqrt{40}$

3.17
a. $\sqrt{54}$
b. $\sqrt{99}$
c. $\sqrt{80}$
d. $\sqrt{96}$
e. $\sqrt{200}$

3.18
a. $\sqrt{147}$
b. $\sqrt{242}$
c. $\sqrt{125}$
d. $\sqrt{216}$
e. $\sqrt{288}$

3.19
a. $\sqrt{675}$
b. $\sqrt{405}$
c. $\sqrt{512}$
d. $\sqrt{338}$
e. $\sqrt{588}$

3.20
a. $\sqrt{1331}$
b. $\sqrt{972}$
c. $\sqrt{2025}$
d. $\sqrt{722}$
e. $\sqrt{676}$

3.21
a. $\sqrt{6} \times \sqrt{3}$
b. $\sqrt{10} \times \sqrt{15}$
c. $2\sqrt{14} \times -3\sqrt{21}$
d. $-4\sqrt{22} \times 5\sqrt{33}$
e. $3\sqrt{30} \times 2\sqrt{42}$

3.22
a. $\sqrt{5} \times \sqrt{3}$
b. $-\sqrt{2} \times \sqrt{7}$
c. $\sqrt{3} \times \sqrt{5} \times \sqrt{2}$
d. $2\sqrt{14} \times 3\sqrt{6}$
e. $3\sqrt{5} \times -2\sqrt{6} \times 4\sqrt{10}$

3.23
a. $3\sqrt{6} \times 2\sqrt{15} \times 4\sqrt{10}$
b. $-5\sqrt{5} \times 10\sqrt{10} \times 2\sqrt{2}$
c. $2\sqrt{21} \times -\sqrt{14} \times -3\sqrt{10}$
d. $\sqrt{15} \times 2\sqrt{3} \times -3\sqrt{35}$
e. $-3\sqrt{30} \times 12\sqrt{14} \times -2\sqrt{21}$

## Square roots of non-negative integers

The *square root* of a number $a \geq 0$ is the number $r$ for which $r \geq 0$ and $r^2 = a$ holds. Notation: $r = \sqrt{a}$.

Example: $\sqrt{25} = 5$ since $5^2 = 25$. Note that $(-5)^2 = 25$ as well, so perhaps you might also call $-5$ a 'square root of 25'. But by definition, $\sqrt{a}$ is solely the *non-negative* number for which the square equals $a$, i.e. $\sqrt{25} = +5$.

The number $\sqrt{20}$ is not an integer, since $4^2 = 16 < 20$ and $5^2 = 25 > 20$ so $4 < \sqrt{20} < 5$. Could $\sqrt{20}$ possibly be a fraction? The answer is no: the square root of a positive integer that itself is not a square of an integer always is *irrational*, which means that it cannot be written as a fraction. However, $\sqrt{20}$ may be simplified, since $20 = 2^2 \times 5$, which gives $\sqrt{20} = \sqrt{2^2 \times 5} = 2 \times \sqrt{5}$. The last expression usually is written as $2\sqrt{5}$.

The square root $\sqrt{a}$ of a positive integer $a$ is called *irreducible* if $a$ cannot be divided by the square of an integer greater than 1. For example, $\sqrt{21} = \sqrt{3 \times 7}$, $\sqrt{66} = \sqrt{2 \times 3 \times 11}$ and $\sqrt{91} = \sqrt{7 \times 13}$ are irreducible square roots, but $\sqrt{63}$ is not, since $\sqrt{63} = \sqrt{7 \times 9} = \sqrt{7 \times 3^2} = 3\sqrt{7}$.

Any square root of a positive integer can be written as an integer or as the product of an integer and an irreducible square root. This notation is called the *standard form* of the square root. You can find the standard form by 'extracting all the squares from the root'. Example: $\sqrt{200} = \sqrt{10^2 \times 2} = 10\sqrt{2}$.

**Why is $\sqrt{20}$ irrational?**

To prove that $\sqrt{20}$ is irrational, we use a *proof by contradiction*: suppose that $\sqrt{20}$ is rational. Then it can be written as a fraction $p/q$ in lowest terms, i.e. with $p$ and $q$ positive integers with $\gcd(p,q) = 1$. From $\sqrt{20} = p/q$ it follows that $20q^2 = p^2$, so $2 \times 2 \times 5 \times q^2 = p^2$. The left-hand side is a multiple of 5, so the same holds for the right-hand side. The prime decomposition of $p$ should then contain at least one prime factor 5, so the prime decomposition of $p^2$ should contain at least *two* factors 5. But $\gcd(p,q) = 1$, so the prime decomposition of $q$ *does not* contain any factor 5. The prime decomposition of $20q^2$ therefore contains exactly one factor 5. As we just showed that the prime decomposition of $p^2$ has at least two factors 5, this is inconsistent with $20q^2 = p^2$. Our presumption that $\sqrt{20}$ is rational has resulted in a contradiction, so it cannot be true. Conclusion: the number $\sqrt{20}$ has to be irrational. A similar proof can be given for any square root of a positive integer that itself is not the square of an integer.

Write the following expressions in the standard form, i.e. in the form $a\sqrt{b}$ where $a$ is an integer or a fraction in lowest terms, and $\sqrt{b}$ is an irreducible square root:

3.24

a. $\left(\frac{\sqrt{3}}{2}\right)^2$

b. $\left(\frac{3}{\sqrt{2}}\right)^2$

c. $\left(\frac{\sqrt{3}}{\sqrt{2}}\right)^2$

d. $\left(\frac{\sqrt{2}}{3}\right)^3$

e. $\left(\frac{2\sqrt{3}}{\sqrt{2}}\right)^3$

3.25

a. $\left(\frac{\sqrt{3}}{\sqrt{6}}\right)^3$

b. $\left(\frac{2\sqrt{3}}{3\sqrt{2}}\right)^3$

c. $\left(\frac{-\sqrt{7}}{2\sqrt{2}}\right)^4$

d. $\left(\sqrt{\frac{3}{2}}\right)^3$

e. $\left(\sqrt{\frac{4}{3}}\right)^5$

3.26

a. $\sqrt{\frac{2}{3}}$

b. $\sqrt{\frac{3}{2}}$

c. $\sqrt{\frac{6}{5}}$

d. $\sqrt{\frac{7}{2}}$

e. $\sqrt{\frac{2}{7}}$

3.27

a. $\sqrt{\frac{5}{12}}$

b. $\sqrt{\frac{4}{27}}$

c. $\sqrt{\frac{9}{20}}$

d. $\sqrt{\frac{6}{15}}$

e. $\sqrt{\frac{7}{32}}$

3.28

a. $\frac{\sqrt{3}}{\sqrt{2}}$

b. $\frac{\sqrt{5}}{\sqrt{3}}$

c. $\frac{\sqrt{7}}{\sqrt{11}}$

d. $\frac{\sqrt{11}}{\sqrt{5}}$

e. $\frac{\sqrt{2}}{\sqrt{11}}$

3.29

a. $\frac{3\sqrt{5}}{\sqrt{6}}$

b. $\frac{2\sqrt{3}}{\sqrt{10}}$

c. $\frac{4\sqrt{12}}{\sqrt{20}}$

d. $\frac{-5\sqrt{2}}{\sqrt{15}}$

e. $\frac{6\sqrt{6}}{3\sqrt{3}}$

## Square roots of fractions in the standard form

The square root of a fraction with a positive numerator and denominator is the quotient of the square root of the numerator and the square root of the denominator. For example, $\sqrt{\frac{4}{9}} = \frac{\sqrt{4}}{\sqrt{9}} = \frac{2}{3}$. Indeed, we have $\left(\frac{2}{3}\right)^2 = \frac{4}{9}$.

The square root of a fraction can always be written as a fraction in lowest terms or as the product of a fraction in lowest terms and an irreducible square root. We call this the *standard form* of such a square root. Examples:

$$\sqrt{\frac{4}{3}} = \sqrt{\frac{4 \times 3}{3 \times 3}} = \frac{2}{3}\sqrt{3} \quad \text{and} \quad \sqrt{\frac{11}{15}} = \sqrt{\frac{11 \times 15}{15 \times 15}} = \frac{1}{15}\sqrt{165}$$

The standard form is obtained by first multiplying the numerator and denominator with a common factor that makes the denominator the square of an integer. Extracting the root in the denominator gives an integer. The numerator is still the square root of an integer, which can be written as the product of an integer and an irreducible square root. The result is the product of a fraction and an irreducible square root. Writing the fraction in lowest terms gives the required standard form.

Similarly, a square root in the denominator of a fractional expression can be removed. For example:

$$\frac{2\sqrt{3}}{\sqrt{7}} = \frac{2\sqrt{3} \times \sqrt{7}}{\sqrt{7} \times \sqrt{7}} = \frac{2\sqrt{21}}{7} = \frac{2}{7}\sqrt{21}$$

Write the following expressions in the standard form:

3.30

a. $\sqrt[3]{8}$
b. $\sqrt[4]{81}$
c. $\sqrt[3]{125}$
d. $\sqrt[5]{1024}$
e. $\sqrt[3]{216}$

3.31

a. $\sqrt[3]{-27}$
b. $\sqrt[4]{16}$
c. $\sqrt[5]{243}$
d. $\sqrt[7]{-128}$
e. $\sqrt[2]{144}$

3.32

a. $\sqrt[3]{16}$
b. $\sqrt[4]{243}$
c. $\sqrt[3]{375}$
d. $\sqrt[5]{96}$
e. $\sqrt[3]{54}$

3.33

a. $\sqrt[3]{-40}$
b. $\sqrt[4]{48}$
c. $\sqrt[5]{320}$
d. $\sqrt[3]{432}$
e. $\sqrt[6]{192}$

3.34

a. $\sqrt[3]{5} \times \sqrt[3]{7}$
b. $\sqrt[4]{4} \times \sqrt[4]{14}$
c. $\sqrt[3]{6} \times \sqrt[3]{4}$
d. $\sqrt[4]{18} \times \sqrt[4]{45}$
e. $\sqrt[5]{16} \times \sqrt[5]{12}$

3.35

a. $\sqrt[4]{24} \times \sqrt[4]{54}$
b. $\sqrt[3]{36} \times \sqrt[3]{12}$
c. $\sqrt[5]{81} \times \sqrt[5]{15}$
d. $\sqrt[6]{288} \times \sqrt[6]{324}$
e. $\sqrt[3]{200} \times \sqrt[3]{35}$

3.36

a. $\sqrt[3]{\frac{1}{343}}$
b. $\sqrt[4]{\frac{-16}{81}}$
c. $\sqrt[5]{\frac{32}{-243}}$
d. $\sqrt[2]{\frac{36}{121}}$
e. $\sqrt[4]{\frac{1296}{625}}$

3.37

a. $\sqrt[3]{\frac{8}{27}}$
b. $\sqrt[4]{\frac{625}{16}}$
c. $\sqrt[5]{\frac{32}{243}}$
d. $\sqrt[3]{\frac{216}{1000}}$
e. $\sqrt[2]{\frac{144}{25}}$

3.38

a. $\sqrt[3]{\frac{1}{4}}$
b. $\sqrt[4]{\frac{2}{27}}$
c. $\sqrt[3]{\frac{3}{25}}$
d. $\sqrt[3]{\frac{5}{9}}$
e. $\sqrt[6]{\frac{3}{8}}$

3.39

a. $\sqrt[3]{\frac{5}{24}}$
b. $\sqrt[4]{\frac{7}{72}}$
c. $\sqrt[5]{\frac{5}{648}}$
d. $\sqrt[3]{\frac{9}{100}}$

3.40

a. $\frac{\sqrt[3]{2}}{\sqrt[3]{3}}$
b. $\frac{\sqrt[4]{3}}{\sqrt[4]{8}}$
c. $\frac{\sqrt[5]{1}}{\sqrt[5]{16}}$
d. $\frac{\sqrt[6]{6}}{\sqrt[6]{81}}$

3.41

a. $\frac{\sqrt[3]{-3}}{\sqrt[3]{2}}$
b. $\frac{\sqrt[4]{3}}{\sqrt[4]{4}}$
c. $\frac{\sqrt[5]{7}}{\sqrt[5]{-27}}$
d. $\frac{\sqrt[3]{35}}{\sqrt[3]{36}}$

## Higher roots in the standard form

Higher roots are treated in a similar way. The *cube root* of a number $a$ is the number $r$ for which $r^3 = a$. Notation: $\sqrt[3]{a}$. Examples: $\sqrt[3]{27} = 3$, since $3^3 = 27$ and $\sqrt[3]{-8} = -2$, since $(-2)^3 = -8$. Note that cube roots can also be defined for negative numbers, and that there are no sign problems for cube roots: there is only one number for which the cube is equal to 27, namely 3, and there is only one number for which the cube equals $-8$, namely $-2$.

In general the $n$-th root $\sqrt[n]{a}$ of $a$ is the number $r$ for which $r^n = a$. If $n$ is even, $a \geq 0$ should hold. In that case, we also have $r^n = (-r)^n$, yielding two possible candidates for the root. Convention dictates you always take the non-negative $r$ for which $r^n = a$.

The $n$-th roots and square roots have much in common:

- The $n$-th root of an integer $a$ is irrational unless $a$ itself is the $n$-th power of an integer.
- The $n$-th root of a positive integer $a$ is called *irreducible* if $a$ does not have any $n$-th power greater than 1 as a divisor.
- The $n$-th root of a fraction can be written as a fraction in lowest terms or as the product of a fraction in lowest terms and an irreducible $n$-th root. This is called the *standard form* of the root.

Examples of cube roots: $\sqrt[3]{24}$ is reducible, since $\sqrt[3]{24} = \sqrt[3]{2^3 \times 3} = 2\sqrt[3]{3}$, but the cube roots $\sqrt[3]{18}$, $\sqrt[3]{25}$ and $\sqrt[3]{450}$ are irreducible.

The standard form of a cube root of a fraction can be found by multiplying the numerator and denominator by a common factor that makes the denominator a cube. For example:

$$\sqrt[3]{\frac{14}{75}} = \sqrt[3]{\frac{2 \times 7}{3 \times 5^2}} = \sqrt[3]{\frac{2 \times 7 \times 3^2 \times 5}{3^3 \times 5^3}} = \frac{1}{15}\sqrt[3]{630}$$

3.42 Write as a root:

a. $2^{\frac{1}{2}}$
b. $3^{\frac{3}{2}}$
c. $7^{\frac{2}{3}}$
d. $5^{\frac{5}{4}}$
e. $4^{\frac{4}{3}}$

3.43 Write as a root:

a. $3^{-\frac{1}{2}}$
b. $7^{-\frac{3}{2}}$
c. $4^{-\frac{1}{3}}$
d. $9^{-\frac{2}{5}}$
e. $2^{-\frac{1}{2}}$

3.44 Write as a power:

a. $\sqrt[3]{5}$
b. $\sqrt[2]{7}$
c. $\sqrt[4]{2}$
d. $\sqrt[6]{12}$
e. $\sqrt[5]{5}$

3.45 Write as a power:

a. $\frac{1}{\sqrt[2]{5}}$
b. $\frac{1}{\sqrt[3]{6}}$
c. $\frac{1}{2\sqrt[4]{2}}$
d. $\frac{3}{\sqrt[2]{3}}$
e. $\frac{7}{\sqrt[5]{7}}$

3.46 Write as a power of 2:

a. $\sqrt[3]{4}$
b. $\sqrt[2]{8}$
c. $\sqrt[4]{32}$
d. $\sqrt[6]{16}$
e. $\sqrt[3]{32}$

3.47 Write as a power of 2:

a. $\frac{4}{\sqrt[2]{2}}$
b. $\frac{1}{2\sqrt[2]{2}}$
c. $\frac{8}{\sqrt[3]{4}}$
d. $\frac{2}{\sqrt[4]{8}}$
e. $\frac{1}{4\sqrt[3]{16}}$

Write the following expressions as a root in the standard form:

3.48

a. $\sqrt[2]{2} \times \sqrt[3]{2}$
b. $\sqrt[3]{3} \times \sqrt[2]{3}$
c. $\sqrt[4]{8} \times \sqrt[3]{16}$
d. $\sqrt[5]{27} \times \sqrt[3]{9}$
e. $\sqrt[3]{16} \times \sqrt[6]{16}$

3.49

a. $\sqrt[2]{7} \times \sqrt[3]{49}$
b. $\sqrt[3]{9} \times \sqrt[2]{3}$
c. $\sqrt[4]{25} \times \sqrt[3]{5}$
d. $\sqrt[5]{81} \times \sqrt[4]{27}$
e. $\sqrt[4]{49} \times \sqrt[2]{7}$

3.50

a. $\sqrt[2]{2} : \sqrt[3]{2}$
b. $\sqrt[3]{9} : \sqrt[2]{3}$
c. $\sqrt[4]{8} : \sqrt[2]{2}$
d. $\sqrt[3]{9} : \sqrt[5]{27}$
e. $\sqrt[2]{2} : \sqrt[3]{4}$

## Fractional powers

In this section, we restrict ourselves to powers with a *positive* base $a$. For any fraction $\frac{m}{n}$ with $n > 1$ we define

$$a^{\frac{m}{n}} = \sqrt[n]{a^m}$$

In particular (take $m = 1$)

$$a^{\frac{1}{n}} = \sqrt[n]{a}$$

giving

$$a^{\frac{1}{2}} = \sqrt{a}, \quad a^{\frac{1}{3}} = \sqrt[3]{a}, \quad a^{\frac{1}{4}} = \sqrt[4]{a} \quad \text{et cetera.}$$

In the same way (take $m = -1$)

$$a^{-\frac{1}{2}} = \sqrt{a^{-1}} = \sqrt{\frac{1}{a}} = \frac{1}{\sqrt{a}}, \quad a^{-\frac{1}{3}} = \sqrt[3]{a^{-1}} = \sqrt[3]{\frac{1}{a}} = \frac{1}{\sqrt[3]{a}} \quad \text{et cetera.}$$

Further examples:

$$7^{\frac{3}{2}} = \sqrt{7^3} = 7\sqrt{7}, \quad 5^{-\frac{2}{7}} = \frac{1}{\sqrt[7]{25}} \quad \text{and} \quad 2^{\frac{5}{3}} = \sqrt[3]{2^5} = 2\sqrt[3]{4}$$

The last example may also be verified using $\frac{5}{3} = 1\frac{2}{3} = 1 + \frac{2}{3}$:

$$2^{\frac{5}{3}} = 2^{1+\frac{2}{3}} = 2^1 \times 2^{\frac{2}{3}} = 2\sqrt[3]{2^2} = 2\sqrt[3]{4}$$

*Calculation rules for powers:*

$$\begin{aligned} a^r \times a^s &= a^{r+s} \\ a^r : a^s &= a^{r-s} \\ (a^r)^s &= a^{r \times s} \\ (a \times b)^r &= a^r \times b^r \\ \left(\frac{a}{b}\right)^r &= \frac{a^r}{b^r} \end{aligned}$$

These rules are valid for all rational numbers $r$ and $s$ and all positive numbers $a$ and $b$.

# II Algebra

$$(a+b)(c+d) = ac + ad + bc + bd$$

Algebra is the art of calculating with letters. These letters usually represent numbers. In the first two chapters of part II, we tackle the basic rules of algebra: priority rules, expanding brackets, factoring out terms, the *banana formula* (see above) and some *special products*. The last chapter is about calculating with fractional expressions in which letters occur, in particular simplifying, splitting and writing such expressions with a common denominator.

# 4 Calculating with letters

Substitute the given numerical values into the algebraic expressions and calculate the result. For example, substituting $a = 5$ into the expression $3a^3 - 2a + 4$ yields $3 \times 5^3 - 2 \times 5 + 4 = 375 - 10 + 4 = 369$.

4.1 Substitute $a = 3$ into

a. $2a^2$
b. $-a^2 + a$
c. $4a^3 - 2a$
d. $-3a^3 - 3a^2$
e. $a(2a - 3)$

4.2 Substitute $a = -2$ into

a. $3a^2$
b. $-a^3 + a$
c. $3(a^2 - 2a)$
d. $-2a^2 + a$
e. $2a(-a + 3)$

4.3 Substitute $a = 4$ into

a. $3a^2 - 2a$
b. $-a^3 + 2a^2$
c. $-2(a^2 - 2a)$
d. $(2a - 4)(-a + 2)$
e. $(3a - 4)^2$

4.4 Substitute $a = -3$ into

a. $-a^2 + 2a$
b. $a^3 - 2a^2$
c. $-3(a^2 - 2a)$
d. $(2a - 1)(-3a + 2)$
e. $(2a + 1)^2$

4.5 Substitute $a = 3$ and $b = 2$ into

a. $2a^2b$
b. $3a^2b^2 - 2ab$
c. $-3a^2b^3 + 2ab^2$
d. $2a^3b - 3ab^3$
e. $-5ab^2 - 2a^2 + 3b^3$

4.6 Substitute $a = -2$ and $b = -3$ into

a. $3ab - a$
b. $2a^2b - 2ab$
c. $-3ab^2 + 3ab$
d. $a^2b^2 - 2a^2b + ab^2$
e. $-a^2 + b^2 + 4ab$

4.7 Substitute $a = 5$ and $b = -2$ into

a. $3(ab)^2 - 2ab$
b. $a(a + b)^2 - (2a)^2$
c. $-3ab(a + 2b)^2$
d. $3a(a - 2b)(a^2 - 2ab)$
e. $(a^2b - 2ab^2)^2$

4.8 Substitute $a = -2$ and $b = -1$ into

a. $-(a^2b)^3 - 2(ab^2)^2$
b. $-b(3a^2 - 2b)^2$
c. $(3a^2b - 2ab^2)(2a^2 - b^2)$
d. $(a^2 + b^2)(a^2 - b^2)$
e. $\left((-a^2b + 2b)(ab^2 - 2a)\right)^2$

## Priority rules

In this part, letters in algebraic expressions always represent numbers. This means that arithmetical operations can also be defined for letters. For instance, $a + b$ is the sum of $a$ and $b$, $a - b$ is the difference of $a$ and $b$ and so on.

The multiplication sign is often replaced by a dot, or omitted completely. For instance, instead of $a \times b$ we write $a \cdot b$ or even $ab$. You can also use mixed forms of numbers and letters: $2ab$ means $2 \times a \times b$. In these cases, the number is usually put first, giving $2ab$ and not $a2b$ or $ab2$.

The following *priority rules* are generally accepted:

a. Addition and subtraction are applied in the order in which they occur, from left to right.
b. Multiplication and division are applied in the order in which they occur, from left to right.
c. Multiplication and division have priority over addition and subtraction.

Below are some examples with numbers. On the right-hand side, we give the order with brackets first, before calculating the answer.

$$\begin{array}{rcrcr} 5-7+8 & = & (5-7)+8 & = & 6 \\ 4-5\times 3 & = & 4-(5\times 3) & = & -11 \\ 9+14:7 & = & 9+(14:7) & = & 11 \\ 12:3\times 4 & = & (12:3)\times 4 & = & 16 \end{array}$$

Below are the same examples with letters. On the right-hand side, the order of calculation is indicated with brackets.

$$\begin{array}{lcl} a-b+c & = & (a-b)+c \\ a-bc & = & a-(b\times c) \\ a+b:c & = & a+(b:c) \\ a:b\times c & = & (a:b)\times c \end{array}$$

Note that if in the last example the left-hand side would be written as $a : bc$, many would interpret this as $a : (b \times c)$, which is not the same as $(a : b) \times c$. If, for instance, $a = 12$, $b = 3$ and $c = 4$, then $(12 : 3) \times 4 = 16$ but $12 : (3 \times 4) = 1$. This explains why you should not write $a : bc$ but $a : (bc)$ when the latter is meant. More generally:

*Use brackets in all cases where the order of calculation might be unclear!*

As a rule of thumb, it is better to use too many brackets than too few!

Write the following expressions as simple as possible as a power or a product of powers:

4.9

a. $a^3 \cdot a^5$
b. $b^3 \cdot b^2$
c. $a^4 \cdot a^7$
d. $b \cdot b^3$
e. $a^7 \cdot a^7$

4.10

a. $(a^2)^3$
b. $(b^3)^4$
c. $(a^5)^5$
d. $(b^4)^2$
e. $(a^6)^9$

4.11

a. $(ab)^4$
b. $(a^2b^3)^2$
c. $(a^4b)^3$
d. $(a^2b^3)^4$
e. $(a^3b^4)^5$

4.12

a. $a^4 \cdot a^3 \cdot a$
b. $2a^5 \cdot 3a^5$
c. $4a^2 \cdot 3a^2 \cdot 5a^2$
d. $5a^3 \cdot 6a^4 \cdot 7a$
e. $a \cdot 2a^2 \cdot 3a^3$

4.13

a. $(2a^2)^3$
b. $(3a^3b^4)^4$
c. $(4a^2b^2)^2$
d. $(5a^5b^3)^3$
e. $(2ab^5)^4$

4.14

a. $3a^2b \cdot 5ab^4$
b. $6a^3b^4 \cdot 4a^6b^2$
c. $3a^2b^2 \cdot 2a^3b^3$
d. $7a^5b^3 \cdot 5a^7b^5$
e. $8a^2b^4 \cdot 3ab^2 \cdot 6a^5b^4$

4.15

a. $3a^2 \cdot -2a^3 \cdot -4a^5$
b. $-5a^3 \cdot 2a^2 \cdot -4a^3 \cdot 3a^2$
c. $4a^2 \cdot -2a^4 \cdot -5a^5$
d. $2a^4 \cdot -3a^5 \cdot -3a^6$
e. $-3a^2 \cdot -2a^4 \cdot -4a$

4.16

a. $(-2a^2)^3$
b. $(-3a^3)^2$
c. $(-5a^4)^4$
d. $(-a^2b^4)^5$
e. $(-2a^3b^5)^7$

4.17

a. $3a^2 \cdot (2a^3)^2$
b. $(-3a^3)^2 \cdot (2a^2)^3$
c. $(3a^4)^3 \cdot -5a^6$
d. $2a^2 \cdot (5a^3)^3 \cdot 3a^5$
e. $-2a^5 \cdot (-2a)^5 \cdot 5a^2$

4.18

a. $2a^3b^4(-3a^2b^3)^2$
b. $(-2a^2b^4)^3(-3a^2b^5)^2$
c. $2a^2b(-2a^2b)^2(-2a^2b)^3$
d. $3a^4b^2(-3a^2b^4)^3(-2a^3b^2)^2$
e. $(2a^3)^4(-3b^2)^2(2a^2b^3)^3$

4.19

a. $(3a^2b^3c^4)^2(2ab^2c^3)^3$
b. $(-2a^3c^4)^2(-a^2b^3)^3(2b^3c^2)^4$
c. $2a^2c^3(3a^3b^2c)^4(-5ab^2c^5)$
d. $(-2a^3c)^6(5a^3b^2)^2(-5b^3c^4)^4$
e. $-(-3a^2b^2c^2)^3(-2a^3b^3c^3)^2$

4.20

a. $((a^3)^4)^3$
b. $((-a^2)^3(2a^3)^2)^2$
c. $((2a^2b^3)^2(-3a^3b^2)^3)^2$
d. $(-2a(-a^3)^2)^5$
e. $(-2(-a^2)^3)^2\,(-3(-a^4)^2)^3$

## Calculating with powers

In the previous chapter, we discussed the following rules for powers:

$$\begin{aligned} a^n \times a^m &= a^{n+m} \\ (a^n)^m &= a^{n\times m} \\ (a \times b)^n &= a^n \times b^n \end{aligned}$$

We calculated with numbers before, but we can also use letters instead. In this section, the exponents are still given integers, but with letters used as bases. Using the rules above, we can simplify complex algebraic expressions with powers.

We will give a few examples, starting with four simple ones:

$$\begin{aligned} &a^4 \cdot a^5 = a^{4+5} = a^9 \\ &(a^2)^4 = a^{2\times 4} = a^8 \\ &(ab)^5 = a^5 \times b^5 = a^5b^5 \\ &(a^2b^4)^3 = (a^2)^3(b^4)^3 = a^6b^{12} \end{aligned}$$

The following four examples also contain numbers:

$$\begin{aligned} &2a^3 \cdot 5a^7 = (2 \times 5)\, a^{3+7} = 10\, a^{10} \\ &(2a)^4 \cdot (5a)^3 = (2^4 \times 5^3)\, a^{4+3} = 2000\, a^7 \\ &(-2a)^7 = (-2)^7\, a^7 = -128\, a^7 \\ &(4a^2)^3 \cdot (-5a)^2 = 64 \cdot 25\, (a^2)^3 \cdot a^2 = 1600\, a^8 \end{aligned}$$

Now try all the exercises on the previous page. Make sure you pay attention to minus signs, if present. Remember the following rules:

*A negative number raised to an* even *power yields a positive number.*

*A negative number raised to an* odd *power yields a negative number.*

Expand all the brackets:

4.21
a. $3(2a+5)$
b. $8(5a-2)$
c. $-5(3a-2)$
d. $12(-5a+1)$
e. $-7(7a+6)$

4.22
a. $2a(a-5)$
b. $7a(2a+12)$
c. $-13a(9a-5)$
d. $8a(8a-15)$
e. $-21a(3a+9)$

4.23
a. $2a(a^2+9)$
b. $3a^2(4a-7)$
c. $-5a^2(2a^2+4)$
d. $9a^2(a^2+2a)$
e. $-3a(a^2-4a)$

4.24
a. $4a^2(3a^2+2a+3)$
b. $-3a^2(2a^3+5a^2-a)$
c. $7a^3(2a^2+3a-6)$
d. $12a^2(-6a^3-2a^2+a-1)$
e. $-5a^2(3a^4+a^2-2)$

4.25
a. $2(3a+4b)$
b. $-5(2a-5b)$
c. $2a(a+2b)$
d. $16a(-4a+6b)$
e. $-22a(8a-11b)$

4.26
a. $3a(9a+5b-12)$
b. $2a^2(7a-6b)$
c. $-8a^2(7a+4b-1)$
d. $6a^2(-2a+2b+2)$
e. $-13a^2(13a+12b-14)$

4.27
a. $2a^2(3a^2+2b-3)$
b. $-5a^3(2a^2+a-2b)$
c. $2b^2(3a^2+2b^2)$
d. $4a^3(-2a^2+5b^2-2b)$
e. $-14b^3(14a^2+2a-5b^2)$

4.28
a. $2a^2(a^2+3ab)$
b. $-5a^2(3a^2+2ab-3b^2)$
c. $2a^3(3a^3+2a^2b^2-b^2)$
d. $-3a^4(2a^3+2a^2b^2+2ab^2)$
e. $7a^3(-7a^3+3a^2b-4ab^2)$

4.29
a. $2ab(a^2+2ab-b^2)$
b. $-5ab(-3a^2b+2ab^2-6b)$
c. $6ab^2(2a^2b-5ab-b^2)$
d. $-12a^2b^2(-12a^2b^2+6ab-12)$
e. $6ab^2(2a^2b+9ab-ab^2)$

4.30
a. $a^3b^2(-5a^2b^3+2a^2b^2-ab^3)$
b. $-a^2b^3(-a^3b^2-a^2b-14)$
c. $15a^4b^3(-a^3b^4-6a^2b^3+ab^4)$
d. $-a^5b^4(13a^4b^5-12a^2b^3+9ab^5)$
e. $7a^2b^2(-7a^3-7ab^2-1)$

4.31
a. $2a(a+6)-4(a+2)$
b. $-4a(3a+6)+2(a-3)$
c. $7a(-2a-1)-2a(-7a+1)$
d. $-8a(a-8)-2(-a+5)$
e. $5a(2a-5)+5(2a-1)$
f. $-2a(a+1)-(a-1)$

4.32
a. $3a(a+2b)-b(-2a+2)$
b. $-a(a-b)+b(-a+1)$
c. $2a(2a+b)-2b(-a+b)-2(a-b)$
d. $-b(-a+2b)+3(2a-b)-a(2a+b)$

## Expanding brackets

The *distributive laws*

$$\begin{aligned} a(b+c) &= ab+ac \\ (a+b)c &= ac+bc \end{aligned}$$

are valid irrespective of the particular values of $a$, $b$ and $c$. Examples:

$$15(3+8) = 15\times 3 + 15\times 8 = 45+120 = 165$$
$$(3-8)(-11) = 3\times(-11) + (-8)\times(-11) = -33+88 = 55$$

By applying the distributive laws, you can 'expand brackets'. For instance:

$$\begin{aligned} 5a^2(4b-2c) &= 20a^2b - 10a^2c \\ 3ab(c+2b) &= 3abc + 6ab^2 \\ (5a-2b)3c^2 &= 15ac^2 - 6bc^2 \end{aligned}$$

Note that the distributive laws are formulated in their most simple form, but that in the given examples we substituted various kinds of algebraic expressions for $a$, $b$ and $c$. It is in fact this possibility to manipulate formulas that makes algebra such a powerful tool.

Also note that in these examples the multiplication sign is omitted. With explicit use of multiplication signs the first example would read

$$5\times a^2\times(4\times b - 2\times c) = 20\times a^2\times b - 10\times a^2\times c$$

which gives a much more cumbersome formula (although more illuminating for beginners).

The above can also be applied in more complicated expressions:

$$3a(4b-2c) + 2b(a-3c) = 12ab - 6ac + 2ab - 6bc = 14ab - 6ac - 6bc$$
$$4a(b+c) - 5a(2b-3c) = 4ab + 4ac - 10ab + 15ac = -6ab + 19ac$$
$$-2a(b-3c) - 5c(a+2b) = -2ab + 6ac - 5ac - 10bc = -2ab + ac - 10bc$$

Note the signs in the last two examples and remember:

| | |
|---|---|
| *Plus times plus equals plus* | *Minus times plus equals minus* |
| *Plus times minus equals minus* | *Minus times minus equals plus* |

Factor out as many integers or algebraic expressions as possible:

4.33

a. $6a + 12$
b. $12a + 16$
c. $9a - 12$
d. $15a - 10$
e. $27a + 81$

4.34

a. $3a - 6b + 9$
b. $12a + 8b - 16$
c. $9a + 12b + 3$
d. $30a - 24b + 60$
e. $24a + 60b - 36$

4.35

a. $-6a + 9b - 15$
b. $-14a + 35b - 21$
c. $-18a - 24b - 12c$
d. $-28a - 70b + 42c$
e. $-45a + 27b - 63c - 18$

4.36

a. $a^2 + a$
b. $a^3 - a^2$
c. $a^3 - a^2 + a$
d. $a^4 + a^3 - a^2$
e. $a^6 - a^4 + a^3$

4.37

a. $3a^2 + 6a$
b. $9a^3 + 6a^2 - 3a$
c. $15a^4 - 10a^3 + 25a^2$
d. $27a^6 - 18a^4 - 36a^2$
e. $48a^4 - 24a^3 + 36a^2 + 60a$

4.38

a. $3a^2b + 6ab$
b. $9a^2b - 9ab^2$
c. $12ab^2 - 4ab$
d. $14a^2b^2 - 21ab^2$
e. $18a^2b^2 - 15a^2b$

4.39

a. $3a^3b^2 + 6a^2b$
b. $6a^4b^3 - 9a^3b^2 + 12a^2b$
c. $10a^3b^2c^2 - 5a^2bc^2 - 15abc$
d. $8a^6b^5c^4 - 12a^4b^4c^3 + 20a^3b^4c^3$
e. $a^3b^3c^3 + a^3b^3c^2 + a^3b^3c$

4.40

a. $-4a^2b^3c^2 + 2a^2b^2c^2 - 6a^2bc^2$
b. $a^6b^5c^4 - a^4b^6c^4 - a^3b^7c^3$
c. $-2a^3c^4 + 2a^2b^2c^3 - 4a^2bc^2$
d. $-a^7b^6 + a^6b^7 - a^5b^6$
e. $-a^8b^7c^6 - a^7b^6c^7 + a^6b^6c^6$

4.41

a. $a(b+3) + 3(b+3)$
b. $a(b-1) - 2(b-1)$
c. $2a(b+4) + 7(b+4)$
d. $a^2(2b-1) + 2(2b-1)$
e. $a(b-2) - (b-2)$

4.42

a. $a^2(b+1) - a(b+1)$
b. $6a(2b+1) + 12(2b+1)$
c. $-2a(b-1) + 4(b-1)$
d. $a^3(4b+3) - a^2(4b+3)$
e. $-6a^2(2b+3) - 9a(2b+3)$

4.43

a. $(a+1)(b+1) + 3(b+1)$
b. $(2a-1)(b+1) + (2a-1)(b-1)$
c. $(a+3)(2b-1) + (2a-1)(2b-1)$
d. $(a-1)(a+3) + (a+2)(a+3)$
e. $(a+1)^2 + (a+1)$

4.44

a. $2(a+3)^2 + 4(a+3)$
b. $(a+3)^2(b+1) - 2(a+3)(b+1)$
c. $(a-1)^2(a+2) - (a-1)(a+2)^2$
d. $3(a+2)^2(a-2) + 9(a+2)(a-2)^2$
e. $-2(a+4)^3 + 6(a+4)^2(a+2)$

## Factoring expressions

The distributive laws can also be read in reverse order:

$$\begin{aligned} ab + ac &= a(b + c) \\ ac + bc &= (a + b)c \end{aligned}$$

This way, they can be used to factor expressions. In the following examples, we only factor out integers:

$$\begin{aligned} 3a + 12 &= 3(a + 4) \\ 27a + 45b - 9 &= 9(3a + 5b - 1) \end{aligned}$$

Similarly, letters or combinations of letters and numbers can also be factored out:

$$\begin{aligned} a^4 - a &= a(a^3 - 1) \\ 15a^2b + 5ab^3 &= 5ab(3a + b^2) \end{aligned}$$

Or even more complicated algebraic expressions:

$$\begin{aligned} (a + 1)b - 3(a + 1) &= (a + 1)(b - 3) \\ 7a^2(b^2 - 3) - 35(b^2 - 3) &= 7(a^2 - 5)(b^2 - 3) \end{aligned}$$

Expand the brackets:

4.45

a. $(a+3)(a+1)$
b. $(2a+3)(a+3)$
c. $(a-6)(3a+1)$
d. $(4a-5)(5a+4)$
e. $(3a+9)(2a-5)$
f. $(6a-12)(4a+10)$

4.46

a. $(-3a+8)(8a-3)$
b. $(7a+12)(8a-11)$
c. $(17a+1)(a-17)$
d. $(-2a+6)(-3a-6)$
e. $(a+3)(b-5)$
f. $(2a+8)(3b+5)$

4.47

a. $(-4a+1)(b-1)$
b. $(3a-1)(-b+3)$
c. $(13a+12)(12b-13)$
d. $(a^2+4)(a-4)$
e. $(a-1)(a^2+7)$
f. $(a^2+3)(a^2+9)$

4.48

a. $(2a^2-7)(a+7)$
b. $(-3a^2+2)(-2a^2+3)$
c. $(a^2+2a)(2a^2-a)$
d. $(3a^2-4a)(-2a^2+5a)$
e. $(-6a^2+5)(a^2+a)$
f. $(9a^2+7a)(2a^2-7a)$

4.49

a. $(-8a^2-3a)(3a^2-8a)$
b. $(2a^3-a)(-5a^2+4)$
c. $(-a^3+a^2)(a^2+a)$
d. $(9a^4-5a^2)(6a^3+2a^2)$
e. $(7a^3-1)(8a^3-5a)$
f. $(-6a^5-5a^4)(-4a^3-3a^2)$

4.50

a. $(2ab+a)(3ab-b)$
b. $(3a^2b+ab)(2ab^2-3ab)$
c. $(-2a^2b^2+3a^2b)(2ab^2-2ab)$
d. $(8a^3b^2-6ab^3)(-4a^2b^3-2ab^2)$
e. $(-a^5b^3+a^3b^5)(a^3b^5-ab^7)$
f. $(2a+3)(a^2+2a-2)$

4.51

a. $(-3a+2)(4a^2-a+1)$
b. $(2a+b)(a+b+4)$
c. $(-3a+3b)(3a-3b-3)$
d. $(9a+2)(2a-9b+1)$
e. $(a^2+a)(a^2-a+1)$
f. $(2a^2+2a-1)(3a+2)$

4.52

a. $(-2a-1)(-a^2-3a-4)$
b. $(a-b-1)(a+b)$
c. $(a^2+ab+b^2)(a^2-b^2)$
d. $(a+1)(a+2)(a+3)$
e. $(a-1)(a+2)(a-3)$
f. $(2a+1)(a-1)(2a+3)$

4.53

a. $(2a+b)(a-b)(2a-b)$
b. $(5a-4b)(4a-3b)(3a-2b)$
c. $-3a(a^2+3)(a-2)$
d. $(-3a+1)(a+3)(-a+1)$
e. $2a^2(a^2-1)(a^2+2)$
f. $(a^2b-ab)(ab^2+ab)(a+b)$

4.54

a. $3a^2b(a^2-b^2)(2a+2b)$
b. $(a+1)(a^3+a^2-a+2)$
c. $(a^2+2a+1)(a^2-a+2)$
d. $(-2a^2+3a+1)(3a^2-2a-1)$
e. $3a(a^2+1)(a^2-2a+4)$
f. $(2a+b-5)(5a-2b+2)$

## The banana formula

The next formula yields the product of two sums of two terms:

$$(a+b)(c+d) = ac + ad + bc + bd$$

As the curved lines indicate, this is the result of applying the distributive law twice:

$$(a+b)(c+d) = a(c+d) + b(c+d) = ac + ad + bc + bd$$

The curved lines act as a nice mnemonic device and look a bit like a banana, which is why this formula also is referred to as the *banana formula*. It also can be used in more complex forms, for instance:

$$(3a^2 + 7bc)(5ab - 2c) = 15a^3b - 6a^2c + 35ab^2c - 14bc^2$$

In some cases, after expanding brackets using the banana formula, certain terms may be combined, for instance:

$$(5a+3b)(2a-7b) = 10a^2 - 35ab + 6ab - 21b^2 = 10a^2 - 29ab - 21b^2$$

When more than two terms are contained in the brackets, the expansion follows the same rules as with the banana formula. For example:

$$\begin{aligned}(3a+2b)(2c-d+8e) &= 3a(2c-d+8e) + 2b(2c-d+8e)\\ &= 6ac - 3ad + 24ae + 4bc - 2bd + 16be\end{aligned}$$

Products with more than two factors are expanded step by step. For example:

$$\begin{aligned}(3a+2b)(a-4b)(2a+c) &= (3a^2 - 12ab + 2ab - 8b^2)(2a+c)\\ &= (3a^2 - 10ab - 8b^2)(2a+c)\\ &= 6a^3 + 3a^2c - 20a^2b - 10abc - 16ab^2 - 8b^2c\end{aligned}$$

# 5 Special products

Expand the brackets:

5.1

a. $(a+6)^2$
b. $(a-2)^2$
c. $(a+11)^2$
d. $(a-9)^2$
e. $(a+1)^2$

5.2

a. $(b+5)^2$
b. $(b-12)^2$
c. $(b+13)^2$
d. $(b-7)^2$
e. $(b+8)^2$

5.3

a. $(a+14)^2$
b. $(-b+5)^2$
c. $(a-15)^2$
d. $(-b-2)^2$
e. $(-a+10)^2$

5.4

a. $(2a+5)^2$
b. $(3a-6)^2$
c. $(11a+2)^2$
d. $(4a-9)^2$
e. $(13a+14)^2$

5.5

a. $(5b+2)^2$
b. $(2a-3)^2$
c. $(9b+7)^2$
d. $(4a-3)^2$
e. $(8b+1)^2$

5.6

a. $(2a+5b)^2$
b. $(3a-13b)^2$
c. $(a+2b)^2$
d. $(2a-b)^2$
e. $(6a+7b)^2$

5.7

a. $(12a-5b)^2$
b. $(-2a+b)^2$
c. $(7a-5b)^2$
d. $(-14a+3)^2$
e. $(a+11b)^2$

5.8

a. $(a^2+5)^2$
b. $(a^2-3)^2$
c. $(b^2-1)^2$
d. $(a^3+2)^2$
e. $(b^4-7)^2$

5.9

a. $(2a+7b)^2$
b. $(3a+8b)^2$
c. $(5a-9b)^2$
d. $(7a-8b)^2$
e. $(6a-11b)^2$

5.10

a. $(a^2+3)^2$
b. $(b^2-4)^2$
c. $(2a^3-13)^2$
d. $(5b^2+14)^2$
e. $(-12a^3-5)^2$

5.11

a. $(2a^2-3b)^2$
b. $(3a^2+2b)^2$
c. $(9a^2-5b^2)^2$
d. $(12a^3+2b^2)^2$
e. $(20a^2-6b^3)^2$

5.12

a. $(2a+3)^2+(a-1)^2$
b. $(a-5)^2-(a+4)^2$
c. $(3a-1)^2-(2a-3)^2$
d. $(2a+b)^2+(a+2b)^2$
e. $(-7a^2+9b^2)^2-(9a^2-7b^2)^2$

## The square of a sum or a difference

Some particular instances of the banana formula are used so often that they are referred to as *special products*.

The first two special products treated here only differ in sign. Actually, the second one is superfluous, as the only difference is $b$ in the first formula being changed into $-b$. However, it is a good idea to memorize both cases.

$$\begin{aligned}(a+b)^2 &= a^2+2ab+b^2 \\ (a-b)^2 &= a^2-2ab+b^2\end{aligned}$$

They are derived from the banana formula as follows:

$$\begin{aligned}(a+b)^2 = (a+b)(a+b) = a^2+ab+ab+b^2 = a^2+2ab+b^2 \\ (a-b)^2 = (a-b)(a-b) = a^2-ab-ab+b^2 = a^2-2ab+b^2\end{aligned}$$

A fun way to apply this formula is to mentally calculate $2003^2$ and $1998^2$:

$$\begin{aligned}2003^2 &= (2000+3)^2 = 2000^2+2\times2000\times3+3^2 \\ &= 4\,000\,000+12\,000+9 = 4\,012\,009\end{aligned}$$

and

$$\begin{aligned}1998^2 &= (2000-2)^2 = 2000^2-2\times2000\times2+2^2 \\ &= 4\,000\,000-8\,000+4 = 3\,992\,004\end{aligned}$$

Algebraic applications are more important, i.e. applications where formulas are rewritten in a different form. Here are some examples:

$(a+4)^2 = a^2+8a+16$
$(a-2b)^2 = a^2-4ab+4b^2$
$(2a+3b)^2 = 4a^2+12ab+9b^2$

Decompose the following expressions:

5.13
a. $a^2 - 16$
b. $a^2 - 1$
c. $a^2 - 144$
d. $a^2 - 81$
e. $a^2 - 121$

5.14
a. $a^2 - 36$
b. $a^2 - 4$
c. $a^2 - 169$
d. $a^2 - 256$
e. $a^2 - 1024$

5.15
a. $4a^2 - 9$
b. $9a^2 - 1$
c. $16a^2 - 25$
d. $25a^2 - 81$
e. $144a^2 - 169$

5.16
a. $36a^2 - 49$
b. $64a^2 - 121$
c. $400a^2 - 441$
d. $196a^2 - 225$
e. $144a^2 - 49$

5.17
a. $a^2 - b^2$
b. $4a^2 - 25b^2$
c. $9a^2 - b^2$
d. $16a^2 - 81b^2$
e. $196a^2 - 169b^2$

5.18
a. $a^2b^2 - 4$
b. $a^2b^2 - 625$
c. $9a^2b^2 - 25c^2$
d. $25a^2 - 16b^2c^2$
e. $100a^2b^2 - 9c^2$

5.19
a. $a^4 - b^2$
b. $25a^4 - 16b^2$
c. $16a^4 - b^4$
d. $81a^4 - 16b^4$
e. $256a^4 - 625b^4$

5.20
a. $a^4b^2 - 1$
b. $a^2b^4 - c^2$
c. $a^4 - 81b^4c^4$
d. $a^8 - b^8$
e. $256a^8 - b^8$

5.21
a. $a^3 - a$
b. $8a^2 - 50$
c. $27a^2 - 12b^2$
d. $125a^3 - 45a$
e. $600a^5 - 24a^3$

5.22
a. $3a^2b^3 - 27b$
b. $128a^3b^3 - 18ab$
c. $a^6b^3 - a^2b$
d. $-5a^3b^3c + 125abc$
e. $3a^2b - 3b$

5.23
a. $a^5 - a$
b. $2a^5 - 32a$
c. $a^5b^5 - 81ab$
d. $-a^7 + 625a$
e. $a^9b - 256ab^9$

5.24
a. $(a+3)^2 - (a+2)^2$
b. $(2a-1)^2 - (a+2)^2$
c. $(a+5)^2 - (2a+3)^2$
d. $(a+1)^2 - (3a-1)^2$
e. $(2a+1)^2 - (3a+2)^2$

Expand the brackets:

5.25
a. $(a-2)(a+2)$
b. $(a+7)(a-7)$
c. $(a-3)(a+3)$
d. $(a+12)(a-12)$
e. $(a-11)(a+11)$

5.26
a. $(2a-5)(2a+5)$
b. $(3a-1)(3a+1)$
c. $(4a+3)(4a-3)$
d. $(9a-12)(9a+12)$
e. $(13a+14)(13a-14)$

## The difference of two squares

The following special product concerns the difference of two squares:

$$a^2 - b^2 = (a + b)(a - b)$$

This formula can also be derived directly from the banana formula:

$$(a + b)(a - b) = a^2 - ab + ab - b^2 = a^2 - b^2$$

As a fun exercise, we mentally calculate:

$$2003 \times 1997 = 2000^2 - 3^2 = 4\,000\,000 - 9 = 3\,999\,991$$

Here again, the algebraic applications are more important, i.e. applications where formulas are rewritten in a different form. Examples:

$$a^2 - 25 = (a + 5)(a - 5)$$
$$4a^2b^2 - 1 = (2ab + 1)(2ab - 1)$$
$$a^6 - 9b^6 = (a^3 + 3b^3)(a^3 - 3b^3)$$

Here, the left-hand side, which in each case is the difference of two squares, is *decomposed into two factors*. This special product can also be used the other way round, to write the product of two terms that only differ in a minus sign as the difference of two squares. In daily mathematical practice, this method is also used very often. Examples:

$$(a + 2b)(a - 2b) = a^2 - 4b^2$$
$$(3a + 5)(3a - 5) = 9a^2 - 25$$
$$(a^2 - b^2)(a^2 + b^2) = a^4 - b^4$$

On the previous page and the following pages, you will find a large number of exercises to practice these skills.

Expand the brackets:

5.27

a. $(6a-9)(6a+9)$
b. $(15a-1)(15a+1)$
c. $(7a-8)(7a+8)$
d. $(16a+5)(16a-5)$
e. $(21a+25)(21a-25)$

5.28

a. $(a^2-5)(a^2+5)$
b. $(a^2+9)(a^2-9)$
c. $(2a^2-3)(2a^2+3)$
d. $(6a^2-5)(6a^2+5)$
e. $(9a^2-11)(9a^2+11)$

5.29

a. $(a^3-4)(a^3+4)$
b. $(a^5+10)(a^5-10)$
c. $(9a^2+2)(9a^2-2)$
d. $(11a^4-3)(11a^4+3)$
e. $(12a^6+13)(12a^6-13)$

5.30

a. $(2a+3b)(2a-3b)$
b. $(6a-10b)(6a+10b)$
c. $(9a+2b)(9a-2b)$
d. $(7a-5b)(7a+5b)$
e. $(a-20b)(a+20b)$

5.31

a. $(a^2+b)(a^2-b)$
b. $(2a^2+3b)(2a^2-3b)$
c. $(5a^2-3b^2)(5a^2+3b^2)$
d. $(6a^2-11b^2)(6a^2+11b^2)$
e. $(13a^2+15b^2)(13a^2-15b^2)$

5.32

a. $(a^3+2b^2)(a^3-2b^2)$
b. $(2a^2+9b^3)(2a^2-9b^3)$
c. $(5a^4+3b^3)(5a^4-3b^3)$
d. $(7a^2-19b^4)(7a^2+19b^4)$
e. $(15a^5-8b^4)(15a^5+8b^4)$

5.33

a. $(2ab+c)(2ab-c)$
b. $(3a^2b+2c)(3a^2b-2c)$
c. $(5ab^2+c^2)(5ab^2-c^2)$
d. $(9a^2b^2-4c^2)(9a^2b^2+4c^2)$
e. $(18a^3b^2-7c^3)(18a^3b^2+7c^3)$

5.34

a. $(2a^2-3bc^2)(2a^2+3bc^2)$
b. $(7a^3b-8c^3)(7a^3b+8c^3)$
c. $(13a^5b^3+14c^5)(13a^5b^3-14c^5)$
d. $(5abc+1)(5abc-1)$
e. $(9a^2bc^3+7)(9a^2bc^3-7)$

## Mixed exercises on expanding brackets

5.35

a. $(a+4)^2$
b. $(a+4)(a-4)$
c. $(a+4)(a+3)$
d. $4(a+3)$
e. $(a-4)(a+3)$

5.36

a. $(a-7)(a+6)$
b. $(a+7)^2$
c. $(a-6)(a+6)$
d. $(a-6)^2$
e. $(2a+6)(a-6)$

5.37

a. $(a+13)^2$
b. $(a-14)^2$
c. $(a+13)(a-14)$
d. $(a-13)(3a+13)$
e. $(13a-14)(14a+13)$

5.38

a. $(2a+8)^2$
b. $(a-8)(a-2)$
c. $2a(a-8)+a(a-2)$
d. $(2a-8)(2a+8)$
e. $(2a+4)(a+2)$

5.39

a. $(a-17)(a+4)$
b. $(a-17)^2$
c. $(a+17)(a-4)$
d. $(4a-17)(4a+17)$
e. $(4a+17)(17a-4)$

5.40

a. $(a+21)^2$
b. $(a+21)(a-12)$
c. $(21a-12)(21a+12)$
d. $(a-12)^2$
e. $(12a-21)(a+12)$

5.41

a. $(a^2-4)(a^2+2a+1)$
b. $(a-2)(a+2)(a+1)^2$
c. $((a-1)(a+1))^2$
d. $(4a^2+24a+9)(a^2-1)$
e. $(a-1)(a+1)(2a+3)^2$

5.42

a. $(a^2+2a+1)(a^2-2a+1)$
b. $(a+1)^2(a-1)^2$
c. $(a^2-1)^2$
d. $(2a+3)^2(2a-3)^2$
e. $(a+1)^4$

5.43

a. $(a^2+1)(a-1)(a+1)$
b. $2a(2a+3)(2a-3)$
c. $(a-2)(a^2+4)(a+2)$
d. $6a^2(3a^2+2)(3a^2-2)$
e. $2a(a-5)(a^2+25)(a+5)$

5.44 Mentally calculate:

a. $17 \cdot 23$
b. $45 \cdot 55$
c. $69 \cdot 71$
d. $93 \cdot 87$
e. $66 \cdot 74$

5.45

a. $(a+1)^2+(a+5)^2$
b. $(a+5)(a-5)+(a-1)^2$
c. $(a+1)(a+5)-(a-1)(a-5)$
d. $(5a+1)(a-1)+(a-5)(a+1)$
e. $(5a-1)(5a+1)-(5a-1)^2$

5.46

a. $(3a-7)(3a+7)-(3a-7)^2$
b. $3a(3a+7)-7a(3a+7)$
c. $(9a+2)^2-(a^2-2)(a^2+2)$
d. $(a^2+2)(a^2+3)-(a^2-2)^2$
e. $(a^2-1)(a^2+1)+(a^2+1)^2$

5.47

a. $(a-1)(a+1)(a+2)(a-2)$
b. $(a+5)(a-4)(a-5)(a+4)$
c. $(a^2+1)(a^2-1)(a^2+2)(a^2-2)$
d. $(a+2)(a+1)^2$
e. $(a+2)^3$

5.48

a. $2a(a+1)^2-3a(a+3)^2$
b. $-a(a+2)(a-2)+a(a+2)^2$
c. $2a(a+2)(a+3)-3a(a-2)(a-3)$
d. $5a(a-5)^2+25(a+5)(a-5)$
e. $a^2(a+3)(a-1)-(a^2+1)(a^2-3)$

# 6 Fractions with letters

Split into fractions with only one term in the numerator (see the first example on the following page):

6.1

a. $\dfrac{a+3}{a-3}$

b. $\dfrac{2a+3b}{a-b}$

c. $\dfrac{a^2+3a+1}{a^2-3}$

d. $\dfrac{2a-b+3}{ab-3}$

e. $\dfrac{2-5a}{b-a^3}$

6.2

a. $\dfrac{a^2+b^2}{a^2-b^2}$

b. $\dfrac{ab+bc-ca}{a-2b}$

c. $\dfrac{b^2-1}{a^2-1}$

d. $\dfrac{4abc+5}{c-ab}$

e. $\dfrac{5ab^2-abc}{ab-c}$

Write the fractions with a common denominator, then perform the addition or subtraction and finally expand all the brackets:

6.3

a. $\dfrac{1}{a-3}-\dfrac{1}{a+3}$

b. $\dfrac{1}{a-3}+\dfrac{1}{a+3}$

c. $\dfrac{2}{a-3}-\dfrac{1}{a+3}$

d. $\dfrac{1}{a-3}+\dfrac{a}{a+3}$

e. $\dfrac{a}{a-3}-\dfrac{a}{a+3}$

6.4

a. $\dfrac{a+1}{a-2}-\dfrac{a-1}{a+3}$

b. $\dfrac{a+1}{a-1}+\dfrac{a-1}{a+1}$

c. $\dfrac{a}{a+4}-\dfrac{a}{a+3}$

d. $\dfrac{3a-5}{a-1}+\dfrac{2a+3}{a-2}$

e. $\dfrac{4-a}{4+a}-\dfrac{2+a}{2-a}$

6.5

a. $\dfrac{a}{a-b}-\dfrac{b}{a-2b}$

b. $\dfrac{1}{a-b}+\dfrac{1}{a+b}$

c. $\dfrac{2}{a-b}-\dfrac{2a}{a-2}$

d. $\dfrac{1}{a-b}+\dfrac{a}{2a+3b}$

e. $\dfrac{a+b}{a-3}-\dfrac{a-b}{a+3}$

6.6

a. $\dfrac{a+b}{a-c}-\dfrac{a-b}{a+c}$

b. $\dfrac{2a+1}{a-b}+\dfrac{a-2}{a+b}$

c. $\dfrac{4-a}{a+4b}-\dfrac{ab}{4a+b}$

d. $\dfrac{a-5c}{b-c}+\dfrac{2b+3}{a-b}$

e. $\dfrac{a}{4+a+b}-\dfrac{2+a}{4-a+b}$

## Splitting and writing with a common denominator

Fractions can also contain letters. Examples:

$$\frac{a+3b}{2a-5c}, \quad \frac{b}{a^2-1}, \quad \frac{a+b}{1+a^2+b^2}$$

They become ordinary fractions as soon as you substitute the letters by numbers. The only thing you have to avoid, is the denominator becoming zero. For example in the first fraction you cannot have $a = 5$ and $c = 2$, and in the second fraction you have to exclude $a = 1$ or $a = -1$. From now on, this precondition will not be mentioned any more, and we assume that the numerical values of the letters, if chosen, stay away from these 'forbidden' areas.

Calculating with fractions which contain letters is not that different from calculating with ordinary fractions. Very often, fractions are split or written with a common denominator as an intermediate step in adding or subtracting fractions. Below are some examples; the first one of splitting:

$$\frac{a+3b}{2a-5c} = \frac{a}{2a-5c} + \frac{3b}{2a-5c}$$

This is true when you substitute the letters by numbers (of course, unless the denominator becomes zero). For instance, if $a = 4$, $b = 3$, $c = 1$ the formula yields

$$\frac{4+3\times 3}{2\times 4-5\times 1} = \frac{4}{2\times 4-5\times 1} + \frac{3\times 3}{2\times 4-5\times 1}$$

which is correct, since $\frac{13}{3} = \frac{4}{3} + \frac{9}{3}$. In the following examples the fractions are written with a common denominator first and then combined. This can again be checked using numerical substitutions.

$$\frac{a}{b} - \frac{b}{a} = \frac{a^2}{ab} - \frac{b^2}{ab} = \frac{a^2-b^2}{ab}$$

$$\frac{1}{a-1} - \frac{1}{a+1} = \frac{a+1}{(a-1)(a+1)} - \frac{a-1}{(a-1)(a+1)} = \frac{2}{a^2-1}$$

$$\begin{aligned}\frac{a+3b}{2a-5} + \frac{b}{a^2-1} &= \frac{(a+3b)(a^2-1)}{(2a-5)(a^2-1)} + \frac{b(2a-5)}{(a^2-1)(2a-5)}\\ &= \frac{(a+3b)(a^2-1)+b(2a-5)}{(2a-5)(a^2-1)}\end{aligned}$$

In the last example, you can also expand the brackets in the numerator and the denominator.

Simplify the following fractions as much as possible:

6.7

a. $\dfrac{3a+18}{9b-6}$

b. $\dfrac{a^2+a}{a+1}$

c. $\dfrac{4a-2}{2a^2-a}$

d. $\dfrac{a+2b}{a^2-4b^2}$

e. $\dfrac{ab+b^3}{b^2-3b}$

6.8

a. $\dfrac{a^2b+ab^2}{3abc}$

b. $\dfrac{a^2-4a}{a+2a^2}$

c. $\dfrac{4ab-3ab^2}{a^2-abc}$

d. $\dfrac{a^2+2ab+b^2}{a^2-b^2}$

e. $\dfrac{a^4-b^2}{a^2-b}$

Write with a common denominator, combine and finally simplify the result as much as possible:

6.9

a. $\dfrac{1}{a-3}-\dfrac{1}{a^2-9}$

b. $\dfrac{1}{a-3}-\dfrac{a}{a^2-9}$

c. $\dfrac{a^2+1}{a-3}-\dfrac{a^2-1}{a+3}$

d. $\dfrac{b}{a-b}+\dfrac{a}{b-a}$

e. $\dfrac{a^2-1}{a-1}-\dfrac{a^2+1}{a+1}$

6.10

a. $\dfrac{a+b}{a-2b}-\dfrac{a-2b}{a+b}$

b. $\dfrac{a^2+ab}{a^2-b^2}+a-1$

c. $\dfrac{a}{a^2-4}-\dfrac{2}{4-a^2}$

d. $\dfrac{3a-2b}{a-b}+\dfrac{2a+3b}{3a}$

e. $\dfrac{4-a}{a}-\dfrac{4+a}{2a}$

## Simplifying fractions

Just as with ordinary fractions, fractions with letters can sometimes be simplified by dividing the numerator and denominator by a common factor:

$$\frac{3a+9b^2}{6a-3} = \frac{a+3b^2}{2a-1}$$

Here, the numerator and denominator are divided by 3. Dividing by a common letter is also possible:

$$\frac{7b}{b+2b^3} = \frac{7}{1+2b^2}$$

However, there is a caveat. We divided the numerator and denominator by $b$, but this is only permitted if $b \neq 0$. The left-hand side is not defined for $b = 0$ (as that would yield $\frac{0}{0}$), while on the right-hand side, $b = 0$ would give 7 as a result. To be accurate, we should say

$$\frac{7b}{b+2b^3} = \frac{7}{1+2b^2} \quad \text{if} \quad b \neq 0$$

Another example:

$$\frac{a^2-4}{a-2} = \frac{(a-2)(a+2)}{a-2} = a+2 \quad \text{if} \quad a \neq 2$$

Here, the numerator is first decomposed using the special product $a^2 - 4 = (a-2)(a+2)$, after which one of the factors is divided away. Of course, on the condition that this factor is not zero, hence $a \neq 2$.

In the following example, the condition is somewhat more complicated as it involves two letters:

$$\frac{a^2-b^2}{a+b} = \frac{(a-b)(a+b)}{a+b} = a-b \quad \text{if} \quad a+b \neq 0$$

The condition $a + b \neq 0$ now yields an infinite number of combinations of $a$ and $b$, for which the left-hand side yields $\frac{0}{0}$ and thus is not defined, while the right-hand side just represents a finite number. For instance, if $a = 1$ and $b = -1$, the left-hand side yields $\frac{0}{0}$, while the right-hand side equals 2. If $a = -137$ and $b = 137$, the right-hand side becomes $-274$, while the left-hand side, again, is $\frac{0}{0}$.

# III Sequences

$$\frac{1}{2} - \frac{1}{4} + \frac{1}{8} - \frac{1}{16} + \frac{1}{32} - \frac{1}{64} + \cdots = \frac{1}{3}$$

When trying to generalize the formula $(a+b)^2 = a^2 + 2ab + b^2$ to formulas for $(a+b)^3$, $(a+b)^4$, ..., a pattern soon emerges, which can be visualized in *Pascal's triangle*. The numbers in this triangle are the *binomial coefficients*. We will show how these can be calculated quickly and easily. *Newton's binomial formula* expresses $(a+b)^n$ in terms of the binomial coefficients. Here, we introduce the frequently used *sigma notation* for sums of sequences. In the next chapter, we will tackle the so-called arithmetic and geometric sequences and their summation formulas. And finally, we will explain what in general is meant by the *limit* of an infinite numerical sequence.

# 7 Factorials and binomial coefficients

Using the formulas on the next page, expand and if possible simplify the following expressions:

7.1

a. $(a+1)^3$
b. $(a-1)^3$
c. $(2a-1)^3$
d. $(a+2)^3$
e. $(2a-3)^3$

7.2

a. $(1-a^2)^3$
b. $(ab+1)^3$
c. $(a+2b)^3$
d. $(a^2-b^2)^3$
e. $(2a-5b)^3$

7.3

a. $(2a-1)^3+(a-2)^3$
b. $(a-2b)^3$
c. $(a+3b)^3$
d. $(5a+2)^3$
e. $(a-7)^3+(a+7)^3$

7.4

a. $(a^2-b)^3$
b. $(a^4+2b^2)^3$
c. $(a+2b)^3+(a-2b)^3$
d. $(a+2b)^3-(a-2b)^3$
e. $(a+2b)^3-(2a+b)^3$

7.5

a. $(a+1)^4$
b. $(a-1)^4$
c. $(2a-1)^4$
d. $(a+2)^4$
e. $(2a-3)^4$

7.6

a. $(1-a^2)^4$
b. $(ab+1)^4$
c. $(a+2b)^4$
d. $(a^2-b^2)^4$
e. $(a-b)^4+(a+b)^4$

## The formulas for $(a+b)^3$ and $(a+b)^4$

The special product $(a+b)^2 = a^2 + 2ab + b^2$ is the starting point for the derivation of formulas for$(a+b)^n$ with $n > 2$. Let's start with $n = 3$:

$$\begin{aligned}
(a+b)^3 &= (a+b)(a+b)^2 \\
&= (a+b)(a^2+2ab+b^2) \quad \text{(special product)} \\
&= a(a^2+2ab+b^2) + b(a^2+2ab+b^2) \\
&= a^3 + 2a^2b + \phantom{2}ab^2 \\
&\phantom{= a^3} + \phantom{2}a^2b + 2ab^2 + b^3 \\
&= a^3 + 3a^2b + 3ab^2 + b^3
\end{aligned}$$

You can see how we used the special product for $(a+b)^2$ and subsequently expanded the brackets step by step. In the fourth and fifth line, we lined up similar terms under each other, which made adding up in the sixth line easier. The result is the formula

$$(a+b)^3 = a^3 + 3a^2b + 3ab^2 + b^3$$

Using this formula, we can calculate for $n = 4$ in a similar way:

$$\begin{aligned}
(a+b)^4 &= (a+b)(a^3+3a^2b+3ab^2+b^3) \quad \text{(formula for } (a+b)^3\text{)} \\
&= a(a^3+3a^2b+3ab^2+b^3) + b(a^3+3a^2b+3ab^2+b^3) \\
&= a^4 + 3a^3b + 3a^2b^2 + \phantom{3}ab^3 \\
&\phantom{= a^4} + \phantom{3}a^3b + 3a^2b^2 + 3ab^3 + b^4 \\
&= a^4 + 4a^3b + 6a^2b^2 + 4ab^3 + b^4
\end{aligned}$$

which yields the formula

$$(a+b)^4 = a^4 + 4a^3b + 6a^2b^2 + 4ab^3 + b^4$$

## Exercises using Pascal's triangle

7.7 Add the next three rows ($n = 8, n = 9, n = 10$) to Pascal's triangle on the next page.

7.8 Use the previous exercise to expand the brackets in $(a+1)^8$, $(a-1)^9$ and $(a-b)^{10}$.

7.9 Adding up all the entries of the $n$-th row of Pascal's triangle yields $2^n$ as a result. Check this for $n = 1$ to $n = 10$ first and then explain this by substituting $a = 1$ and $b = 1$ in the formula for $(a+b)^n$.

7.10 If the entries in one of the rows of Pascal's triangle are alternately given plus and minus signs and added up, the result will always be 0. First, check this for $n = 1$ to $n = 10$, and then explain this by substituting $a = 1$ and $b = -1$ in the formula for $(a+b)^n$.

7.11 In Pascal's triangle, replace every even number by 0 and every odd number by 1. Draw the first 20 rows of the resulting 'binary' Pascal triangle and explain the patterns that appear. You will get a nice looking 'sieve pattern' if all the zeros are left open and all the ones are replaced by an asterisk.

## Binomial coefficients and Pascal's triangle

Up to now we have derived the following formulas:

$$\begin{aligned}(a+b)^2 &= a^2+2ab+b^2\\ (a+b)^3 &= a^3+3a^2b+3ab^2+b^3\\ (a+b)^4 &= a^4+4a^3b+6a^2b^2+4ab^3+b^4\end{aligned}$$

We have arranged the terms in descending powers of $a$ and ascending powers of $b$. Step by step, the power of $a$ is decreased, while the power of $b$ is increased by 1. The integer coefficients are called *binomial coefficients*. For $n = 2$ they are 1, 2 and 1, for $n = 3$ they are 1, 3, 3 and 1, and for $n = 4$ they are 1, 4, 6, 4 and 1. Of course, the coefficients equal to 1 are omitted from the formulas: we write $a^2$ instead of $1a^2$ and so on. But they do appear in *Pascal's triangle*, which gives all the binomial coefficients:

$$\begin{array}{ccccccccccccccc|l}
 & & & & & & & 1 & & & & & & & & \leftarrow n=0\\
 & & & & & & 1 & & 1 & & & & & & & \leftarrow n=1\\
 & & & & & 1 & & 2 & & 1 & & & & & & \leftarrow n=2\\
 & & & & 1 & & 3 & & 3 & & 1 & & & & & \leftarrow n=3\\
 & & & 1 & & 4 & & 6 & & 4 & & 1 & & & & \leftarrow n=4\\
 & & 1 & & 5 & & 10 & & 10 & & 5 & & 1 & & & \leftarrow n=5\\
 & 1 & & 6 & & 15 & & 20 & & 15 & & 6 & & 1 & & \leftarrow n=6\\
1 & & 7 & & 21 & & 35 & & 35 & & 21 & & 7 & & 1 & \leftarrow n=7\\
 & & & & & & & \dots & & & & & & & & \dots
\end{array}$$

*Pascal's triangle*

This gives e.g. the following formula:

$$(a+b)^6 = a^6+6a^5b+15a^4b^2+20a^3b^3+15a^2b^4+6ab^5+b^6$$

At the top we also included $n = 0$ and $n = 1$, corresponding with $(a+b)^0 = 1$ and $(a+b)^1 = a+b$, respectively. Pascal's triangle is constructed according to the following rule:

*The left and right borders are filled with 1-s and every other entry equals the sum of its right and left upper neighbour.*

If you have followed the derivations on page 53 for $n = 3$ and $n = 4$, you will understand why this generic rule is valid. It allows you to expand Pascal's triangle as far as you like.

Calculate the given binomial coefficients $\binom{n}{k}$ using the formula on the next page. Start by applying the simplifications described underneath the formula and then also strike out all the factors in the denominator against corresponding factors in the numerator (this is always possible, since binomial coefficients are integers!). The final calculation then should be very easy. Check your results for $n \leq 10$ using Pascal's triangle.

7.12

a. $\binom{4}{2}$

b. $\binom{5}{0}$

c. $\binom{4}{4}$

d. $\binom{5}{3}$

e. $\binom{6}{3}$

7.13

a. $\binom{7}{1}$

b. $\binom{6}{4}$

c. $\binom{7}{5}$

d. $\binom{7}{2}$

e. $\binom{7}{7}$

7.14

a. $\binom{8}{2}$

b. $\binom{9}{3}$

c. $\binom{9}{8}$

d. $\binom{8}{4}$

e. $\binom{8}{5}$

7.15

a. $\binom{8}{3}$

b. $\binom{9}{4}$

c. $\binom{9}{7}$

d. $\binom{7}{3}$

e. $\binom{9}{6}$

7.16

a. $\binom{12}{0}$

b. $\binom{15}{14}$

c. $\binom{13}{5}$

d. $\binom{21}{2}$

e. $\binom{18}{14}$

7.17

a. $\binom{12}{7}$

b. $\binom{11}{5}$

c. $\binom{48}{2}$

d. $\binom{49}{3}$

e. $\binom{50}{48}$

7.18

a. $\binom{17}{3}$

b. $\binom{51}{50}$

c. $\binom{12}{9}$

7.19

a. $\binom{42}{3}$

b. $\binom{13}{6}$

c. $\binom{27}{5}$

7.20

a. $\binom{78}{75}$

b. $\binom{14}{5}$

c. $\binom{28}{4}$

## Calculating binomial coefficients

Using Pascal's triangle, you can find all the binomial coefficients. For big values of $n$, however, this might be very time consuming. Since binomial coefficients are important for many applications (e.g. probability theory and statistics), it is useful to have a formula to calculate them in a more direct way. But, before we look at the formula, we need to establish some notations first.

The horizontal row for $n = 2$ of Pascal's triangle contains three coefficients: 1, 2 and 1. In general, on the $n$-th row there are $n + 1$ coefficients. We number these consecutively from left to right, from 0 to $n$. The $k$-th coefficient on the $n$-th row is denoted by $\binom{n}{k}$, pronounced as '$n$ over $k$'. Examples:

$$\binom{3}{0} = 1, \quad \binom{3}{1} = 3, \quad \binom{4}{2} = 6, \quad \binom{6}{6} = 1 \quad \text{and} \quad \binom{7}{4} = 35$$

Another notation that is frequently used, is $k!$, pronounced as '*k-factorial*'. We define:

$$\begin{aligned} 0! &= 1 \\ k! &= 1 \times \cdots \times k \quad \text{for each positive integer } k \end{aligned}$$

Examples: $1! = 1, 2! = 1 \times 2 = 2, 3! = 1 \times 2 \times 3 = 6$ and $7! = 1 \times 2 \times 3 \times 4 \times 5 \times 6 \times 7 = 5040$.

With these notations the following formula is valid:

$$\binom{n}{k} = \frac{n!}{k!(n-k)!}$$

*Warning:* when you calculate binomial coefficients, you *never* explicitly calculate these factorials! In the fraction on the right-hand side, you can always strike out the greatest factorial in the denominator against the first part of the factorial $n!$ in the numerator. Since $k!$ increases quickly with $k$, it is advisable to apply this simplification, even for small values of $n$. Examples:

$$\binom{7}{3} = \frac{7!}{3!\,4!} = \frac{1 \times 2 \times \cdots \times 7}{(1 \times 2 \times 3) \times (1 \times 2 \times 3 \times 4)} = \frac{7 \times 6 \times 5}{1 \times 2 \times 3} = 35$$

This simplification is easy to remember: the numerator contains just as many terms as the denominator (three, in this case). In the denominator, you end up with one factorial (3! in this case) and in the numerator the terms count down from $n$ (in this case $7 \times 6 \times 5$).
You should memorize the following special cases:

$$\binom{n}{0} = \binom{n}{n} = 1, \quad \binom{n}{1} = \binom{n}{n-1} = n, \quad \binom{n}{2} = \binom{n}{n-2} = \frac{1}{2}n(n-1)$$

Write in the sigma notation using Newton's binomial formula:

7.21

a. $(a+1)^7$
b. $(a-1)^{12}$
c. $(a+10)^{12}$
d. $(2a-1)^9$
e. $(2a+b)^{10}$

7.22

a. $(a+5)^7$
b. $(1-a)^5$
c. $(ab+1)^{18}$
d. $(a+2b)^9$
e. $(a-b)^8$

Calculate the following sums using Newton's binomial formula, in each case choosing suitable values for $a$ and $b$.
Example:

$$\sum_{k=0}^{5} \binom{5}{k} = \binom{5}{0} + \binom{5}{1} + \cdots + \binom{5}{5} = (1+1)^5 = 2^5 = 32$$

(Binomial formula for $(a+b)^5$ with $a = b = 1$.)

7.23

a. $\sum_{k=0}^{8} \binom{8}{k}$
b. $\sum_{k=0}^{8} \binom{8}{k} (-1)^k$
c. $\sum_{k=0}^{8} \binom{8}{k} 2^k$

7.24

a. $\sum_{k=0}^{8} \binom{8}{k} (-2)^k$
b. $\sum_{k=0}^{n} \binom{n}{k}$
c. $\sum_{k=0}^{n} \binom{n}{k} (-1)^k$

The following exercises are meant to familiarize you with the sigma notation. Calculate the given sums:

7.25

a. $\sum_{k=0}^{6} k^2$
b. $\sum_{k=-4}^{4} k^3$
c. $\sum_{k=3}^{7} (2k+4)$

7.26

a. $\sum_{j=-1}^{4} (j^2-1)$
b. $\sum_{j=1}^{3} (j+\frac{1}{j})$
c. $\sum_{j=2}^{5} j^4$

## Newton's binomial formula and the sigma notation

The $n$-th row of Pascal's triangle contains the binomial coefficients $\binom{n}{0}, \binom{n}{1}, \ldots, \binom{n}{n}$, which gives:

$$(a+b)^n = \binom{n}{0}a^n + \binom{n}{1}a^{n-1}b + \cdots + \binom{n}{n-1}ab^{n-1} + \binom{n}{n}b^n$$

This formula is known as *Newton's binomial formula*.

Literally, binomium means a set of *two terms*. This refers to the two terms $a$ and $b$ between the brackets on the left-hand side. The binomial coefficients can be calculated using Pascal's triangle or using the formula on page 57.

The $n+1$ terms on the right-hand side of the formula above all have the same form, namely the product of a binomial coefficient $\binom{n}{k}$, a power of $a$ and a power of $b$. Even the first term $\binom{n}{0}a^n$ and the last term $\binom{n}{n}b^n$ have this form, since the 'missing' powers of $b$ and $a$ can be written as $b^0$ and $a^0$, respectively.

All terms, therefore, have the form $\binom{n}{k}a^{n-k}b^k$, in which $k$ takes all values from 0 up to $n$. In such cases, where a number of similar terms have to be added up, and where only an 'index' $k$ changes from term to term, a notation with a Greek capital $\Sigma$ (sigma) is often used. In this notation, Newton's binomial formula can be written as

$$(a+b)^n = \sum_{k=0}^{n} \binom{n}{k} a^{n-k}b^k$$

Next to the capital sigma, the general expression of a term is written involving the letter $k$ (the so-called *summation index*). Underneath and above the sigma, the first and last values of $k$ are given. Of course, you can use any other letter instead of $k$.

Here are two more examples of the use of the sigma notation for a sum:

$$\sum_{k=1}^{5} k^2 = 1^2 + 2^2 + 3^2 + 4^2 + 5^2$$

$$\sum_{j=-2}^{2} 3^j = 3^{-2} + 3^{-1} + 3^0 + 3^1 + 3^2$$

# 8 Sequences and limits

8.1 Calculate the sum of the following sequences:

a. All positive integers from 1 up to 2003.
b. All positive three-digit integers.
c. All odd integers between 1000 and 2000.
d. All positive integers with a maximum of three digits, ending in 3.
e. All positive four-digit integers ending in 2 or 7.
f. All positive four-digit integers ending in 6 or 7.

8.2 Calculate the following sums:

a. $\sum_{k=1}^{20}(3k+2)$

b. $\sum_{k=10}^{70}(7k-2)$

c. $\sum_{k=3}^{30}(8k+7)$

d. $\sum_{k=0}^{14}(5k+3)$

e. $\sum_{k=-2}^{22}(100k+10)$

8.3 The runners in a marathon wear serial numbers running from 1 to 97. One of the runners notices that the sum of all the even serial numbers equals the sum of all the odd serial numbers. However, in this sums he didn't include his own number. What is his serial number? (Dutch mathematical olympiad, first round, 1997.)

## Arithmetic sequences

An *arithmetic sequence* is a sequence of numbers $a_1, a_2, a_3, \ldots$ where the difference $a_{k+1} - a_k$ between consecutive terms is constant. The most simple example is the sequence $1, 2, 3, 4, 5, \ldots$ of all the positive integers. Here, the difference is always equal to 1. Another example is given by the sequence $3, 8, 13, 18, 23, 28, \ldots$. In this case, the common difference is 5.

Legend has it that, when he was still in school, the great mathematician Gauss was asked to add up all integers from 1 up to 100. He amazed his teacher by instantly giving the answer: 5050. His idea was to mentally write down the sum twice, first from 1 up to 100, and then in reverse order. Vertically, you always have two numbers that add up to 101:

$$\begin{array}{rcrcccrcr} 1 & + & 2 & + & \cdots & + & 99 & + & 100 \\ 100 & + & 99 & + & \cdots & + & 2 & + & 1 \\ \hline 101 & + & 101 & + & \cdots & + & 101 & + & 101 \end{array}$$

There are 100 terms in the sequence, giving you $100 \times 101 = 10100$, which is twice the required sum, so this should be halved, which gives 5050. This is also how you prove that $1 + 2 + 3 + \cdots + n = \frac{1}{2}n(n+1)$ for any $n$.

Gauss's trick can be applied to any arithmetic sequence. To calculate the sum $a_1 + a_2 + \cdots + a_{n-1} + a_n$ of the first $n$ terms of such a sequence, you mentally write down the sum twice, the second time in reverse order. Added up vertically, you always get the same answer, namely $a_1 + a_n$, which means that the required sum is $\frac{1}{2}n(a_1 + a_n)$. Using the sigma notation (see page 59) we can write this result as follows:

If $a_1, a_2, a_3, \ldots$ is an arithmetic sequence, then

$$\sum_{k=1}^{n} a_k = \frac{1}{2}n(a_1 + a_n)$$

This is the *summation formula for an arithmetic sequence.*

8.4 Calculate the sum of the following geometric sequences:

a. $2 + 4 + 8 + 16 + \cdots + 256$
b. $1 + \frac{1}{2} + \frac{1}{4} + \frac{1}{8} + \cdots + \frac{1}{256}$
c. $2 + 6 + 18 + 54 + \cdots + 1458$
d. $\frac{2}{3} + \frac{4}{9} + \cdots + \frac{64}{729}$
e. $\frac{3}{10} + \frac{3}{100} + \cdots + \frac{3}{10\,000\,000}$

8.5 Calculate the sum of the following infinite geometric sequences:

a. $4 + 2 + 1 + \frac{1}{2} + \frac{1}{4} + \frac{1}{8} + \cdots$
b. $1 + \frac{2}{3} + \frac{4}{9} + \cdots$
c. $1 - \frac{7}{8} + \frac{49}{64} - \frac{343}{512} + \cdots$
d. $7 + \frac{7}{10} + \frac{7}{100} + \cdots$
e. $1 - \frac{9}{10} + \frac{81}{100} - \frac{729}{1000} + \cdots$

8.6 Calculate the sum of the following infinite geometric sequences. In each case also give the ratio $r$:

a. $0.1 - 0.01 + 0.001 - 0.0001 + \cdots$
b. $0.3 + 0.03 + 0.003 + 0.0003 + \cdots$
c. $0.9 + 0.09 + 0.009 + 0.0009 + \cdots$
d. $0.12 + 0.0012 + 0.000012 + \cdots$
e. $0.98 - 0.0098 + 0.000098 - \cdots$

8.7 Using the preceding exercise, show that

a. $0.3333\ldots = \frac{1}{3}$
b. $0.9999\ldots = 1$
c. $0.12121212\ldots = \frac{4}{33}$
d. $0.0012121212\ldots = \frac{4}{3300}$ (*Hint:* $0.0012121212\ldots = \frac{1}{100} \times 0.12121212\ldots$)
e. $10.3333\ldots = \frac{31}{3}$ (*Hint:* $10.33333\ldots = 10 + 0.33333\ldots$)

8.8 Using a calculator or a computer, for the following values of $r$ calculate the numbers $r^{100}$ and $r^{1000}$. Write your answers in the form $m \times 10^k$ with $0.1 \leq m < 1$ and integer $k$. Round off $m$ to five decimals places.

a. $r = 0.2$
b. $r = 0.5$
c. $r = 0.7$
d. $r = 0.9$
e. $r = 0.99$

## Geometric sequences

Sometimes, it is a good idea not to start the numbering of the terms of a sequence at 1, but at, for instance, 0, because this makes certain formulas become more simple. We will do this for geometric sequences.

A sequence $a_0, a_1, a_2, \ldots$ is called a *geometric sequence* with *ratio* $r$ if $a_{n+1} = a_n r$ for all $n$. Every term, except the first one, is the result of multiplication of its predecessor by $r$. An example is the sequence $1, 2, 4, 8, 16, 32, 64, \ldots$, in which each term is double its predecessor. This means that here $r = 2$.

If we call $a_0$ simply $a$, we have $a_1 = ar$, $a_2 = a_1 r = ar^2$ and so on. In general, for each $n$ we have $a_n = ar^n$. Thus, every geometric sequence can be written as

$$a, ar, ar^2, ar^3, ar^4, \ldots$$

For geometric sequences, there is also a simple formula for the sum $s_n = a + ar + ar^2 + \cdots + ar^{n-1}$ of the first $n$ terms. To find it, we start with noting that $rs_n = ar + ar^2 + ar^3 + \cdots + ar^n$ so $s_n - rs_n = a - ar^n$ (all intermediate terms cancel out in the subtraction). When $r \neq 1$, we can solve $s_n$ from this equation: $s_n = a(1 - r^n)/(1 - r)$.

Using the sigma notation, this yields the *summation formula for a (finite) geometric sequence*:

$$\sum_{k=0}^{n-1} ar^k = \frac{a(1 - r^n)}{1 - r} \quad \text{if } r \neq 1$$

Suppose that the ratio $r$ satisfies $-1 < r < 1$. As $n$ increases without bounds, the number $r^n$ gets ever closer to 0. Example: if $r = 0.95$ then $r^{100} \approx 0.0059205$ and $r^{1000} \approx 0.52918 \times 10^{-22}$. This can be written as $\lim_{n\to\infty} r^n = 0$. In words: *the limit of $r^n$ for $n$ to infinity is* 0. If we apply this to the summation formula for a geometric sequence above, this gives us the *summation formula for an infinite geometric sequence*:

$$\sum_{k=0}^{\infty} ar^k = \frac{a}{1 - r} \quad \text{if } -1 < r < 1$$

Notice the symbol $\infty$ ('infinity') above the summation sign.
Example: take $a = \frac{1}{2}$ and $r = -\frac{1}{2}$, then

$$\sum_{k=0}^{\infty} \frac{1}{2}\left(-\frac{1}{2}\right)^k = \frac{\frac{1}{2}}{1 + \frac{1}{2}} = \frac{1}{3}$$

or, written out explicitly:

$$\frac{1}{2} - \frac{1}{4} + \frac{1}{8} - \frac{1}{16} + \frac{1}{32} - \frac{1}{64} + \cdots = \frac{1}{3}$$

In the following exercises, write the given repeating decimal fraction as an ordinary irreducible fraction, assuming that the regularity shown continues indefinitely: all given decimal numbers are periodic from a certain decimal.

8.9

a. $0.222222222222\ldots$

b. $0.313131313131\ldots$

c. $1.999999999999\ldots$

d. $0.123123123123\ldots$

e. $0.123333333333\ldots$

8.10

a. $0.101010101010\ldots$

b. $0.330330330330\ldots$

c. $1.211211211211\ldots$

d. $0.000111111111\ldots$

e. $3.091919191919\ldots$

8.11

a. $22.24444444444\ldots$

b. $0.700700700700\ldots$

c. $0.699699699699\ldots$

d. $8.124444444444\ldots$

e. $1.131313131313\ldots$

8.12

a. $0.111109999999\ldots$

b. $0.365656565656\ldots$

c. $3.141514151415\ldots$

d. $2.718281828182\ldots$

e. $0.090909090909\ldots$

8.13 Using a computer or a calculator, approximate the numbers $r^{101}$ and $r^{1001}$ for the given values of $r$. Write your answers in the form $\pm m \times 10^k$ with $0.1 \leq m < 1$ and integer $k$. Round off $m$ to five decimal places.

a. $r = 1.02$

b. $r = -2$

c. $r = 10.1$

d. $r = -0.999$

e. $r = 9.99$

Using a computer or a calculator, approximate the terms $a_{101}$ and $a_{1001}$ for the given sequences $a_1, a_2, a_3, \ldots$. Write your answers in the form $m \times 10^k$ with $0.1 \leq m < 1$ and integer $k$. Round off $m$ to five decimal places. In each case, the general term $a_n$ of the sequence is given by a formula.

8.14

a. $a_n = n^2$

b. $a_n = n^{-3}$

c. $a_n = n^{-1.1}$

d. $a_n = n^{1000.1}$

e. $a_n = \sqrt{n}$

8.15

a. $a_n = n^{-0.333}$

b. $a_n = \sqrt[n]{10}$

c. $a_n = \sqrt[n]{1000}$

d. $a_n = \sqrt[n]{0.01}$

e. $a_n = \sqrt[n]{0.9}$

## Repeating decimal fractions

On page 62, we saw that repeating decimal fractions such as 0.333333... or 0.12121212... can be considered as the sum of an infinite geometric sequence. When calculating such a sum, you get a fraction, in the given examples $\frac{1}{3}$ and $\frac{4}{33}$, respectively. Even when such a decimal fraction only starts repeating from a certain moment on, it still represents a fraction. *Irrational numbers*, i.e. numbers that cannot be written as a fraction, such as $\sqrt{2}$ or $\pi$, therefore have a decimal expansion that *never* becomes periodic.

## Special limits

For many sequences $a_0, a_1, a_2, \ldots$ it is important to know what happens to the terms $a_n$ if $n$ becomes bigger and bigger, in other words, if $n$ goes to infinity. For the geometric sequences $1, r, r^2, r^3, \ldots$ with $-1 < r < 1$ we already know that $\lim_{n\to\infty} r^n = 0$. Below is a complete overview of the behaviour of $r^n$ if $n \to \infty$ for all values of $r$.

$$\begin{array}{ll} \lim\limits_{n\to\infty} r^n = \infty & \text{if } r > 1 \\ r^n = 1 & \text{if } r = 1 \\ \lim\limits_{n\to\infty} r^n = 0 & \text{if } -1 < r < 1 \\ r^n = (-1)^n = \pm 1 & \text{if } r = -1 \\ \lim\limits_{n\to\infty} |r|^n = \infty & \text{if } r < -1 \end{array}$$

Below, we list some more types of sequences whose limits you should know.

The sequence $1, 4, 9, 16, 25 \ldots$ with $a_n = n^2$ as its $n$-th term has limit infinity. In formula: $\lim\limits_{n\to\infty} n^2 = \infty$. In general

$$\lim_{n\to\infty} n^p = \infty \quad \text{for every } p > 0$$

The sequence $1, \frac{1}{4}, \frac{1}{9}, \frac{1}{16}, \frac{1}{25} \ldots$ with $a_n = \dfrac{1}{n^2}$ as its $n$-th term has limit 0. In formula: $\lim\limits_{n\to\infty} \dfrac{1}{n^2} = 0$. In general

$$\lim_{n\to\infty} \frac{1}{n^p} = 0 \quad \text{for every } p > 0$$

The sequence $2, \sqrt{2}, \sqrt[3]{2}, \sqrt[4]{2}, \sqrt[5]{2}, \ldots$ with $a_n = \sqrt[n]{2}$ as its $n$-th term has limit 1. In formula: $\lim\limits_{n\to\infty} \sqrt[n]{2} = 1$. In general

$$\lim_{n\to\infty} \sqrt[n]{p} = 1 \quad \text{for every } p > 0$$

Calculate the following limits:

8.16

a. $\lim_{n\to\infty} \frac{n+1}{n}$

b. $\lim_{n\to\infty} \frac{2n}{n+12}$

c. $\lim_{n\to\infty} \frac{n^2-1}{n^2+1}$

d. $\lim_{n\to\infty} \frac{n^2+2n+5}{3n^2-2}$

e. $\lim_{n\to\infty} \frac{n^3+3n^2}{3n^4+4}$

f. $\lim_{n\to\infty} \frac{2n^4-6n^2}{5n^4+4}$

8.17

a. $\lim_{n\to\infty} \frac{2n+\sqrt{n}}{n-\sqrt{n}}$

b. $\lim_{n\to\infty} \frac{10n+4}{n^2+4}$

c. $\lim_{n\to\infty} \frac{2n^5}{4n^4+5n^2-3}$

d. $\lim_{n\to\infty} \frac{8n^3+4n+7}{8n^3-4n-7}$

e. $\lim_{n\to\infty} \frac{n+\frac{1}{n}}{n-\frac{2}{n}}$

f. $\lim_{n\to\infty} \frac{3n^3-5n^2}{n^4+n^2-2}$

8.18

a. $\lim_{n\to\infty} \frac{n^3-1}{n^3+n^2}$

b. $\lim_{n\to\infty} \frac{2n^2}{n\sqrt{n+2}}$

c. $\lim_{n\to\infty} \frac{4n^2+5n+n\sqrt{n}}{3n^2-2n-1}$

d. $\lim_{n\to\infty} \frac{n^2+1}{n\sqrt{n^2+n}}$

e. $\lim_{n\to\infty} \frac{4n^4+1}{5n^4+1000n^3}$

8.19

a. $\lim_{n\to\infty} \left(\frac{8}{9}\right)^n$

b. $\lim_{n\to\infty} \left(-\frac{9}{8}\right)^n$

c. $\lim_{n\to\infty} \sqrt[n]{\frac{8}{9}}$

d. $\lim_{n\to\infty} \sqrt[n]{\frac{9}{8}}$

e. $\lim_{n\to\infty} 1.0001^n$

8.20

a. $\lim_{n\to\infty} \frac{n^3}{3^n}$

b. $\lim_{n\to\infty} \frac{2n^2}{1+2^n}$ (*Hint:* divide the numerator and denominator by $2^n$)

c. $\lim_{n\to\infty} \frac{n!}{3^n}$

d. $\lim_{n\to\infty} \frac{n^n}{n!}$

e. $\lim_{n\to\infty} \frac{n^n}{n^n+n!}$ (*Hint:* divide the numerator and denominator by $n^n$)

## Limits of quotients

When calculating the limits of expressions in the form of a fraction where both the numerator and the denominator go to infinity, it is often useful to divide the numerator and denominator by a 'dominating term' in the denominator (for instance the highest power). Example:

$$\lim_{n\to\infty} \frac{2n^2 - 3n + 4}{3n^2 + 4n + 9} = \lim_{n\to\infty} \frac{2 - \frac{3}{n} + \frac{4}{n^2}}{3 + \frac{4}{n} + \frac{9}{n^2}}$$

Since $\frac{3}{n} \to 0$, $\frac{4}{n^2} \to 0$, $\frac{4}{n} \to 0$ and $\frac{9}{n^2} \to 0$, we immediately see that the required limit equals $\frac{2}{3}$.

In these cases it is easy to find a dominating term: simply take the highest power. But, apart from powers, fractions can also contain other terms such as $n!$ or $2^n$. What is the dominating term in those cases? Which one 'wins' as $n$ goes to infinity? That question will be discussed now.

## Fast growing terms

Compare the following sequences:

a. $1, 2^{100}, 3^{100}, 4^{100}, 5^{100}, \ldots$ with $a_n = n^{100}$ as general term
b. $100, 100^2, 100^3, 100^4, 100^5, \ldots$ with $b_n = 100^n$ as general term
c. $1, 2, 6, 24, 120, \ldots$ with $c_n = n!$ as general term
d. $1, 2^2, 3^3, 4^4, 5^5, \ldots$ with $d_n = n^n$ as general term

All have infinity as their limit, but which sequence grows fastest? For $n = 100$, the terms $a_n$, $b_n$ and $d_n$ are equal, namely, $100^{100} = 10^{200}$, which written out is a 1 followed by 200 zeroes, while $c_n = 100! = 1 \times 2 \times 3 \times \cdots \times 100$ is clearly much smaller (it 'only' counts 158 digits). But, for $n = 1000$ matters are completely different: $a_n = 10^{300}$, $b_n = 10^{2000}$, $c_n \approx 0.40 \times 10^{2568}$, $d_n = 10^{3000}$. This pattern remains unchanged: as $n$ grows endlessly, $b_n$ grows much faster than $a_n$, $c_n$ grows much faster than $b_n$, and $d_n$ grows much faster than $c_n$. In general we have:

$$\lim_{n\to\infty} \frac{n^p}{a^n} = 0 \text{ if } a > 1, \qquad \lim_{n\to\infty} \frac{a^n}{n!} = 0, \qquad \lim_{n\to\infty} \frac{n!}{n^n} = 0$$

Above, we illustrated these limits for $p = 100$ and $a = 100$. By the way, do you see why the first limit is self-evident if $p \leq 0$? For $p > 0$ this is not the case: the numerator $n^p$ and the denominator $a^n$ both go to infinity. However, the denominator wins.

Calculate the following limits.
*Note:* for the limits on this page, it is often possible to divide both the numerator and the denominator by a suitably chosen 'dominating term' in the denominator. The solution then follows if we apply one of the three standard limits on the previous page. Example:

$$\lim_{n\to\infty} \frac{n^2+2^n}{n^2-2^n} = \lim_{n\to\infty} \frac{(n^2/2^n)+1}{(n^2/2^n)-1} = -1$$

since $\lim_{n\to\infty} \frac{n^2}{2^n} = 0$, according to the first standard limit (with $p = a = 2$).

8.21

a. $\lim_{n\to\infty} \frac{2^n+1}{3^n+1}$

b. $\lim_{n\to\infty} \frac{2^n-2^{-n}}{2^n-1}$

c. $\lim_{n\to\infty} \frac{2^{n+1}+1}{2^n+1}$

d. $\lim_{n\to\infty} \frac{3^n-2^n}{3^n+2^n}$

e. $\lim_{n\to\infty} \frac{2^{3n}-1}{2^{3n}+3^{2n}}$

8.22

a. $\lim_{n\to\infty} \frac{n^3-3^n}{n^3+3^n}$

b. $\lim_{n\to\infty} \frac{2n^2}{n+2^{-n}}$

c. $\lim_{n\to\infty} \frac{n^2+n!}{3^n-n!}$

d. $\lim_{n\to\infty} \frac{n^n+3n!}{n^n+(3n)!}$

e. $\lim_{n\to\infty} \frac{n!+3n^9-7}{n^n+3n^9+7}$

8.23

a. $\lim_{n\to\infty} \frac{n^3+\sqrt[3]{n}}{3^n}$

b. $\lim_{n\to\infty} \frac{2^n+1000}{n^{1000}}$

c. $\lim_{n\to\infty} \frac{2^n+4^n}{3^n}$

d. $\lim_{n\to\infty} \frac{2^n}{n^n+2^n}$

e. $\lim_{n\to\infty} \frac{1.002^n}{n^{1000}+1.001^n}$

8.24

a. $\lim_{n\to\infty} n^{10}\, 0.9999^n$

b. $\lim_{n\to\infty} (n^2+3n-7)\left(\frac{1}{2}\right)^{n-1}$

c. $\lim_{n\to\infty} \frac{(n+3)!}{n!+2^n}$

d. $\lim_{n\to\infty} \frac{2^n}{1+(n+1)!}$

e. $\lim_{n\to\infty} \frac{2^n+(n+1)!}{3^n+n!}$

## What exactly is the limit of a sequence?

On the previous pages, we have seen limits of many types of sequences. We will now define exactly what is meant by the limit of a sequence $a_1, a_2, a_3, \ldots$ For each case, we will give two possible notations: a notation with the letters 'lim' and an 'arrow notation', which is also frequently used. We also use the symbol $\infty$ ('infinity'). This is not a real number, but a symbol, the meaning of which is explained in the last column.

| *'lim' notation:* | *arrow notation:* | *description:* |
|---|---|---|
| $\lim\limits_{n\to\infty} a_n = L$ | $a_n \to L$ if $n \to \infty$ | For each positive number $p$ (no matter how small), there is a term $a_N$ in the sequence for which all terms $a_n$ with $n > N$ satisfy $\lvert a_n - L \rvert < p$. |
| $\lim\limits_{n\to\infty} a_n = \infty$ | $a_n \to \infty$ if $n \to \infty$ | For each positive number $P$ (no matter how big), there is a term $a_N$ in the sequence for which all terms $a_n$ with $n > N$ satisfy $a_n > P$. |
| $\lim\limits_{n\to\infty} a_n = -\infty$ | $a_n \to -\infty$ if $n \to \infty$ | This simply means that $\lim\limits_{n\to\infty} (-a_n) = \infty$. |

However, not all sequences have a limit in the above sense. For example, the sequence where the $n$-th term equals $(-1)^n$ (the geometric sequence with ratio$(-1)$) does not have a finite or infinite limit, since the terms alternate between $+1$ and $-1$. The geometric sequence $(-2)^n$ also does not have a limit, although in absolute value its terms grow endlessly. However, they alternate in sign, which means that for any number $M$ there are infinitely many terms greater than $M$ and infinitely many terms less than $M$.

We again want to stress that $\infty$ and $-\infty$ are not real numbers, but symbols, used to explain the limit behaviour of certain sequences. However, with some care, it is still possible to use these symbols in calculations. For instance, if for a certain sequence $a_1, a_2, a_3, \ldots$ it is known that $\lim\limits_{n\to\infty} a_n = \infty$, then certainly for the sequence $\frac{1}{a_1}, \frac{1}{a_2}, \frac{1}{a_3}, \ldots$ it is true that $\lim\limits_{n\to\infty} \frac{1}{a_n} = 0$. The reason is clear: if the numbers $a_n$ grow endlessly, the numbers $\frac{1}{a_n}$ will get closer and sloser to 0. Page 65 and the exercises contain plenty of examples.

# IV Equations

$$x_{1,2} = \frac{-b \pm \sqrt{b^2 - 4ac}}{2a}$$

Many mathematical applications require the solving of equations or inequalities. All numbers that satisfy one or more given equations or inequalities must be found. In this part, we will tackle the most elementary solution methods. In particular, we will show you how to solve linear and quadratic equations and inequalities in one variable. The famous formula for solving quadratic equations shown above is an important instance. In the last chapter we will show you a solution method for simple systems of linear equations.

# 9 Linear equations

Find the solution $x$ for each of the following equations:

9.1

a. $x+7=10$
b. $x-12=4$
c. $x+3=-10$
d. $x-10=-7$
e. $x+8=0$

9.2

a. $-x+15=6$
b. $-x-7=10$
c. $-x+17=-10$
d. $-x-8=-9$
e. $-x-19=0$

9.3

a. $2x+7=9$
b. $3x-8=7$
c. $4x+3=11$
d. $9x-10=17$
e. $6x+6=0$

9.4

a. $-3x+15=21$
b. $-2x-7=11$
c. $-5x+17=32$
d. $-4x-8=16$
e. $-6x-18=0$

9.5

a. $2x+9=12$
b. $3x-12=9$
c. $-4x+3=-11$
d. $5x-12=17$
e. $-6x+9=0$

9.6

a. $-x-15=6$
b. $-9x-7=-10$
c. $6x+17=12$
d. $-9x-18=-6$
e. $5x-19=0$

9.7

a. $x+7=10-2x$
b. $x-12=4+5x$
c. $2x+3=-10+x$
d. $3x-10=2x-7$
e. $5x+9=2x$

9.8

a. $-x+15=6-4x$
b. $-2x-7=2x-10$
c. $3x+17=-11+x$
d. $-x-8=-9x-4$
e. $2x-19=19-2x$

9.9

a. $x-12=3-4x$
b. $-3x+5=2x-8$
c. $-x+7=-12-x$
d. $4x-1=-7x+4$
e. $2x+12=9+4x$

In the following exercises, first remove the fractions by multiplying both sides by a suitable number (rule R2):

9.10

a. $\frac{1}{2}x+\frac{3}{2}=1+\frac{5}{2}x$
b. $-\frac{1}{3}x-\frac{2}{3}=\frac{4}{3}x-1$
c. $\frac{2}{5}x+\frac{3}{5}=-\frac{3}{5}-\frac{1}{5}x$
d. $-\frac{3}{7}x-\frac{3}{7}=-\frac{6}{7}-\frac{1}{7}x$
e. $\frac{2}{9}x-\frac{1}{9}=x-\frac{2}{9}$

9.11

a. $\frac{1}{3}x+\frac{3}{2}=1+\frac{1}{6}x$
b. $-\frac{2}{3}x-\frac{3}{4}=\frac{4}{3}x-1$
c. $\frac{2}{5}x+\frac{5}{3}=-\frac{5}{6}-\frac{2}{3}x$
d. $-\frac{2}{9}x-\frac{1}{4}=-\frac{3}{2}-\frac{1}{6}x$
e. $\frac{1}{8}x-\frac{5}{6}=x-\frac{3}{4}$

9.12

a. $3(x+4)=-2(x+8)$
b. $-2(x-3)+1=-3(-x+7)+2$
c. $2-(x+4)=-2(x+1)-3$

9.13

a. $6(-x+2)-(x-3)=3(-x+1)$
b. $2x-(-x+1)=-3(-x+1)$
c. $5(-2x+3)+(2x-5)=4(x-4)$

## General solution rules

Suppose that a number $x$ satisfies the equation:

$$3x + 7 = -2x + 1$$

and that you are asked to find $x$.

Solution:

1. add $2x$ on both sides: $5x + 7 = 1$,
2. add $-7$ on both sides: $5x = -6$,
3. divide both sides by 5: $x = -\dfrac{6}{5}$.

It has taken three steps to find the unknown number $x$. To verify the answer, you can substitute this value $x = -\frac{6}{5}$ into the original equation.

To solve the equation above, the following general rules have been used:

> R1. *The validity of an equation doesn't change if you add the same number on both sides of the equation.*
>
> R2. *The validity of an equation doesn't change if you multiply or divide both sides by the same number, provided that this number is not equal to 0.*

The first two steps of the solution in the given example can also be considered as

> *moving a term from one side of the equality sign to the other side, while flipping the sign of the term that is moved (from plus to minus or vice versa).*

This is how most people memorize rule R1. In step 1, we moved the term $-2x$ from the right-hand side to the left-hand side, and in step 2, we moved the term $+7$ from the left-hand side to the right-hand side.

Write the given inequalities in one of the following forms:
$x < a,\ x \leq a,\ x > a$ or $x \geq a$.

*Example:* $-3x + 7 > 5$. Subtracting 7 yields $-3x > -2$ and dividing by $-3$ subsequently yields $x < \frac{2}{3}$.

9.14

a. $x + 6 < 8$
b. $x - 8 > 6$
c. $x + 9 \leq 7$
d. $x - 1 \geq -3$
e. $x + 6 > 7$

9.15

a. $-2x + 4 < 8$
b. $-3x - 8 > 7$
c. $-5x + 9 \leq -6$
d. $-4x + 1 \geq -3$
e. $-2x + 6 > 5$

9.16

a. $2x + 6 < x - 8$
b. $3x - 8 > 7 - 2x$
c. $x + 9 \leq 7 - 3x$
d. $2x - 1 \geq x - 3$
e. $5x + 6 > 3x + 7$

9.17

a. $-2x + 6 < x + 9$
b. $x - 8 > 3x + 6$
c. $2x + 9 \leq 3x + 1$
d. $-3x - 1 \geq 3 - x$
e. $5x + 6 > 7x + 2$

9.18

a. $\frac{1}{2}x + 1 < 2 - \frac{1}{3}x$
b. $\frac{2}{3}x - \frac{1}{2} > 1 + \frac{1}{3}x$
c. $\frac{3}{4}x + \frac{1}{2} \leq \frac{1}{2}x - \frac{1}{4}$
d. $\frac{1}{6}x - \frac{1}{3} \geq \frac{2}{3}x - \frac{1}{6}$
e. $\frac{2}{5}x - \frac{5}{2} > \frac{1}{2}x - \frac{2}{5}$

9.19

a. $-\frac{3}{2}x - 1 < 2 - \frac{1}{4}x$
b. $\frac{1}{5}x - \frac{1}{2} > 1 + \frac{2}{5}x$
c. $-\frac{3}{4}x + \frac{1}{3} \leq \frac{1}{2}x - \frac{5}{6}$
d. $\frac{2}{7}x - \frac{1}{2} \geq \frac{1}{2}x - \frac{3}{7}$
e. $-\frac{3}{5}x - \frac{5}{2} > -\frac{1}{2}x + \frac{2}{5}$

Write the following inequalities in one of the following forms:
$a < x < b,\ a \leq x < b,\ a < x \leq b$ or $a \leq x \leq b$.

*Example:* $-2 \leq 3 - 6x < 4$. Subtracting 3 yields $-5 \leq -6x < 1$ and subsequently dividing by $-6$ yields $\frac{5}{6} \geq x > -\frac{1}{6}$, so $-\frac{1}{6} < x \leq \frac{5}{6}$.

9.20

a. $-3 < x + 1 < 4$
b. $2 < 2x + 4 < 6$
c. $0 \leq 3x + 6 < 9$
d. $-6 < 4x - 2 \leq 4$
e. $1 \leq 1 + 2x \leq 2$

9.21

a. $-3 < -x + 1 < 2$
b. $2 < 2x - 4 < 4$
c. $0 \leq -3x + 9 < 6$
d. $-6 < -4x + 2 \leq 4$
e. $-1 \leq 1 - 2x \leq 0$

## Inequalities

The manipulation of inequalities requires more care than the manipulation of equations. There are, however, a number of common features. Inequalities occur in four forms:

$$a < b, \qquad a \leq b, \qquad a > b, \qquad a \geq b,$$

meaning, respectively, '$a$ is less than $b$','$a$ is less than or equal to $b$', '$a$ is greater than $b$' and '$a$ is greater than or equal to $b$'. Of course, $a > b$ means the same as $b < a$, and $a \geq b$ means the same as $b \leq a$. Furthermore, the following general rule holds true:

> R3. *The validity of an inequality doesn't change if the same number is added to both sides of the inequality.*

Just as with equations, this rule implies that you can move a term from one side of the inequality sign to the other, provided you flip the sign of the moved term (from plus to minus or vice versa).

When multiplying or dividing both sides of an inequality by the same number (not equal to 0), you have to be careful:

> R4. *The validity of an inequality doesn't change if both sides are multiplied or divided by the same positive number.*
>
> R5. *When multiplying or dividing both sides of an inequality by a negative number, the inequality sign should be reversed.*

Sometimes similar inequalities are combined. For instance, $a < b \leq c$ means that $b$ is greater than $a$ and less than or equal to $c$. However, you should *never* combine dissimilar inequalities: combinations of 'greater' and 'less' never occur in one chain. For instance, it is allowed to write $a > b > c$, but not $a < b > c$, even if the separate inequalities $a < b$ and $b > c$ are valid. The reason is that in such a case, you know that $a$ and $c$ are both less than $b$, but you cannot draw any conclusions about the mutual relationship between $a$ and $c$.

Find all the solutions $x$ for the following equations:

9.22

a. $\frac{1}{x+1} = 5$

b. $\frac{x}{x-4} = 2$

c. $\frac{2x+1}{x} = -3$

d. $\frac{4x-1}{x-3} = -2$

e. $\frac{x+7}{-3x+8} = 1$

9.23

a. $\frac{2x}{3x-4} = -1$

b. $\frac{8x}{4x-4} = 2$

c. $\frac{4-4x}{x-1} = -3$

d. $\frac{2x+3}{4x} = 6$

e. $\frac{x-5}{x-4} = 1$

9.24

a. $(x+1)^2 = 1$

b. $(x-4)^2 = 9$

c. $(1-x)^2 = 25$

d. $(2x+1)^2 = 4$

e. $(-3x+1)^2 = 16$

9.25

a. $(x+2)^2 = 3$

b. $(x-1)^2 = 2$

c. $(3-x)^2 = 5$

d. $(2x+1)^2 = 6$

e. $(6-2x)^2 = 8$

9.26

a. $(x-1)^3 = 1$

b. $(x+4)^3 = -8$

c. $(1-x)^3 = 1$

d. $(2x-1)^3 = 27$

e. $(-4x-1)^3 = 64$

9.27

a. $(x-2)^4 = 1$

b. $(x+1)^4 = 16$

c. $(3-2x)^4 = 4$

d. $(2x+3)^4 = 81$

e. $(4-3x)^4 = 625$

9.28

a. $(x+1)^2 = (2x-1)^2$

b. $(3x-1)^2 = (x-1)^2$

c. $(x+1)^2 = (-2x+1)^2$

d. $(2x+5)^2 = (3-x)^2$

e. $(4x+3)^2 = x^2$

9.29

a. $(x+2)^2 = 4x^2$

b. $(2x+1)^2 = 4(x+1)^2$

c. $(-x+2)^2 = 9(x+2)^2$

d. $4(x+1)^2 = 25(x-1)^2$

e. $9(2x+1)^2 = 4(1-2x)^2$

## Reducing an equation to a linear equation

An equation of the form

$$ax + b = 0$$

where $x$ is an unknown number and $a$ and $b$ are given (known) numbers with $a \neq 0$, is called a *linear equation* in $x$, or also an *equation of degree* 1 in $x$. The equations on page 72 can all be written in this form. These equations can be solved using rules R1 and R2 on page 73. The solution is

$$x = -\frac{b}{a}$$

In some cases, more complicated equations can be reduced to linear equations.

**Example 1:**

$$\frac{3x+2}{4x-5} = 2$$

Multiplying both sides by $4x - 5$ yields the equation

$$3x + 2 = 2(4x - 5)$$

which can be solved using the methods of page 73. The result is $x = \frac{12}{5}$, as you can check for yourself.

In the first step, we used rule R2. This, however, is only allowed if the number $4x - 5$, which we used to multiply both sides, does not equal 0. Since the value of $x$ is still unknown in this stage of the solution, we don't know whether $4x - 5 \neq 0$ is true. We can only check this at the end, after solving the new equation for $x$. Such a *final check* is not superfluous, as can be seen in some of the exercises on the previous page.

**Example 2:**

$$(3x - 1)^2 = 4$$

If $x$ satisfies this equation, the square of $3x - 1$ should be 4. This means that $3x - 1$ is equal to $+2$ or $-2$. This gives two possibilities:

$$3x - 1 = 2 \quad \text{and} \quad 3x - 1 = -2$$

with solutions $x = 1$ and $x = -\frac{1}{3}$, respectively (check this for yourself).

# 10 Quadratic equations

Find all $x$ satisfying the following equations:

10.1

a. $x^2 = 9$
b. $4x^2 = 16$
c. $3x^2 + 1 = 13$
d. $-2x^2 + 21 = 3$
e. $2x^2 - 48 = 50$

10.2

a. $3x^2 - 2 = x^2 + 2$
b. $x^2 - 15 = 2x^2 - 2$
c. $12 - x^2 = x^2 - 4$
d. $3(2 - x^2) = x^2 + 6$
e. $-2(1 - x^2) = x^2$

10.3

a. $\frac{1}{2}x^2 = 2$
b. $\frac{2}{3}x^2 = \frac{1}{2}$
c. $\frac{3}{2}x^2 = \frac{2}{3}$
d. $\frac{4}{5}x^2 = \frac{5}{4}$
e. $2x^2 = \frac{9}{4}$

10.4

a. $\frac{1}{2}x^2 + \frac{2}{3} = \frac{5}{6}$
b. $\frac{1}{3}x^2 - \frac{1}{2} = \frac{1}{4}$
c. $-\frac{2}{5}x^2 - \frac{3}{7} = \frac{4}{3}$
d. $\frac{1}{8}x^2 + \frac{3}{4} = \frac{5}{2}$
e. $\frac{1}{3}(x^2 - \frac{1}{2}) = \frac{1}{4}$

10.5

a. $x(x+3) = 0$
b. $(x+1)(x-5) = 0$
c. $(x-1)(x+1) = 0$
d. $(x+7)(x-2) = 0$
e. $(x-3)(x+9) = 0$

10.6

a. $x(2x-1) = 0$
b. $(2x+1)(x-3) = 0$
c. $(3x+2)(2x-3) = 0$
d. $(5x+3)(3x-5) = 0$
e. $(2-3x)(3x-2) = 0$

10.7

a. $3(x-1)(x+3) = 0$
b. $5(x-1)(x+5) = 0$
c. $-2(2x+1)(3x-4) = 0$
d. $4(3x+2)(6x+3) = 0$
e. $-5(3x-2)(3x+2) = 0$

10.8

a. $(\frac{1}{2}x + 3)(x - \frac{2}{3}) = 0$
b. $(\frac{2}{3}x - \frac{4}{5})(\frac{1}{3}x - \frac{2}{7}) = 0$
c. $\frac{1}{2}(\frac{3}{4}x - \frac{4}{3})(\frac{1}{3}x - \frac{1}{2}) = 0$

## Quadratic equations

An equation of the form

$$ax^2 + bx + c = 0$$

where $x$ is an unknown number and $a$, $b$ and $c$ are given (known) numbers with $a \neq 0$, is called a *quadratic equation* in $x$, or also an *equation of degree* 2 in $x$.

These equations have 0, 1 or 2 solutions, i.e. there are 0, 1 or 2 numbers $x$ satisfying the equation. These solutions are often called the *roots* of the equation, although the notation of these numbers does not need to contain roots in the sense of chapter 3. We give examples of each of the three possible cases.

1. The equation $x^2 + 1 = 0$ has no solutions, since the left-hand side is greater than or equal to 1 for each choice of $x$ (a square is always greater than or equal to 0).
2. The equation $x^2 + 2x + 1 = 0$ has one solution, since the left-hand side can be written as $(x + 1)^2$, and this is only 0 if $x + 1 = 0$, i.e. if $x = -1$.
3. The equation $x^2 - 1 = 0$ has two solutions, namely $x = 1$ and $x = -1$.

In some cases, the solution of a quadratic equation doesn't require special techniques. For instance,

$$x^2 - 3x = 0$$

can be written as

$$x(x - 3) = 0$$

When two numbers are both unequal to 0, their product must also be unequal to 0, so in the equation above either $x$ or $x - 3$ has to be 0. The solutions here are $x = 0$ and $x = 3$.

This is an application of a general rule that is used frequently:

$$a \cdot b = 0 \quad \Longleftrightarrow \quad a = 0 \text{ or } b = 0$$

where 'or' has to be taken in the sense that $a$ and $b$ can also both be zero. Indeed, if at least one of the numbers $a$ and $b$ is zero, then $a \cdot b$ is also zero, and if both are unequal to zero, then their product cannot be zero either.

Solve the following equations by completing the square:

10.9

a. $x^2 + 4x + 1 = 0$
b. $x^2 + 6x - 2 = 0$
c. $x^2 + 8x + 3 = 0$
d. $x^2 - 2x - 1 = 0$
e. $x^2 + 10x + 5 = 0$

10.10

a. $x^2 - 12x + 6 = 0$
b. $x^2 - 13x - 7 = 0$
c. $x^2 + x - 42 = 0$
d. $x^2 - 12x + 27 = 0$
e. $x^2 + 6x - 12 = 0$

10.11

a. $x^2 + 7x - 1 = 0$
b. $x^2 + 3x - 4 = 0$
c. $x^2 + 4x + 4 = 0$
d. $x^2 - 4x - 4 = 0$
e. $x^2 - 11x + 7 = 0$

10.12

a. $x^2 + 20x + 60 = 0$
b. $x^2 - 18x - 80 = 0$
c. $x^2 + 13x - 42 = 0$
d. $x^2 - 15x + 56 = 0$
e. $x^2 + 60x + 800 = 0$

10.13

a. $x^2 + \frac{1}{2}x - \frac{3}{4} = 0$
b. $x^2 + \frac{4}{3}x - \frac{5}{9} = 0$
c. $x^2 - \frac{1}{3}x - \frac{1}{9} = 0$
d. $x^2 + \frac{3}{2}x - \frac{5}{8} = 0$
e. $x^2 - \frac{2}{5}x - \frac{1}{5} = 0$

10.14

a. $x^2 + \frac{3}{4}x - \frac{3}{8} = 0$
b. $x^2 + \frac{5}{2}x + \frac{3}{2} = 0$
c. $x^2 - \frac{2}{3}x + \frac{1}{9} = 0$
d. $x^2 - \frac{3}{2}x - \frac{3}{4} = 0$
e. $x^2 + \frac{4}{5}x - \frac{4}{5} = 0$

Occasionally an equation that doesn't look like a quadratic equation can be reduced to a quadratic equation by using a clever trick.
*Example:* $x^4 - 6x^2 - 16 = 0$.

This is an equation of degree 4, but taking $y = x^2$, it becomes a quadratic equation: $y^2 - 6y - 16 = 0$. Completing the square yields $(y-3)^2 = 25$, so $y - 3 = \pm 5$, with solutions $y = 8$ and $y = -2$. Since $y = x^2$, $y = -2$ doesn't yield solutions in $x$ (a square cannot be negative). The solutions here are $x = \pm\sqrt{8} = \pm 2\sqrt{2}$.

10.15

a. $x^4 + 4x^2 - 5 = 0$
b. $x^4 - 6x^2 = 7$
c. $x^4 + 4x^2 + 4 = 0$
d. $x^4 - 4x^2 + 4 = 0$
e. $x^6 - 11x^3 = 12$

10.16

a. $x - 2\sqrt{x} = 3$ (put $y = \sqrt{x}$)
b. $x - 18\sqrt{x} + 17 = 0$
c. $x + 4\sqrt{x} = 21$
d. $x - 15\sqrt{x} + 26 = 0$
e. $x + 6\sqrt{x} = 7$

## Completing the square

To solve the equation

$$x^2 - 6x + 3 = 0$$

we write it as

$$x^2 - 6x + 9 = 6$$

resulting in the left-hand side becoming a square, namely the square of $x - 3$ (remember that $(x-3)^2 = x^2 - 6x + 9$). Solving the resulting equation is simple:

$$(x-3)^2 = 6$$

which gives

$$x - 3 = \sqrt{6} \quad \text{or} \quad x - 3 = -\sqrt{6}$$

Hence the solutions are

$$x = 3 + \sqrt{6} \quad \text{and} \quad x = 3 - \sqrt{6}$$

This method of *completing the square* always works when the coefficient of $x^2$ is equal to 1. Divide the coefficient of $x$ by 2 to produce a complete square on the left-hand side and a constant on the right-hand side, i.e. a number that doesn't depend on $x$. If this constant is positive or zero, taking square roots on both sides solves the equation. If the constant is negative, there are no solutions, since the left-hand side is a square and a square cannot be negative.

Another example:

$$x^2 + 10x + 20 = 0$$

Halving the coefficient 10 yields 5. Since $(x+5)^2 = x^2 + 10x + 25$, the equation can be written as

$$x^2 + 10x + 25 = 5$$

giving

$$(x+5)^2 = 5$$

which leads to

$$x + 5 = \sqrt{5} \quad \text{or} \quad x + 5 = -\sqrt{5}$$

Hence the solutions are

$$x = -5 + \sqrt{5} \quad \text{and} \quad x = -5 - \sqrt{5}$$

Solve the following equations using the general formula:

10.17

a. $x^2 + 5x + 1 = 0$
b. $x^2 - 3x + 2 = 0$
c. $x^2 + 7x + 3 = 0$
d. $x^2 - x + 1 = 0$
e. $x^2 + 11x + 11 = 0$

10.18

a. $x^2 + 3x + 1 = 0$
b. $x^2 - 4x + 3 = 0$
c. $x^2 + 9x - 2 = 0$
d. $x^2 - 12x + 3 = 0$
e. $x^2 - 5x + 1 = 0$

10.19

a. $2x^2 + 4x + 3 = 0$
b. $2x^2 - 12x + 9 = 0$
c. $3x^2 + 12x - 8 = 0$
d. $4x^2 + 12x + 1 = 0$
e. $6x^2 - 12x - 1 = 0$

10.20

a. $2x^2 + x - 1 = 0$
b. $3x^2 + 2x + 1 = 0$
c. $2x^2 + 8x - 2 = 0$
d. $6x^2 + 18x + 7 = 0$
e. $4x^2 - 8x + 1 = 0$

10.21

a. $-x^2 + 2x + 1 = 0$
b. $-2x^2 + 8x - 3 = 0$
c. $-3x^2 + 9x - 1 = 0$
d. $-4x^2 - 12x + 9 = 0$
e. $-x^2 + x + 1 = 0$

10.22

a. $3x^2 - 4x + 3 = 0$
b. $-2x^2 + 3x + 2 = 0$
c. $-4x^2 + 6x + 5 = 0$
d. $6x^2 + 18x - 1 = 0$
e. $-x^2 - x - 1 = 0$

10.23

a. $\frac{1}{2}x^2 + x - 1 = 0$
b. $\frac{2}{3}x^2 + 2x - 3 = 0$
c. $\frac{1}{2}x^2 - x - 1 = 0$
d. $\frac{4}{5}x^2 + 3x - 2 = 0$
e. $\frac{5}{2}x^2 + 5x - 2 = 0$

10.24

a. $\frac{1}{2}x^2 + \frac{3}{2}x - \frac{1}{4} = 0$
b. $-\frac{2}{3}x^2 + \frac{1}{3}x - \frac{1}{2} = 0$
c. $\frac{3}{4}x^2 + \frac{3}{8}x - \frac{3}{4} = 0$
d. $\frac{2}{5}x^2 + \frac{3}{5}x - \frac{5}{4} = 0$
e. $-\frac{3}{2}x^2 + \frac{1}{4}x - \frac{1}{8} = 0$

10.25

a. $x(1 - x) = -2$
b. $(3x + 1)(x + 3) = 1$
c. $(x - 2)(2 - 3x) = x$
d. $(5 - x)(5 + x) = 5$
e. $(1 - x)(2 - x) = 3 - x$

10.26

a. $(x^2 - 4)(x^2 - 1) = 5$
b. $(1 - x^2)(1 + 2x^2) = x^2$
c. $(\sqrt{x} - 1)(\sqrt{x} - 3) = 1$
d. $\sqrt{x}(1 + \sqrt{x}) = 1 - \sqrt{x}$
e. $(1 - x^3)(2 - x^3) = x^3$

## The general solution formula for quadratic equations

The method of completing the square can also be applied to more general quadratic equations. To avoid fractions as much as possible, we will slightly modify the recipe.

Suppose a quadratic equation

$$ax^2 + bx + c = 0$$

with $a \neq 0$ has to be solved. Multiply both sides by $4a$ to get

$$4a^2x^2 + 4ab\,x + 4ac = 0$$

and write this as

$$4a^2x^2 + 4ab\,x + b^2 = b^2 - 4ac$$

The left-hand side has now become a complete square, namely the square of $2ax + b$, since $(2ax + b)^2 = 4a^2x^2 + 4abx + b^2$. Hence, the equation can be written as

$$(2ax + b)^2 = b^2 - 4ac$$

The right-hand side $b^2 - 4ac$ is called the *discriminant*. If the discriminant is negative, the equation has no solutions, since the left-hand side is a square.

If the discriminant is positive or zero, taking square roots yields

$$2ax + b = \sqrt{b^2 - 4ac} \quad \text{or} \quad 2ax + b = -\sqrt{b^2 - 4ac}$$

which gives the following solutions:

$$x = \frac{-b + \sqrt{b^2 - 4ac}}{2a} \quad \text{and} \quad x = \frac{-b - \sqrt{b^2 - 4ac}}{2a}$$

If the discriminant is zero, both solutions coincide.

The solutions are often written together as one general solution formula:

$$x_{1,2} = \frac{-b \pm \sqrt{b^2 - 4ac}}{2a}$$

This is sometimes called the *abc-formula*. It is best memorized in this form.

# 11 Systems of linear equations

Solve the following systems of linear equations:

11.1

a. $\begin{cases} 2x + 3y = 4 \\ 3x - 2y = 6 \end{cases}$

b. $\begin{cases} 3x + 5y = 8 \\ -x + 6y = 5 \end{cases}$

c. $\begin{cases} 5x + 2y = 3 \\ 2x - 4y = 6 \end{cases}$

d. $\begin{cases} 2x + 5y = 9 \\ -3x + 4y = -2 \end{cases}$

e. $\begin{cases} 2x - 4y = 3 \\ 4x - 2y = 3 \end{cases}$

11.2

a. $\begin{cases} 2x + 5y = 1 \\ 5x - 4y = 0 \end{cases}$

b. $\begin{cases} 7x - 5y = 1 \\ 4x - 7y = 13 \end{cases}$

c. $\begin{cases} x - 3y = 7 \\ -5x + 2y = 4 \end{cases}$

d. $\begin{cases} 2x - 2y = 6 \\ 3x + 4y = -5 \end{cases}$

e. $\begin{cases} 7x + 5y = 1 \\ 3x - 4y = 25 \end{cases}$

11.3

a. $\begin{cases} 2x + 3y = 2 \\ 3x + 4y = 2 \end{cases}$

b. $\begin{cases} 2x - 3y = 1 \\ 3x - 4y = 0 \end{cases}$

c. $\begin{cases} 2x + 3y = 4 \\ 3x + 5y = 1 \end{cases}$

d. $\begin{cases} 2x - 7y = 5 \\ x - 4y = -1 \end{cases}$

e. $\begin{cases} 4x - 7y = 8 \\ 3x - 5y = 4 \end{cases}$

11.4

a. $\begin{cases} x - 5y = 4 \\ 3x + 4y = 1 \end{cases}$

b. $\begin{cases} 3x - 7y = 2 \\ 3x - 4y = -2 \end{cases}$

c. $\begin{cases} 2x + 9y = 5 \\ -x - 4y = 2 \end{cases}$

d. $\begin{cases} 6x + 5y = 1 \\ 7x + 6y = 2 \end{cases}$

e. $\begin{cases} 5x - 3y = 7 \\ 8x - 5y = 2 \end{cases}$

## Two linear equations with two unknown variables

Suppose that two unknown numbers $x$ and $y$ satisfy each of the following two linear equations:

$$\begin{cases} 2x + 5y = 9 \\ 3x - 4y = 2 \end{cases}$$

This is called a *system of two linear equations with two variables.* The numbers $x$ and $y$ satisfying this system can be found as follows.

In the first equation, multiply both sides by 3, and in the second equation, multiply both sides by 2, making the coefficients of $x$ in both equations equal to 6:

$$\begin{cases} 6x + 15y = 27 \\ 6x - 8y = 4 \end{cases}$$

Now, subtract the second equation from the first. This results in an equation containing only the unknown $y$:

$$23y = 23$$

with $y = 1$ as its solution. Substituting this value into one of the original equations yields an equation from which $x$ can be solved. We choose the first equation:

$$2x + 5 \times 1 = 9$$

or $2x = 4$ so $x = 2$. In this way, the numbers $x$ and $y$ have been found. You can verify that the combination $x = 2$ and $y = 1$ indeed satisfies the two original equations.

In general, this methods works as follows: multiply the equations with factors that make the coefficients of $x$ (or, if preferred, $y$) equal. Subtraction then yields an equation which only contains $y$ (or $x$, respectively). With this equation, the number $y$ (or $x$) can be found, and substituting the value found into one of the original equations yields an equation which can be used to find the remaining variable.

Solve the following systems of linear equations:

11.5

a.
$$\begin{aligned} x + 3y + z &= 1 \\ 2x - y - 3z &= -8 \\ -3x + 2y + 2z &= 7 \end{aligned}$$

b.
$$\begin{aligned} x - 4y + z &= -2 \\ -2x + 3y - 2z &= -1 \\ -4x + y + z &= -2 \end{aligned}$$

c.
$$\begin{aligned} -2x + 2y + 3z &= -3 \\ x - 2y + 4z &= 8 \\ -3x + y &= -7 \end{aligned}$$

d.
$$\begin{aligned} 4x - 3y + z &= 2 \\ -2x - y - 2z &= 2 \\ -x + 2y + 4z &= -9 \end{aligned}$$

e.
$$\begin{aligned} x - 3y + z &= -9 \\ x - y - 2z &= -6 \\ -4x + 3z &= 7 \end{aligned}$$

11.6

a.
$$\begin{aligned} x - 5y + z &= -2 \\ x - 3y - 2z &= 1 \\ -3x + 5y + 7z &= -4 \end{aligned}$$

b.
$$\begin{aligned} -2x - y + 2z &= 5 \\ x + y - z &= -3 \\ -3x + 2y - 6z &= -5 \end{aligned}$$

c.
$$\begin{aligned} x - 6y + z &= -8 \\ - y - 2z &= -1 \\ -3x + 2y + 4z &= 8 \end{aligned}$$

d.
$$\begin{aligned} x - 2y + z &= -5 \\ -3x - y - 3z &= 1 \\ -2x - 3y + 2z &= -8 \end{aligned}$$

e.
$$\begin{aligned} x - 8y + 3z &= -9 \\ - 2y - 3z &= 1 \\ -4x + 5y &= -3 \end{aligned}$$

When you try to solve the following systems of linear equations, strange things happen. Investigate what happens and try to explain it. What does 'solving' such a system tell you about the possible values (if any) of $x$, $y$ and $z$ satisfying the system of equations?

11.7

a.
$$\begin{aligned} x - 2y + z &= 0 \\ x - y - 3z &= 4 \\ -4x + 6y + 4z &= -8 \end{aligned}$$

b.
$$\begin{aligned} x - 2y + z &= 1 \\ x - y - 3z &= 4 \\ -4x + 6y + 4z &= -9 \end{aligned}$$

c.
$$\begin{aligned} x - 3y + z &= -1 \\ -2x + y &= 5 \\ 5y - 2z &= -3 \end{aligned}$$

11.8

a.
$$\begin{aligned} x - 3y + z &= -2 \\ -2x + y &= 4 \\ 5y - 2z &= -1 \end{aligned}$$

b.
$$\begin{aligned} x + 5y - 2z &= 5 \\ 2x - 4z &= 1 \\ -x + 5y + 2z &= -4 \end{aligned}$$

c.
$$\begin{aligned} x + 5y - 2z &= 4 \\ 2x - 4z &= -2 \\ -x + 5y + 2z &= 6 \end{aligned}$$

## Three linear equations with three unknown variables

For a system of three linear equations that is satisfied by three unknown numbers $x$, $y$ and $z$, we can proceed as follows. For instance, take the system

$$\left\{\begin{array}{rcrcrcr} x & - & 2y & + & z & = & 0 \\ x & - & y & - & 3z & = & 4 \\ -4x & + & 5y & + & 9z & = & -9 \end{array}\right.$$

The variable $x$ can be eliminated from the first two equations by adding $-1$ times the first to the second:

$$\begin{array}{rcrcrcrl} x & - & 2y & + & z & = & 0 & (\times -1) \\ x & - & y & - & 3z & = & 4 & (\times 1) \\ \hline & & y & - & 4z & = & 4 & \end{array}$$

The variable $x$ can also be eliminated from the first and third equation by adding 4 times the first to the third:

$$\begin{array}{rcrcrcrl} x & - & 2y & + & z & = & 0 & (\times 4) \\ -4x & + & 5y & + & 9z & = & -9 & (\times 1) \\ \hline & & -3y & + & 13z & = & -9 & \end{array}$$

This way, the following system of two linear equations with the two unknown variables $y$ and $z$ is obtained

$$\left\{\begin{array}{rcrcr} y & - & 4z & = & 4 \\ -3y & + & 13z & = & -9 \end{array}\right.$$

This can be solved in the manner explained before. In this case, it is easy to add three times the first to the second, which yields $z = 3$. Substituting this value into the first equation yields

$$y - 4 \times 3 = 4$$

giving $y = 16$. Substituting $y = 16$ and $z = 3$ into the first equation of the original system yields

$$x - 2 \times 16 + 3 = 0$$

with $x = 29$ as its solution. This solves the original system. The solution values are $x = 29$, $y = 16$, $z = 3$, which can be verified by substituting them into the original system.

This method generally works for linear $3 \times 3$ systems: eliminate one of the unknown variables from two pairs of equations, solve the resulting system of two equations with two unknown variables and finally find the third unknown variable via substitution.

# V Geometry

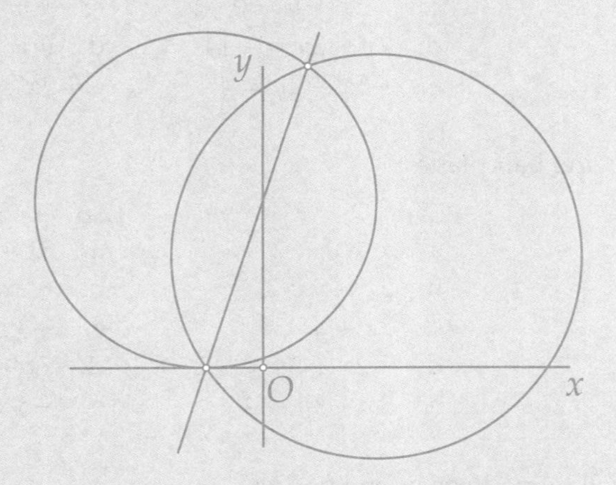


More than two thousand years ago, the ancient Greeks established a firm foundation of classical geometry with their axiomatic method. In the seventeenth century, however, Fermat and Descartes came up with a new concept: geometry with coordinates. Gradually, this approach became the basis of almost all geometry and its many applications. In part V, we will discuss the most important properties of distances, angles, lines and circles using coordinates. And in the last chapter of this part, we will see how the same methods can be applied to planes and spheres in space.

# 12 Lines in the plane

In the following exercises, we assume that a rectangular coordinate system $Oxy$ is chosen. This is easy to visualize on graph paper.

Sketch out the following lines on graph paper. Always find the intersection points (if present) with the $x$-axis and the $y$-axis.

12.1

a. $x + y = 1$
b. $x - y = 0$
c. $2x + y = 2$
d. $-x + 2y = -2$
e. $x + 3y = 4$

12.2

a. $x - 4y = -3$
b. $2x + 8y = -10$
c. $-3x + y = 0$
d. $7x - 2y = -14$
e. $-5x - 2y = 4$

12.3

a. $x = 0$
b. $x = -3$
c. $x = 2y$
d. $y = -1$
e. $3x = 2y + 1$

Draw the following half planes:

12.4

a. $x < 0$
b. $x > -3$
c. $x > y$
d. $y < -2$
e. $3x < y$

12.5

a. $x + y < 2$
b. $2x - y > 0$
c. $2x + y < 2$
d. $-2x + 3y < -2$
e. $3x + 3y > 4$

12.6

a. $5x - 4y > 3$
b. $-2x + 7y < -9$
c. $-3x > y + 2$
d. $7x + 2 < y$
e. $-5 < x + 2y$

12.7 Plot the following lines in one figure:

a. $x + y = -1,\ x + y = 0,\ x + y = 1$ and $x + y = 2$
b. $x - y = -1,\ x - y = 0,\ x - y = 1$ and $x - y = 2$
c. $x + 2y = -1,\ x + 2y = 0,\ x + 2y = 1$ and $x + 2y = 2$
d. $2x + 2y = -1,\ 2x + 2y = 0,\ 2x + 2y = 1$ and $2x + 2y = 2$
e. $x = -2y,\ x = -y,\ x = 0,\ x = y$ and $x = 2y$

## The equation of a line in the plane

The equation

$$4x + 3y = 12$$

contains the two unknown variables $x$ and $y$. A solution of this equation thus is a *pair* $(x, y)$ which satisfies the equation. For instance, $(1, \frac{8}{3})$ is a solution, since substituting $x = 1, y = \frac{8}{3}$ into the equation yields equality. But, there are other solutions as well, e.g. $(3, 0)$, $(0, 4)$ or $(-1, \frac{16}{3})$.

Actually, you are free to choose the value of one of the variables $x$ and $y$, enabling you to solve the equation for the other variable. By selecting a rectangular coordinate system $Oxy$ in the plane, the solutions for the equation are the points on a straight line, in this case the line through $(0, 4)$ and $(3, 0)$. The points $(x, y)$ on this line are exactly the points for which the coordinates satisfy the equation $4x + 3y = 12$. From now on, by 'line' we always mean a straight line.

The line with equation $4x + 3y = 12$ divides the plane into two half planes. For the points in one half plane $4x + 3y > 12$ is true, while the points in the other half plane satisfy $4x + 3y < 12$. Which inequality holds where, can easily be determined by substituting any point not on the line. The origin $O$ satisfies $4 \times 0 + 3 \times 0 < 12$, meaning that the half plane containing the origin is given by $4x + 3y < 12$.

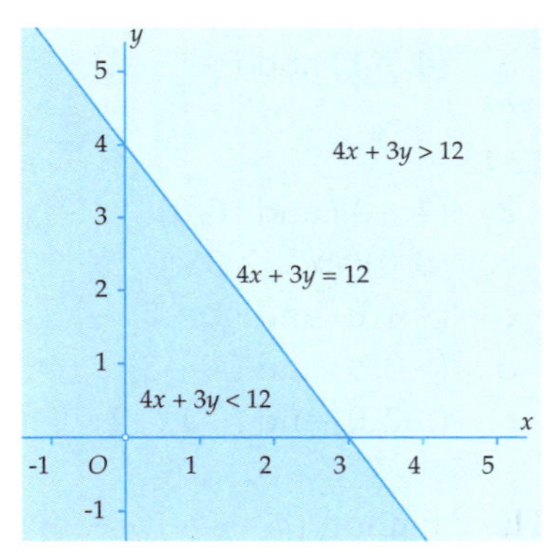


In general, each equation of the form

$$ax + by = c$$

represents a line. The only condition is that $a$ and $b$ should not both be zero. Such an equation is not determined uniquely: multiplying both sides by the same number (not equal to zero), yields another equation of the same line. For instance: $8x + 6y = 24$ represents the same line as $4x + 3y = 12$.

The intersection points of the line with equation $ax + by = c$ with the $y$-axis can be found by putting $x$ equal to 0. This yields $y = \frac{c}{b}$. Therefore, the intersection point is $(0, \frac{c}{b})$. The intersection point with the $x$-axis is found by putting $y$ equal to 0. This yields $(\frac{c}{a}, 0)$ as the intersection point. The line is horizontal if $a = 0$ and vertical if $b = 0$.

A line is completely determined by two distinct points on it. In the next section, we will give a simple method to find an equation of a line through two given points.

Find an equation of the line through the two given points and plot the line on graph paper.

12.8
a. $(3,0)$ and $(0,3)$
b. $(3,0)$ and $(2,0)$
c. $(-1,0)$ and $(0,5)$
d. $(-2,0)$ and $(0,5)$
e. $(-2,-1)$ and $(-2,-2)$

12.9
a. $(3,0)$ and $(0,-2)$
b. $(3,1)$ and $(3,-1)$
c. $(2,0)$ and $(0,5)$
d. $(-2,2)$ and $(2,-2)$
e. $(1,-1)$ and $(2,0)$

12.10
a. $(2,1)$ and $(1,2)$
b. $(2,2)$ and $(-2,0)$
c. $(-1,1)$ and $(1,5)$
d. $(-3,-1)$ and $(-1,5)$
e. $(4,-1)$ and $(-1,-2)$

12.11
a. $(1,-2)$ and $(3,5)$
b. $(7,1)$ and $(5,-1)$
c. $(-1,1)$ and $(4,5)$
d. $(3,-2)$ and $(2,-6)$
e. $(4,-1)$ and $(-1,-3)$

12.12
a. $(4,-1)$ and $(0,0)$
b. $(0,0)$ and $(2,3)$
c. $(-1,0)$ and $(1,-5)$
d. $(-3,4)$ and $(4,-3)$
e. $(-2,0)$ and $(-1,-2)$

12.13
a. $(10,0)$ and $(0,10)$
b. $(3,-1)$ and $(-3,-1)$
c. $(5,-2)$ and $(1,3)$
d. $(-2,-8)$ and $(8,-2)$
e. $(1,-1)$ and $(2,7)$

The equation $(a_1 - b_1)(y - b_2) = (a_2 - b_2)(x - b_1)$ is an equation of the line passing through $(a_1, a_2)$ and $(b_1, b_2)$. Each point $(x, y)$ satisfying the equation is on this line, and, conversely, each point on the line satisfies the equation. Hence, a point $(c_1, c_2)$ is on the line passing through $(a_1, a_2)$ and $(b_1, b_2)$ if

$$(a_1 - b_1)(c_2 - b_2) = (a_2 - b_2)(c_1 - b_1)$$

Use this relation to investigate whether the following three points are on the same line. In each case, verify your result by plotting the points on graph paper.

12.14
a. $(2,1)$, $(3,0)$ and $(1,2)$
b. $(2,2)$, $(0,1)$ and $(-2,0)$
c. $(-1,1)$, $(3,9)$ and $(1,5)$
d. $(-3,-1)$, $(0,2)$ and $(-1,1)$
e. $(4,-1)$, $(1,1)$ and $(-1,2)$

12.15
a. $(1,-2)$, $(0,-5)$ and $(3,4)$
b. $(7,1)$, $(1,-5)$ and $(5,-1)$
c. $(-1,1)$, $(1,3)$ and $(4,5)$
d. $(3,2)$, $(-1,-10)$ and $(2,-1)$
e. $(4,1)$, $(0,-2)$ and $(-1,-3)$

## The equation of a line through two given points

The equation

$$ax + by = c$$

with $a$, $b$ and $c$ given, represents a *straight line* in the $Oxy$-plane (provided $a$ and $b$ are not both equal to zero). Therefore, such an equation is called a *linear equation* in $x$ and $y$. Conversely, with any straight line corresponds a linear equation in $x$ and $y$ in which the coefficients of $x$ and $y$ are not both equal to zero.

To find an equation of a line through two given points, the following formula can be used:

*An equation of the line through the points $(a_1, a_2)$ en $(b_1, b_2)$ is*
$$(a_1 - b_1)(y - b_2) = (a_2 - b_2)(x - b_1)$$

Example: an equation of the line through $A = (-2, 2)$ and $B = (3, -2)$ is

$$(-2 - 3)(y + 2) = (2 - (-2))(x - 3)$$

or, after expanding the brackets and reordering:

$$4x + 5y = 2$$

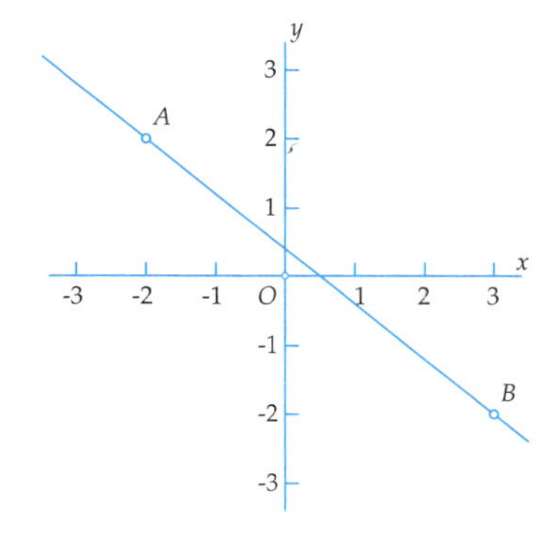


Indeed, substituting $(-2, 2)$ or $(3, -2)$ verifies this equation, and since a straight line is determined by two points, this has to be the required equation.

To verify the general formula, it also is sufficient to check whether $(a_1, a_2)$ and $(b_1, b_2)$ both satisfy the equation

$$(a_1 - b_1)(y - b_2) = (a_2 - b_2)(x - b_1)$$

Substituting $x = a_1$ and $y = a_2$ yields $(a_1 - b_1)(a_2 - b_2) = (a_2 - b_2)(a_1 - b_1)$, which is true, and substituting $x = b_1$ and $y = b_2$ makes both sides equal to zero.

The nice thing about this formula is that it *always* works, even if the line is vertical. For instance, for the points $(3, 5)$ and $(3, 7)$, the formula yields

$$(3 - 3)(y - 7) = (5 - 7)(x - 3)$$

which, after expanding the brackets and simplifying, gives the expected vertical line $x = 3$.

Find the point of intersection of the two given lines, provided they are not parallel or coincident:

12.16

a. $x + y = 2$
$x - y = 1$

b. $x + y = 3$
$2x + y = 6$

c. $-5x + 2y = 4$
$x - 3y = 0$

d. $x + y = 3$
$-x - y = 7$

e. $8x + 3y = 7$
$7y = -4$

12.17

a. $x + 2y = -8$
$3x - 8y = 5$

b. $-2x + 7y = 3$
$-5x - 2y = 6$

c. $5x = 14$
$3x - 2y = 7$

d. $4x = -17$
$9y = 11$

e. $8x - 5y = 1$
$-2x - 11y = 0$

12.18

a. $x + y = 3$
$x - y = 5$

b. $2x + y = 3$
$-x - 2y = 6$

c. $-3x + 2y = 4$
$x - 2y = 2$

d. $4x - 7y = -2$
$5x + 4y = 11$

e. $x + 3y = 6$
$3x + 9y = -2$

12.19

a. $-x + 2y = 9$
$13x - 8y = 15$

b. $12x - 7y = 13$
$-5x - y = 8$

c. $5x + 8y = 14$
$9x - 12y = 5$

d. $4x - 6y = -12$
$-6x + 9y = 18$

e. $-8x + 3y = 5$
$3x - 7y = -12$

12.20 Find an equation of:

a. the line through $(0, 0)$ parallel to $x + y = 4$
b. the line through $(1, 0)$ parallel to $2x - y = -2$
c. the line through $(0, 3)$ parallel to $-x + 4y = 5$
d. the line through $(1, -1)$ parallel to $-5x + 2y = -7$
e. the line through $(-2, 5)$ parallel to $8x + 7y = 14$

## The intersection point of two lines

Two distinct lines in the plane either intersect in one point or are parallel. If they intersect, how can we find the intersection point? For example, suppose that the lines are given by the equations

$$2x + 5y = 9 \qquad \text{and} \qquad 3x - 4y = 2$$

Their intersection point $(x, y)$ satisfies both equations, in other words, it is a solution of the *system of equations*

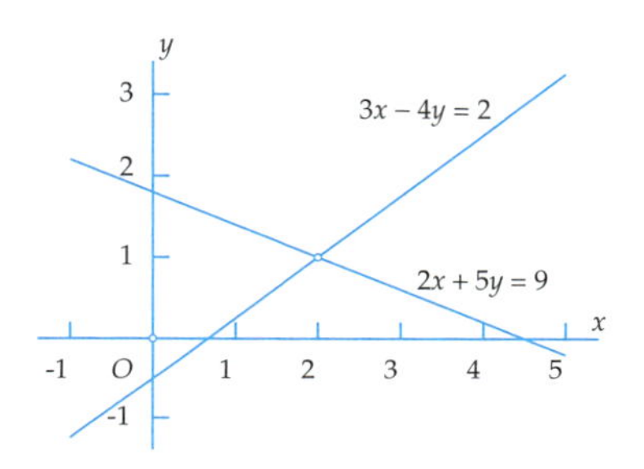


$$\begin{cases} 2x + 5y = 9 \\ 3x - 4y = 2 \end{cases}$$

In chapter 11, we learned how to solve such a system. The intersection point turns out to be $(x, y) = (2, 1)$.

Two lines do not always have exactly one point of intersection. They can be parallel, in which case they don't intersect, or they may coincide, and then one could say that they have infinitely many intersection points. How can this be seen from the equations?

For two coinciding lines, the equations are equal or a multiple of each other, which is immediately obvious. For parallel lines in the standard form $ax + by = c$, only the left-hand sides are equal or a multiple of each other. For instance, take the lines $-6x + 8y = 1$ and $3x - 4y = 2$. In the system

$$\begin{cases} -6x + 8y = 1 \\ 3x - 4y = 2 \end{cases}$$

after multiplying the second equation by $-2$, we get

$$\begin{cases} -6x + 8y = 1 \\ -6x + 8y = -4 \end{cases}$$

This is an *inconsistent system*, i.e. there are no solutions $(x, y)$. Indeed, the expression $-6x + 8y$ cannot be equal to 1 and $-4$ at the same time. Lines without an intersection point are parallel lines.

# 13 Distances and angles

Calculate the distance between the following pairs of points:

13.1

a. $(0,0)$ and $(0,-3)$
b. $(2,0)$ and $(-2,0)$
c. $(0,0)$ and $(1,-5)$
d. $(-1,1)$ and $(-3,3)$
e. $(2,2)$ and $(-4,0)$

13.2

a. $(1,2)$ and $(1,-2)$
b. $(3,-1)$ and $(4,-2)$
c. $(-1,-3)$ and $(3,1)$
d. $(-1,0)$ and $(0,-2)$
e. $(1,1)$ and $(-2,2)$

13.3

a. $(3,0)$ and $(0,3)$
b. $(3,0)$ and $(2,1)$
c. $(-1,0)$ and $(1,5)$
d. $(-2,1)$ and $(3,5)$
e. $(-2,-1)$ and $(-4,-2)$

13.4

a. $(3,2)$ and $(1,-2)$
b. $(3,1)$ and $(4,-1)$
c. $(-2,3)$ and $(3,5)$
d. $(-1,2)$ and $(2,-2)$
e. $(1,-1)$ and $(2,2)$

Find the equation of the perpendicular bisector of the following pairs of points and plot the lines on graph paper:

13.5

a. $(3,0)$ and $(0,3)$
b. $(0,0)$ and $(2,1)$
c. $(-2,0)$ and $(0,0)$
d. $(-2,1)$ and $(2,5)$
e. $(-2,-1)$ and $(-4,-2)$

13.6

a. $(3,2)$ and $(1,-2)$
b. $(3,1)$ and $(4,-1)$
c. $(-2,3)$ and $(3,5)$
d. $(-1,2)$ and $(2,-2)$
e. $(1,-1)$ and $(2,2)$

13.7 In the following exercises, first take $a = 2$, $b = 3$ and plot the lines on graph paper. Then solve the general case in letters.

a. Find an equation of the perpendicular bisector of $(a,b)$ and $(a,-b)$, and also an equation of the line through $(a,b)$ and $(a,-b)$.
b. Find an equation of the perpendicular bisector of $(a,b)$ and $(b,a)$, and also an equation of the line through $(a,b)$ and $(b,a)$.
c. Find an equation of the perpendicular bisector of $(a,b)$ and $(-a,-b)$, and also an equation of the line through $(a,b)$ and $(-a,-b)$.
d. Find an equation of the line through $(1,1)$ perpendicular to the line connecting $(0,0)$ to $(a,b)$.
e. Find an equation of the line through $(a,b)$ perpendicular to the line connecting $(0,0)$ to $(a,b)$.

## Distance and perpendicular bisector

A rectangular coordinate system $Oxy$ is called *orthonormal* if the scales on both axes are the same. In geometric applications, we almost always use orthonormal coordinate systems.

Pythagoras' theorem implies that in an orthonormal coordinate system $Oxy$, the distance $d(A, B)$ between the points $A = (a_1, a_2)$ and $B = (b_1, b_2)$ is given by

$$d(A, B) = \sqrt{(a_1 - b_1)^2 + (a_2 - b_2)^2}$$

For instance, if $A = (4, 9)$ and $B = (8, 1)$, then $d(A, B) = \sqrt{(4-8)^2 + (9-1)^2} = \sqrt{80} = 4\sqrt{5}$.

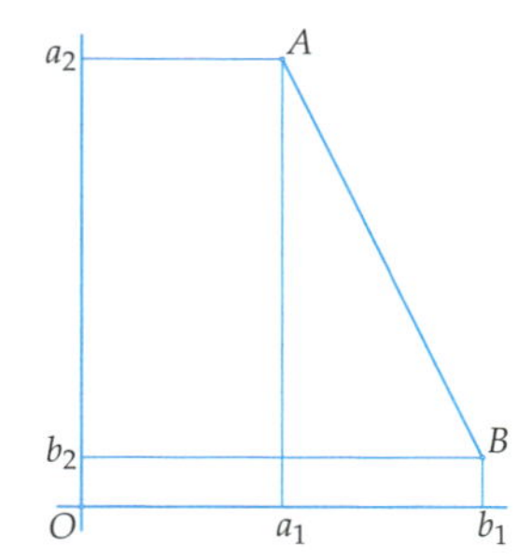


All points $P$ for which $d(P, A) = d(P, B)$ form the *perpendicular bisector* of $A$ and $B$. It is the line bisecting the segment $AB$ perpendicularly into two equal subsegments.

In the figure to the right, the perpendicular bisector of $A = (3, 1)$ and $B = (1, 5)$ is drawn. If $P = (x, y)$ is on the perpendicular bisector, then $d(P, A) = d(P, B)$, giving

$$\sqrt{(x-3)^2 + (y-1)^2} = \sqrt{(x-1)^2 + (y-5)^2}$$

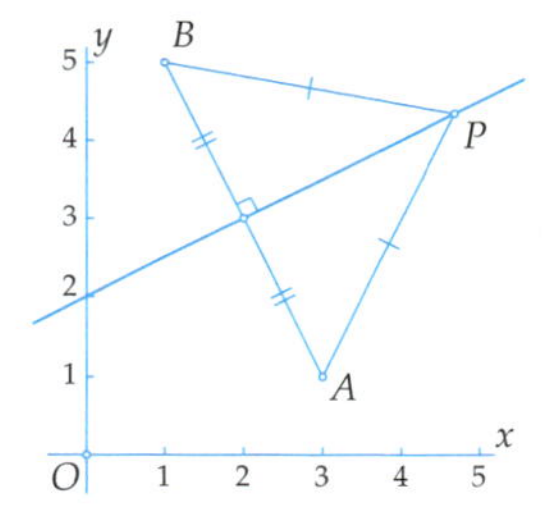


Squaring and expanding the brackets yields the equation

$$x^2 - 6x + 9 + y^2 - 2y + 1 = x^2 - 2x + 1 + y^2 - 10y + 25$$

which can be simplified to a *linear* equation, since the squares can be removed. The result is $-4x + 8y = 16$, or

$$-x + 2y = 4$$

This way, the equation of the perpendicular bisector of $A$ and $B$ can be found. The midpoint $(2, 3)$ of the segment $AB$ is on this line, and, indeed, at this point the perpendicular bisector intersects the segment perpendicularly.

Perpendicular lines are often called *orthogonal*, and a rectangular coordinate system is often called an *orthogonal* coordinate system. Remember that it is called an ortho*normal* coordinate system if the scales on both axes are equal.

In each of the following exercises, a vector **n** and a point $A$ are given. Find the equation of the line through $A$ with **n** as a normal vector.

*Example:* $\mathbf{n} = \begin{pmatrix} 3 \\ 2 \end{pmatrix}$, $A = (1, 1)$.
The equation then has the form

$$3x + 2y = c$$

and substituting the coordinates of the point $A$ yields $3 \times 1 + 2 \times 1 = c$ so $c = 5$. The required equation thus is

$$3x + 2y = 5$$

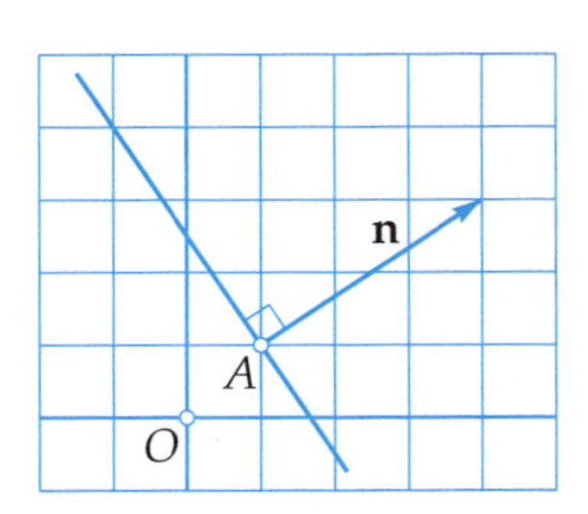


13.8

a. $\mathbf{n} = \begin{pmatrix} 1 \\ 0 \end{pmatrix}$, $A = (0, -3)$

b. $\mathbf{n} = \begin{pmatrix} 1 \\ 2 \end{pmatrix}$, $A = (4, 3)$

c. $\mathbf{n} = \begin{pmatrix} 1 \\ -1 \end{pmatrix}$, $A = (-1, 2)$

d. $\mathbf{n} = \begin{pmatrix} 4 \\ -3 \end{pmatrix}$, $A = (5, 0)$

e. $\mathbf{n} = \begin{pmatrix} 7 \\ -6 \end{pmatrix}$, $A = (1, -2)$

13.9

a. $\mathbf{n} = \begin{pmatrix} 1 \\ 1 \end{pmatrix}$, $A = (2, -3)$

b. $\mathbf{n} = \begin{pmatrix} -1 \\ 3 \end{pmatrix}$, $A = (4, 7)$

c. $\mathbf{n} = \begin{pmatrix} 5 \\ -8 \end{pmatrix}$, $A = (5, 8)$

d. $\mathbf{n} = \begin{pmatrix} -2 \\ 9 \end{pmatrix}$, $A = (2, 4)$

e. $\mathbf{n} = \begin{pmatrix} 5 \\ -7 \end{pmatrix}$, $A = (8, 8)$

13.10

a. $\mathbf{n} = \begin{pmatrix} 2 \\ 2 \end{pmatrix}$, $A = (1, 3)$

b. $\mathbf{n} = \begin{pmatrix} 0 \\ 3 \end{pmatrix}$, $A = (-4, 2)$

c. $\mathbf{n} = \begin{pmatrix} 4 \\ -2 \end{pmatrix}$, $A = (-3, 0)$

d. $\mathbf{n} = \begin{pmatrix} 5 \\ -1 \end{pmatrix}$, $A = (-5, 2)$

e. $\mathbf{n} = \begin{pmatrix} 2 \\ -3 \end{pmatrix}$, $A = (2, -1)$

13.11

a. $\mathbf{n} = \begin{pmatrix} 1 \\ -2 \end{pmatrix}$, $A = (2, 3)$

b. $\mathbf{n} = \begin{pmatrix} -3 \\ 4 \end{pmatrix}$, $A = (-3, 8)$

c. $\mathbf{n} = \begin{pmatrix} -2 \\ 6 \end{pmatrix}$, $A = (-4, 7)$

d. $\mathbf{n} = \begin{pmatrix} 2 \\ 11 \end{pmatrix}$, $A = (-2, 7)$

e. $\mathbf{n} = \begin{pmatrix} -7 \\ -5 \end{pmatrix}$, $A = (5, -3)$

## The normal vector of a line

In many mathematical applications, *vectors* are used to represent quantities that have both magnitude and direction. A vector is an arrow with the corresponding magnitude and direction. Arrows with the same direction and length represent *the same* vector. The starting point of a vector thus can be chosen at will. Vectors are represented in boldface type.

If in a plane an orthonormal coordinate system $Oxy$ is given, a vector $\mathbf{v}$ in that plane can be given coordinates by choosing the origin as its starting point. The coordinates of the end point then are the coordinates of the vector $\mathbf{v}$. To distinguish them from point coordinates, we place the coordinates of a vector above each other. The vector $\mathbf{v} = \begin{pmatrix} a \\ b \end{pmatrix}$ is the vector represented by an arrow starting at the origin $O$ with the point $(a, b)$ as its endpoint, or by any other arrow with the same length and direction.

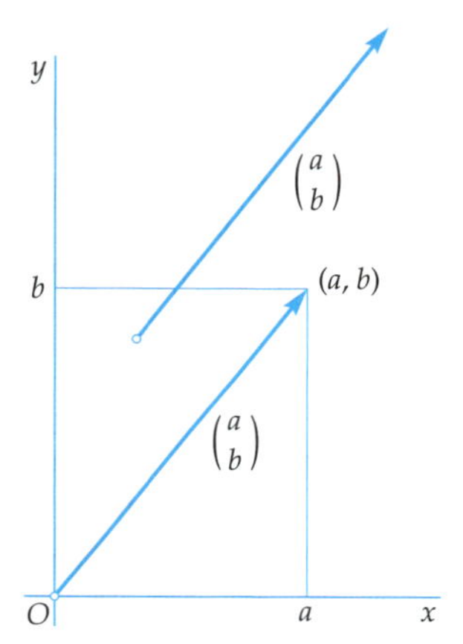


The perpendicular bisector of $(a, b)$ and $(-a, -b)$ is the set of all points with equal distances to $(a, b)$ and $(-a, -b)$. It consists of the points $(x, y)$ satisfying $\sqrt{(x-a)^2 + (y-b)^2} = \sqrt{(x+a)^2 + (y+b)^2}$. Squaring and simplifying yields the equation $ax + by = 0$.

This is the equation of a line through the origin. The vector $\begin{pmatrix} a \\ b \end{pmatrix}$ clearly is orthogonal to this line. Any vector orthogonal to a line is called a *normal vector* of that line. Since the equation $ax + by = c$ for any choice of $c$ represents a line parallel to the line $ax + by = 0$, we have:

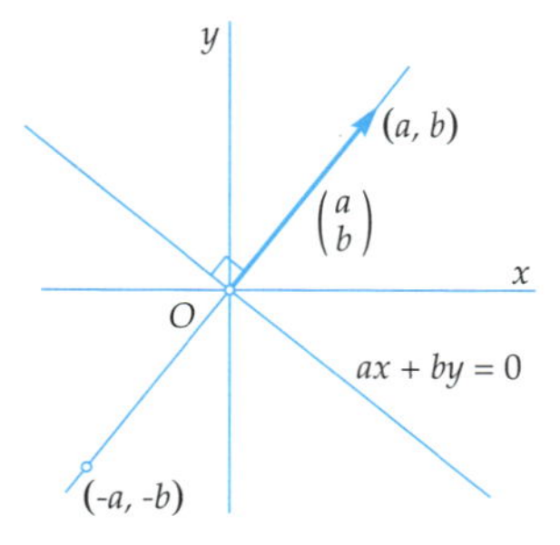


*The vector* $\begin{pmatrix} a \\ b \end{pmatrix}$ *is a normal vector to the line* $ax + by = c$ *for any* $c$.

In each of the following exercises, a point $A$ and the equation of a line are given. Find the equation of the line through $A$ intersecting the given line orthogonally.

*Example:* $A = (1, 2)$ and $3x + 4y = 5$. Each line perpendicular to this line has an equation of the form $4x - 3y = c$ (interchange the coefficients of $x$ and $y$ and add one minus sign). Substituting the coordinates of $A$ yields $4 \times 1 - 3 \times 2 = c$, giving $c = -2$. Hence, the equation of the required line is $4x - 3y = -2$.

13.12

a. $A = (2, 0)$, $2x - 3y = 4$
b. $A = (3, -2)$, $4x + 5y = -1$
c. $A = (-1, 1)$, $x - 7y = 2$
d. $A = (8, -6)$, $4x + 3y = 5$
e. $A = (-2, 1)$, $3x - 3y = 1$

13.13

a. $A = (0, 0)$, $4x - 9y = 1$
b. $A = (0, -3)$, $2x + 7y = -2$
c. $A = (-2, 1)$, $-x + 5y = 3$
d. $A = (4, 6)$, $4x + 5y = 8$
e. $A = (-4, 1)$, $2x - 7y = 6$

In each of the following exercises, a point $A$ and the equation of a line are given. Find the foot of the line through $A$ perpendicular to the given line (in other words, find the orthogonal projection of $A$ onto the given line).

*Example:* $A = (1, 2)$ and $3x + 4y = 5$. The perpendicular line through $A$ has the equation $4x - 3y = -2$ (see above). The intersection of the two lines is easily found as the point $(\frac{7}{25}, \frac{26}{25})$.

13.14

a. $A = (1, -2)$, $2x - 3y = 0$
b. $A = (1, 1)$, $x + y = -1$
c. $A = (2, 0)$, $2x - y = 1$
d. $A = (1, -1)$, $2x + y = -2$
e. $A = (-2, 2)$, $x - 3y = 3$

13.15

a. $A = (0, 5)$, $x - 4y = 1$
b. $A = (1, -3)$, $x + 2y = -2$
c. $A = (2, -1)$, $-x + y = 3$
d. $A = (-2, 2)$, $3x + y = 1$
e. $A = (4, 0)$, $2x - y = 6$

## Orthogonality of lines and vectors

We have seen that the vector $\begin{pmatrix} a \\ b \end{pmatrix}$ is a normal vector of each line of the form $ax + by = c$, and in particular of the line $ax + by = 0$ through the origin.

For each point $(c, d)$ on the line $ax + by = 0$ we have $ac + bd = 0$. Hence, the corresponding vector $\begin{pmatrix} c \\ d \end{pmatrix}$ is perpendicular to the vector $\begin{pmatrix} a \\ b \end{pmatrix}$. We denote this with the sign $\perp$, giving

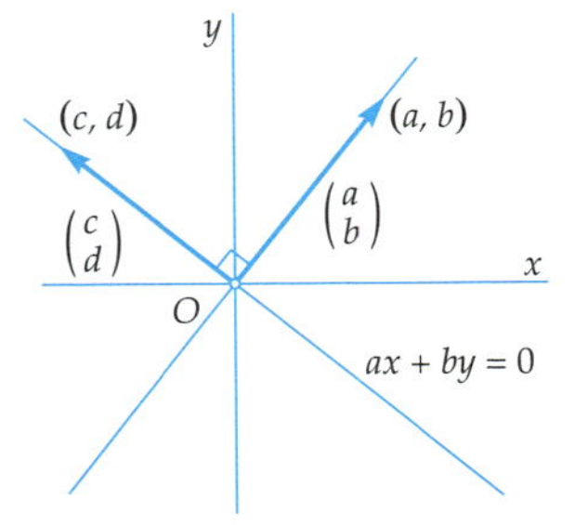


$$\begin{pmatrix} a \\ b \end{pmatrix} \perp \begin{pmatrix} c \\ d \end{pmatrix} \iff ac + bd = 0$$

In particular, the vectors $\begin{pmatrix} -b \\ a \end{pmatrix}$ and $\begin{pmatrix} b \\ -a \end{pmatrix}$ are perpendicular to the vector $\begin{pmatrix} a \\ b \end{pmatrix}$, since $(-b) \times a + a \times b = 0$ and $b \times a + (-a) \times b = 0$. These vectors all have the same length; the first results from $\begin{pmatrix} a \\ b \end{pmatrix}$ after an anticlockwise rotation over 90 degrees, the second after a clockwise rotation over 90 degrees.

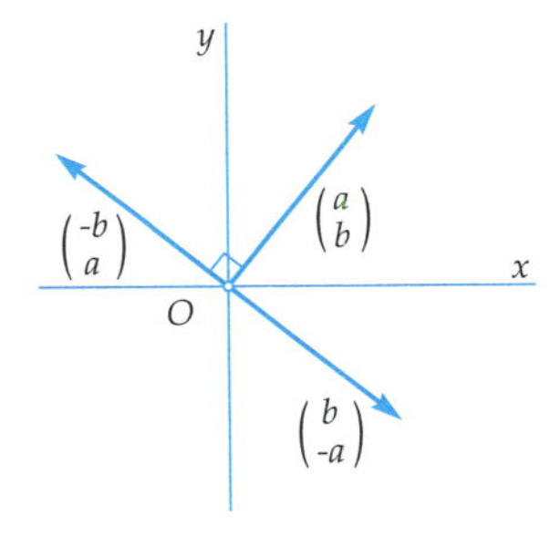


If we say that $\begin{pmatrix} a \\ b \end{pmatrix}$ is a normal vector of the line $ax + by = c$, we silently assumed that $a$ and $b$ are not both equal to zero, since otherwise $ax + by = c$ would not be the equation of a line. However, we can define the *null vector* by $\mathbf{0} = \begin{pmatrix} 0 \\ 0 \end{pmatrix}$. It is an 'arrow' without direction and with length zero. Since $a \times 0 + b \times 0 = 0$, by convention we say that the null vector is orthogonal to *any* vector, including itself.

In each of the following exercises, two vectors **a** and **b** are given. Calculate the cosine of the angle between these vectors and, using a calculator, the angle itself, rounded off to degrees.

*Example:* $\mathbf{a} = \begin{pmatrix}1\\1\end{pmatrix}$, $\mathbf{b} = \begin{pmatrix}1\\-2\end{pmatrix}$. Then $|\mathbf{a}| = \sqrt{2}$, $|\mathbf{b}| = \sqrt{5}$, $\langle \mathbf{a}, \mathbf{b} \rangle = -1$, giving $\cos\varphi = \frac{-1}{\sqrt{2}\sqrt{5}} = -\frac{1}{10}\sqrt{10} \approx -0.31623$, which yields $\varphi \approx 108°$.

13.16

a. $\mathbf{a} = \begin{pmatrix}1\\0\end{pmatrix}$, $\mathbf{b} = \begin{pmatrix}-1\\1\end{pmatrix}$

b. $\mathbf{a} = \begin{pmatrix}2\\-1\end{pmatrix}$, $\mathbf{b} = \begin{pmatrix}1\\2\end{pmatrix}$

c. $\mathbf{a} = \begin{pmatrix}3\\1\end{pmatrix}$, $\mathbf{b} = \begin{pmatrix}3\\2\end{pmatrix}$

d. $\mathbf{a} = \begin{pmatrix}4\\-2\end{pmatrix}$, $\mathbf{b} = \begin{pmatrix}0\\1\end{pmatrix}$

e. $\mathbf{a} = \begin{pmatrix}-2\\-1\end{pmatrix}$, $\mathbf{b} = \begin{pmatrix}2\\1\end{pmatrix}$

13.17

a. $\mathbf{a} = \begin{pmatrix}2\\-2\end{pmatrix}$, $\mathbf{b} = \begin{pmatrix}1\\2\end{pmatrix}$

b. $\mathbf{a} = \begin{pmatrix}-1\\0\end{pmatrix}$, $\mathbf{b} = \begin{pmatrix}1\\-1\end{pmatrix}$

c. $\mathbf{a} = \begin{pmatrix}4\\0\end{pmatrix}$, $\mathbf{b} = \begin{pmatrix}2\\3\end{pmatrix}$

d. $\mathbf{a} = \begin{pmatrix}5\\1\end{pmatrix}$, $\mathbf{b} = \begin{pmatrix}-1\\5\end{pmatrix}$

e. $\mathbf{a} = \begin{pmatrix}6\\-7\end{pmatrix}$, $\mathbf{b} = \begin{pmatrix}1\\1\end{pmatrix}$

In each of the following exercises, the equations of two lines are given. Calculate, using a calculator, the intersection angle of the lines, rounded off to degrees. Always take the angle less than or equal to 90 degrees.

*Example:* $x + y = -1$, $x - 2y = 4$. The angle between the lines is equal to the angle between the normal vectors $\begin{pmatrix}1\\1\end{pmatrix}$ and $\begin{pmatrix}1\\-2\end{pmatrix}$. Rounded off to degrees, this is 108° (see the example above). This, however, is an obtuse angle; the acute angle between the lines is, rounded off to degrees, $180° - 108° = 72°$.

13.18

a. $x + y = 3$, $2x - 3y = 4$

b. $x - 2y = 5$, $4x + 5y = -1$

c. $2x - 2y = 1$, $x + y = -3$

d. $2x - y = 3$, $x - y = 1$

e. $x - 2y = -1$, $x + 3y = -3$

13.19

a. $x + 2y = 0$, $2x + 3y = 1$

b. $-2x - y = 5$, $4x = -1$

c. $3x + y = 1$, $-4x + y = -2$

d. $6x - 7y = 1$, $-2x - 3y = 0$

e. $-3x - 2y = 2$, $3x + y = 2$

## The inner product

For each pair of vectors $\mathbf{a} = \begin{pmatrix} a_1 \\ a_2 \end{pmatrix}$ and $\mathbf{b} = \begin{pmatrix} b_1 \\ b_2 \end{pmatrix}$ we define the *inner product*, notation $\langle \mathbf{a}, \mathbf{b} \rangle$, by

$$\langle \mathbf{a}, \mathbf{b} \rangle = a_1 b_1 + a_2 b_2$$

Other names frequently used for the inner product are *scalar product* and *dot product*. In the latter case, the dot product of $\mathbf{a}$ and $\mathbf{b}$ usually is indicated as $\mathbf{a} \bullet \mathbf{b}$.

In the preceding section, we saw that $\mathbf{a} \perp \mathbf{b}$ if and only if $\langle \mathbf{a}, \mathbf{b} \rangle = 0$. Furthermore, fom Pythagoras' theorem it follows that the length $|\mathbf{a}|$ of a vector $\mathbf{a}$ is given by

$$|\mathbf{a}| = \sqrt{\langle \mathbf{a}, \mathbf{a} \rangle} = \sqrt{a_1^2 + a_2^2}$$

In general, the inner product has a geometric interpretation from which these properties follow as special cases. Indeed, it can be proven that

$$\langle \mathbf{a}, \mathbf{b} \rangle = |\mathbf{a}||\mathbf{b}| \cos \varphi$$

where $\varphi$ denotes the angle between the two vectors taken at the origin. (See pages 141 and 143 for the definition of the cosine of an angle and page 261 for a proof of this formula.)

Here, $|\mathbf{b}| \cos \varphi$ is the length of the projection of the vector $\mathbf{b}$ onto the line supporting the vector $\mathbf{a}$ (with a minus sign if $\varphi$ is obtuse, since then the cosine is negative). Of course, you can switch the roles of $\mathbf{a}$ and $\mathbf{b}$: the inner product is also the length of $\mathbf{b}$ times the length of the projection of the vector $\mathbf{a}$ onto the line supporting the vector $\mathbf{b}$.

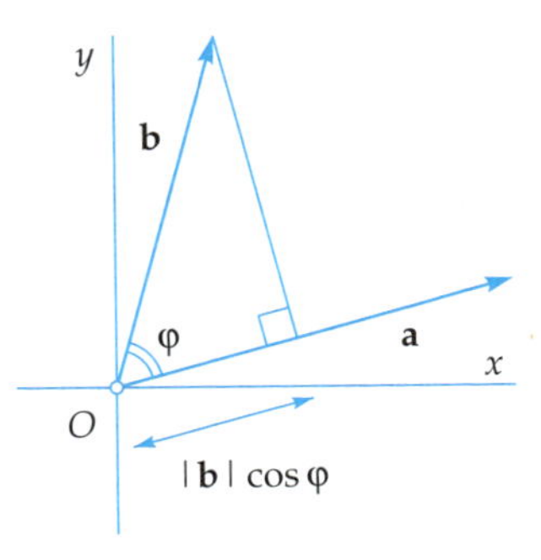


Geometrically seen, this also explains why the inner product is 0 when the vectors are perpendicular (since then $\cos \varphi = 0$), and why it is equal to the square of the length if both vectors are equal (since then $\varphi = 0$ and $\cos \varphi = 1$).

# 14 Circles

In each of the following exercises, a centre $M$ and a radius $r$ are given. Find the equation of the circle with this centre and radius and write it in the form $x^2 + y^2 + ax + by + c = 0$:

14.1

a. $M = (0,0)$ and $r = 2$
b. $M = (2,0)$ and $r = 2$
c. $M = (0,-3)$ and $r = 5$
d. $M = (1,2)$ and $r = 4$
e. $M = (-2,2)$ and $r = 2\sqrt{2}$

14.2

a. $M = (4,0)$ and $r = 1$
b. $M = (3,-2)$ and $r = \sqrt{13}$
c. $M = (2,-1)$ and $r = 5$
d. $M = (1,7)$ and $r = 7$
e. $M = (-5,12)$ and $r = 13$

Investigate which of the next equations represents a circle. If so, find its centre and radius:

14.3

a. $x^2 + y^2 + 4x - 2y + 1 = 0$
b. $x^2 + y^2 + x - y - 1 = 0$
c. $x^2 + y^2 + 2x + 2y = 0$
d. $x^2 + y^2 - 8x + 12 = 0$
e. $x^2 + y^2 + 4x - 2y + 6 = 0$

14.4

a. $x^2 + y^2 = 4x - 5$
b. $x^2 + y^2 = 4x + 5$
c. $x^2 + y^2 = 4y - 4$
d. $3x^2 + 3y^2 = 2y$
e. $4x^2 + 4y^2 - 16x - 8y + 19 = 0$

Find an equation of the circle through the three given points. First, draw a figure on graph paper. The points have been chosen in such a way that it is easy to find the centre and the radius from the figure:

14.5

a. $(0,0)$, $(2,0)$ and $(0,2)$
b. $(0,0)$, $(2,0)$ and $(0,4)$
c. $(0,0)$, $(6,0)$ and $(0,8)$
d. $(0,0)$, $(2,2)$ and $(2,-2)$
e. $(3,4)$, $(3,0)$ and $(0,4)$

14.6

a. $(1,1)$, $(1,5)$ and $(5,1)$
b. $(-2,0)$, $(-2,2)$ and $(2,2)$
c. $(1,-2)$, $(1,0)$ and $(-1,-2)$
d. $(3,3)$, $(3,1)$ and $(1,3)$
e. $(-1,-2)$, $(-1,0)$ and $(0,-1)$

## Circle equations

The circle with centre $M = (m, n)$ and radius $r$ is the set of all points with distance $r$ to $M$. If such a point $P$ has coordinates $(x, y)$, then $d(P, M) = r$, i.e.

$$\sqrt{(x-m)^2 + (y-n)^2} = r$$

or, after squaring both sides:

$$(x-m)^2 + (y-n)^2 = r^2$$

*This is the equation of the circle with centre $(m, n)$ and radius $r$.* All points $(x, y)$ satisfying this equation are on the circle, and, conversely, all points on the circle satisfy this equation.

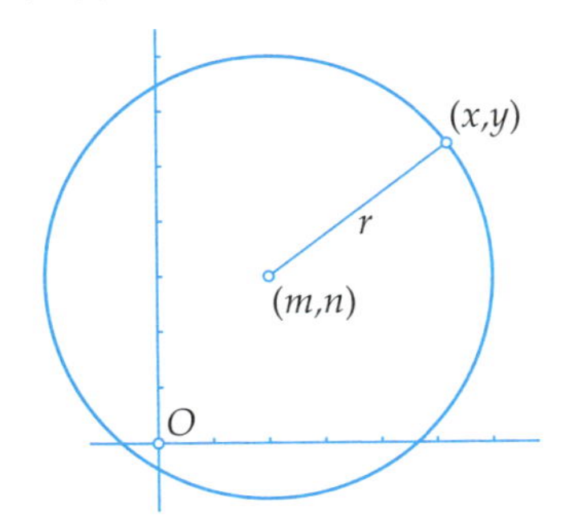


All points $(x, y)$ *inside* the circle satisfy $(x-m)^2 + (y-n)^2 < r^2$ and all points *outside* the circle satisfy $(x-m)^2 + (y-n)^2 > r^2$.

*Example:* in the figure above the circle with centre $(2, 3)$ and radius 4 has been drawn. Its equation is $(x-2)^2 + (y-3)^2 = 16$. After expanding the brackets and reordering, this can be written as

$$x^2 + y^2 - 4x - 6y - 3 = 0$$

In general, any circle equation can be written in the form

$$x^2 + y^2 + ax + by + c = 0$$

but not every equation of this form represents a circle. For instance, take

$$x^2 + y^2 - 4x - 6y + 14 = 0$$

After completing the squares (see page 81) this can be written as

$$(x-2)^2 + (y-3)^2 = -1$$

since $(x-2)^2 = x^2 - 4x + 4$ and $(y-3)^2 = y^2 - 6y + 9$. The left-hand side is the sum of two squares, so it is greater than or equal to zero, while the right-hand side is negative. This means that there cannot be any point $(x, y)$ satisfying this equation, and, consequently, it cannot represent a circle.

You can check that the equation $x^2 + y^2 - 4x - 6y + 13 = 0$ is satisfied by just one point. It is a 'circle' with radius 0.

Find the intersection points of the following circles with the coordinate axes, if any:

14.7

a. $x^2 + y^2 + 4x - 2y + 1 = 0$
b. $x^2 + y^2 + x - y - 1 = 0$
c. $x^2 + y^2 + 2x + 2y = 0$
d. $x^2 + y^2 - 8x + 12 = 0$
e. $x^2 + y^2 + 3x - 4y + 1 = 0$

14.8

a. $x^2 + y^2 = 4x + 5$
b. $x^2 + y^2 = 4x + 6y - 5$
c. $x^2 + y^2 = 4y - 2$
d. $3x^2 + 3y^2 = 2y$
e. $4x^2 + 4y^2 - 16x - 8y + 19 = 0$

In each of the following exercises, a circle and a line are given. Find their intersection points, if any:

14.9

a. $x^2 + y^2 = 9$ and $x = 2$
b. $x^2 + y^2 = 9$ and $x = 2y$
c. $x^2 + y^2 = 9$ and $x + y = 3$
d. $x^2 + y^2 = 9$ and $x + 2y = -3$
e. $x^2 + y^2 = 9$ and $x - y = 3\sqrt{2}$

14.10

a. $x^2 + y^2 = 16$ and $y = -2$
b. $x^2 + y^2 = 16$ and $3x = 4y$
c. $x^2 + y^2 = 16$ and $x + y = -4$
d. $x^2 + y^2 = 16$ and $x - 2y = 4$
e. $x^2 + y^2 = 16$ and $x = \sqrt{3}y$

14.11

a. $x^2 + y^2 + 10x - 8y = 0$ and $x = y$
b. $x^2 + y^2 - 6x - 8y + 21 = 0$ and $x + y = 7$
c. $x^2 + y^2 + 12x + 11 = 0$ and $x - y = -1$
d. $x^2 + y^2 - 16x - 4y + 4 = 0$ and $3x + y = 2$
e. $x^2 + y^2 + 4x - 10y + 20 = 0$ and $-2x + y = 3$

14.12 In each of the following exercises, a line $L$, a point $P$ and a distance $d$ are given. Find all points on $L$ with distance $d$ to $P$. It is a good idea to plot a figure first.

a. $L:\ x = 1,\quad P = (-3, 1),\quad d = 5$
b. $L:\ -x + 4y = 13,\quad P = (2, -2),\quad d = 2$
c. $L:\ x + y = 1,\quad P = (0, 0),\quad d = 5$
d. $L:\ -x + 3y = 4,\quad P = (-4, -1),\quad d = \sqrt{13}$
e. $L:\ 2x - y = 1,\quad P = (1, -1),\quad d = 2$

## The intersection points of a circle and a line

A circle and a line may have two, one or no points in common. If there is only one intersection point, then the line is tangent to the circle in that point. We will give an example of how intersection points can be found. Suppose that the circle and the line are given by the equations

$$x^2 + y^2 - 4x - 6y - 3 = 0 \qquad \text{and} \qquad x + 2y = 3$$

The equation of the line can be written as $x = 3 - 2y$. Substituting this into the circle equation yields

$$(3 - 2y)^2 + y^2 - 4(3 - 2y) - 6y - 3 = 0$$

This is a quadratic equation in $y$ that, by expanding the brackets and reordering, can be written as

$$5y^2 - 10y - 6 = 0$$

(check this!).

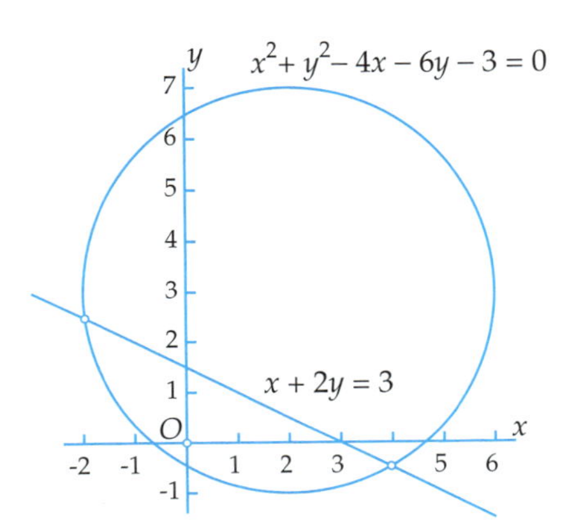


The *abc*-formula (see page 83) yields the solutions:

$$y_1 = \frac{10 + \sqrt{220}}{10} = 1 + \frac{1}{5}\sqrt{55} \quad \text{and} \quad y_2 = \frac{10 - \sqrt{220}}{10} = 1 - \frac{1}{5}\sqrt{55}$$

Substituting this into the equation $x = 3 - 2y$ of the line yields

$$x_1 = 1 - \frac{2}{5}\sqrt{55} \qquad \text{and} \qquad x_2 = 1 + \frac{2}{5}\sqrt{55}$$

The two intersection points thus are

$$\left(1 - \frac{2}{5}\sqrt{55}, 1 + \frac{1}{5}\sqrt{55}\right) \qquad \text{and} \qquad \left(1 + \frac{2}{5}\sqrt{55}, 1 - \frac{1}{5}\sqrt{55}\right)$$

Find the intersection points of the following pairs of circles, if any:

14.13

a. $x^2 + y^2 = 4$
$x^2 + y^2 - 4x = 0$

b. $x^2 + y^2 = 9$
$x^2 + y^2 - 4x + 2y = 3$

c. $x^2 + y^2 = 25$
$x^2 + y^2 + 6x + 2y + 1 = 0$

d. $x^2 + y^2 = 4$
$x^2 + y^2 - 2x - 2y = 0$

e. $x^2 + y^2 = 36$
$x^2 + y^2 - 4x - 4y + 4 = 0$

14.14

a. $x^2 + y^2 + 2x + 2y = 0$
$x^2 + y^2 - 4x = 0$

b. $x^2 + y^2 - 2x - 4y = 0$
$x^2 + y^2 - 4x + 2y = 0$

c. $x^2 + y^2 - 6x + 2y + 6 = 0$
$x^2 + y^2 + 6x + 2y + 6 = 0$

d. $x^2 + y^2 - 2x - 8y + 8 = 0$
$x^2 + y^2 - 4x - 4y + 6 = 0$

e. $x^2 + y^2 + 4x + 4y + 4 = 0$
$x^2 + y^2 - 8x - 2y + 12 = 0$

14.15

a. $x^2 + y^2 + 4x - 2y = 5$
$x^2 + y^2 - 2x + 4y = 11$

b. $x^2 + y^2 - 2x - 8y + 8 = 0$
$x^2 + y^2 - 3x + 2y = 1$

c. $x^2 + y^2 - 5x - y - 6 = 0$
$x^2 + y^2 + 3x + 2y + 2 = 0$

d. $x^2 + y^2 - x - 5y + 2 = 0$
$x^2 + y^2 - 4x - 4y - 2 = 0$

e. $x^2 + y^2 - 4x + 4y + 3 = 0$
$x^2 + y^2 - 8x - 2y + 15 = 0$

14.16

a. $x^2 + y^2 + 2x = 3$
$x^2 + y^2 - 6x + 5 = 0$

b. $x^2 + y^2 - 3x - y = 1$
$x^2 + y^2 + 4x + 3y + 4 = 0$

c. $x^2 + y^2 - 6x - 2y + 8 = 0$
$x^2 + y^2 + 3x + 2y - 7 = 0$

d. $x^2 + y^2 - x + y = 0$
$x^2 + y^2 - 4x + 4y + 6 = 0$

e. $x^2 + y^2 + 4y - 1 = 0$
$x^2 + y^2 - 8x + 3 = 0$

## The intersection points of two circles

Two distinct circles have two, one or no common points. If they have one common point, they touch each other in this point. Of course, concentric circles (circles with the same centre) never intersect. From now on, we will always take circles with distinct centres. We will give an example of how the intersection points, if any, can be found.

Suppose that the circles are given by the equations

$$\begin{aligned} x^2 + y^2 + 2x - 6y + 1 &= 0 \\ x^2 + y^2 - 4x - 4y - 5 &= 0 \end{aligned}$$

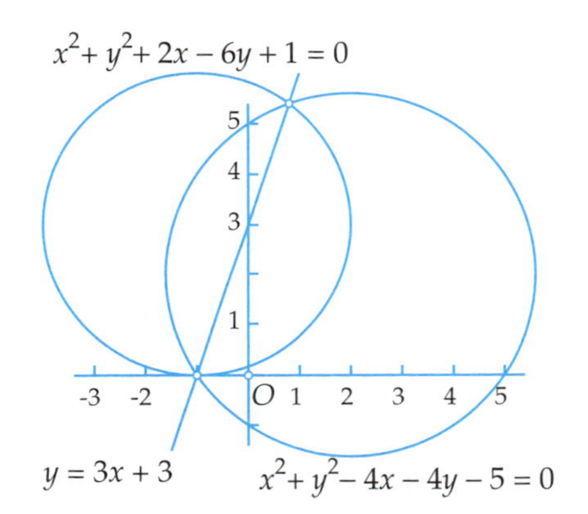


Each intersection point $(x, y)$ is a solution of this system of two equations. By subtracting the second equation from the first, we get the *linear* equation

$$6x - 2y + 6 = 0$$

which can be written as $y = 3x + 3$.

Substituting this into one of the two circle equations, for instance into the first one, results in

$$x^2 + (3x + 3)^2 + 2x - 6(3x + 3) + 1 = 0$$

or

$$10x^2 + 2x - 8 = 0$$

The *abc*-formula yields $x_1 = \frac{4}{5}$ and $x_2 = -1$. Since every solution $(x, y)$ also should satisfy the linear equation $y = 3x + 3$, we get $y_1 = 3 \times \frac{4}{5} + 3 = \frac{27}{5}$ and $y_2 = 3 \times (-1) + 3 = 0$. The intersection points thus are

$$\left(\frac{4}{5}, \frac{27}{5}\right) \qquad \text{and} \qquad (-1, 0)$$

You can check this by substituting these points into the original circle equations.

14.17 Find an equation of:

a. the circle with centre $(0,0)$ tangent to the line $x = 4$.
b. the circle with centre $(2,0)$ tangent to the line $x = y$.
c. the circle with centre $(0,2)$ tangent to the line $4x = 3y$.
d. the circle with centre $(-1,-1)$ tangent to the line $x + 2y = 0$.
e. the circle with centre $(1,2)$ tangent to the line $x + y = -1$.

Hint: plot out a figure first!

14.18 Find an equation of the tangent line to the given circle in the given point $A$:

a. $x^2 + y^2 = 5, \quad A = (1,2)$
b. $x^2 + y^2 = 2, \quad A = (1,-1)$
c. $x^2 + y^2 - 2x - 4y + 4 = 0, \quad A = (1,1)$
d. $x^2 + y^2 + 2x + 6y - 8 = 0, \quad A = (2,0)$
e. $x^2 + y^2 + 6x - 8 = 0, \quad A = (1,-1)$

14.19

a. Find all the intersection points of the circle in the figure on the next page with the two coordinate axes.
b. Find an equation of the tangent line to the circle in each of these intersection points.

14.20 For each of the following circles, find equations of the horizontal and vertical tangent lines:

a. $x^2 + y^2 + 2x = 2$
b. $x^2 + y^2 + 4x - 6y = 20$
c. $x^2 + y^2 - 2x - 4y - 12 = 0$
d. $x^2 + y^2 + 2x + 8y = 0$
e. $x^2 + y^2 - 6y - 2x - 2 = 0$

## Tangent lines to a circle

The figure to the right shows the circle

$$x^2 + y^2 - 4x - 4y - 5 = 0$$

with centre $M = (2,2)$. In the point $A = (-1,0)$ on this circle, the tangent line is drawn. How can we find an equation of this tangent line?

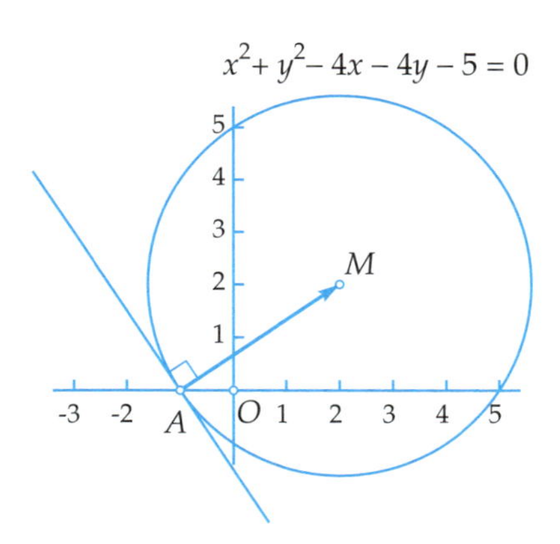


Since the radius $MA$ is orthogonal to the tangent, the vector $\mathbf{r}$ from $A$ to $M$ is a normal vector to the tangent line. Its coordinates are $\begin{pmatrix} 2-(-1) \\ 2-0 \end{pmatrix} = \begin{pmatrix} 3 \\ 2 \end{pmatrix}$.

Now that we know a normal vector of the tangent line, we know that its equation must be $3x + 2y = c$ for some $c$. Since $A = (-1,0)$ is on the tangent line, we can find $c$ by substituting $A$, which yields $c = -3$. The equation of the tangent line thus is

$$3x + 2y = -3$$

The method always works:

*If $M = (m_1, m_2)$ is the centre of a circle and $A = (a_1, a_2)$ is a point on the circle, the equation of the tangent line to the circle in $A$ is given by*

$$(m_1 - a_1)x + (m_2 - a_2)y = (m_1 - a_1)a_1 + (m_2 - a_2)a_2$$

Note that this equation can also be written as

$$(m_1 - a_1)(x - a_1) + (m_2 - a_2)(y - a_2) = 0$$

# 15 Geometry in space

Calculate the distance between the following pairs of points:

15.1

a. $(0,0,0)$ and $(1,0,-3)$
b. $(2,0,1)$ and $(-2,1,0)$
c. $(0,0,0)$ and $(-1,1,-5)$
d. $(3,-1,1)$ and $(2,-3,3)$
e. $(1,2,2)$ and $(-4,0,0)$

15.2

a. $(1,2,-1)$ and $(0,1,-2)$
b. $(3,2,-1)$ and $(4,-1,-2)$
c. $(-1,-3,0)$ and $(3,0,1)$
d. $(-1,1,0)$ and $(0,0,-2)$
e. $(1,1,1)$ and $(-2,2,2)$

Calculate the coordinates of the vector (arrow) that starts in the given point $A$ and ends in the given point $B$:

15.3

a. $A = (3,1,0)$, $B = (0,0,3)$
b. $A = (1,3,0)$, $B = (-1,2,1)$
c. $A = (2,-1,0)$, $B = (1,5,-1)$
d. $A = (0,-2,1)$, $B = (3,0,5)$
e. $A = (2,-1,1)$, $B = (1,-4,2)$

15.4

a. $A = (0,3,2)$, $B = (1,1,-2)$
b. $A = (3,2,1)$, $B = (0,4,-1)$
c. $A = (-2,3,-1)$, $B = (1,3,5)$
d. $A = (-1,2,2)$, $B = (0,2,-2)$
e. $A = (-1,1,-1)$, $B = (0,2,2)$

Calculate the cosine of $\angle AOB$ ($O$ is the origin) and, using a calculator, calculate the angle rounded off to degrees.

15.5

a. $A = (0,1,0)$, $B = (0,2,3)$
b. $A = (1,-3,1)$, $B = (0,2,1)$
c. $A = (2,-1,2)$, $B = (1,3,-1)$
d. $A = (-1,-2,0)$, $B = (3,0,1)$
e. $A = (0,-1,1)$, $B = (0,4,-4)$

15.6

a. $A = (0,1,2)$, $B = (1,-1,1)$
b. $A = (0,2,1)$, $B = (0,1,-2)$
c. $A = (-2,3,1)$, $B = (1,-3,5)$
d. $A = (-1,2,1)$, $B = (0,1,-2)$
e. $A = (-2,1,-1)$, $B = (0,2,1)$

15.7 In a cube, calculate the cosine of the angle between an edge and a solid diagonal (a diagonal connecting opposite vertices). Then, using a calculator, calculate the angle, rounded off to degrees.
*Hint:* choose a suitable coordinate system.

## Coordinates and the inner product in space

In space, we use three coordinates. An orthonormal coordinate system $Oxyz$ is a system with three mutually perpendicular coordinate axes with equal scales.

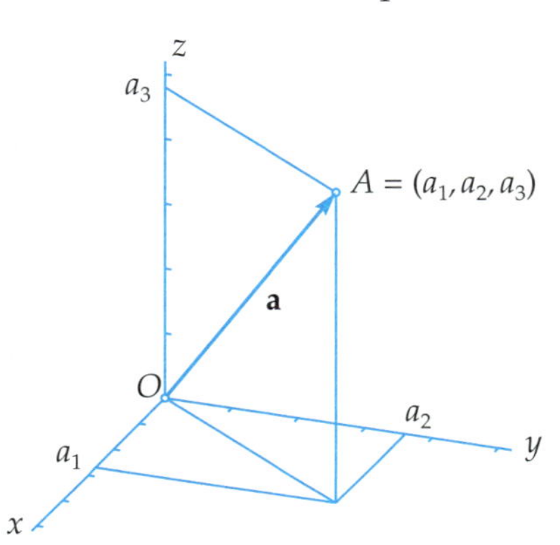


By Pythagoras' theorem (applied twice) for any point $A$ with coordinates $(a_1, a_2, a_3)$, the distance to the origin is equal to $\sqrt{a_1^2 + a_2^2 + a_3^2}$. This is also the length $|\mathbf{a}|$ of the vector $\mathbf{a} = \begin{pmatrix} a_1 \\ a_2 \\ a_3 \end{pmatrix}$, i.e. the arrow connecting the origin to the point $A$. In general, the distance between two points $A = (a_1, a_2, a_3)$ and $B = (b_1, b_2, b_3)$ is given by

$$d(A, B) = \sqrt{(a_1 - b_1)^2 + (a_2 - b_2)^2 + (a_3 - b_3)^2}$$

In space, the inner product of the vectors $\mathbf{a} = \begin{pmatrix} a_1 \\ a_2 \\ a_3 \end{pmatrix}$ and $\mathbf{b} = \begin{pmatrix} b_1 \\ b_2 \\ b_3 \end{pmatrix}$ is given by

$$\langle \mathbf{a}, \mathbf{b} \rangle = a_1 b_1 + a_2 b_2 + a_3 b_3$$

Just as in the plane, the following formula holds

$$\langle \mathbf{a}, \mathbf{b} \rangle = |\mathbf{a}||\mathbf{b}| \cos \varphi$$

where $\varphi$ is the angle between $\mathbf{a}$ and $\mathbf{b}$. The inner product is zero if the vectors are orthogonal and conversely:

$$\mathbf{a} \perp \mathbf{b} \quad \iff \quad \langle \mathbf{a}, \mathbf{b} \rangle = 0$$

When $\mathbf{a} = \mathbf{b}$, the inner product is equal to the square of the length of $\mathbf{a}$:

$$\langle \mathbf{a}, \mathbf{a} \rangle = |\mathbf{a}|^2$$

Just as in the plane, the inner product can be used to calculate the angle between two vectors. First, the cosine is calculated using

$$\cos \varphi = \frac{\langle \mathbf{a}, \mathbf{b} \rangle}{|\mathbf{a}||\mathbf{b}|}$$

and next, using a calculator, the angle itself (in degrees or in radians).

15.8 Find the intersection points with the coordinate axes of each of the following planes:

a. $3x + y - z = 3$
b. $4x + 2y + 3z = 1$
c. $ax + by + cz = 1$ (with $a, b, c, \neq 0$)

15.9 Find an equation of the plane through the three given points:

a. $(1,0,0)$, $(0,1,0)$ and $(0,0,1)$
b. $(2,0,0)$, $(0,3,0)$ and $(0,0,4)$
c. $(1,0,0)$, $(0,-1,0)$ and $(0,0,-3)$
d. $(-1,0,0)$, $(0,1,0)$ and $(0,0,0)$ *Hint:* plot a figure!
e. $(1,0,0)$, $(1,1,0)$ and $(1,1,1)$ *Hint:* plot a figure!

Find an equation of the perpendicular bisecting plane of the given pairs of points:

15.10

a. $(1,1,0)$ and $(0,1,1)$
b. $(2,1,0)$ and $(1,0,4)$
c. $(1,0,1)$ and $(0,1,-3)$
d. $(1,1,1)$ and $(0,0,0)$
e. $(2,1,1)$ and $(1,1,2)$

15.11

a. $(1,-1,2)$ and $(1,1,1)$
b. $(3,1,-1)$ and $(1,5,1)$
c. $(0,0,1)$ and $(2,1,-3)$
d. $(1,1,-1)$ and $(4,0,0)$
e. $(2,2,1)$ and $(1,2,2)$

Find an equation of the plane through $A$ with normal vector $\mathbf{n}$:

15.12

a. $A = (1,1,0), \mathbf{n} = \begin{pmatrix} 1 \\ 0 \\ 0 \end{pmatrix}$

b. $A = (0,1,2), \mathbf{n} = \begin{pmatrix} 1 \\ 0 \\ 1 \end{pmatrix}$

c. $A = (2,1,-1), \mathbf{n} = \begin{pmatrix} -2 \\ 3 \\ 1 \end{pmatrix}$

d. $A = (0,5,5), \mathbf{n} = \begin{pmatrix} 1 \\ 1 \\ -1 \end{pmatrix}$

e. $A = (3,1,3), \mathbf{n} = \begin{pmatrix} 0 \\ 1 \\ 2 \end{pmatrix}$

15.13

a. $A = (-2,0,0), \mathbf{n} = \begin{pmatrix} 3 \\ -2 \\ 0 \end{pmatrix}$

b. $A = (1,4,1), \mathbf{n} = \begin{pmatrix} 0 \\ 3 \\ -1 \end{pmatrix}$

c. $A = (5,1,-2), \mathbf{n} = \begin{pmatrix} 3 \\ -1 \\ 1 \end{pmatrix}$

d. $A = (6,0,0), \mathbf{n} = \begin{pmatrix} 1 \\ 0 \\ 0 \end{pmatrix}$

e. $A = (4,0,4), \mathbf{n} = \begin{pmatrix} 0 \\ 1 \\ 0 \end{pmatrix}$

## Planes and normal vectors

A linear equation in $x$, $y$ and $z$, such as

$$15x + 20y + 12z = 60$$

represents a plane in three-dimensional space. The intersection with the $x$-axis can be found by putting $y = z = 0$, from which it follows that $15x = 60$, so $x = 4$. The intersection point therefore is $(4, 0, 0)$. In a similar way, you can find the intersection point $(0, 3, 0)$ with the $y$-axis and the intersection point $(0, 0, 5)$ with the $z$-axis.

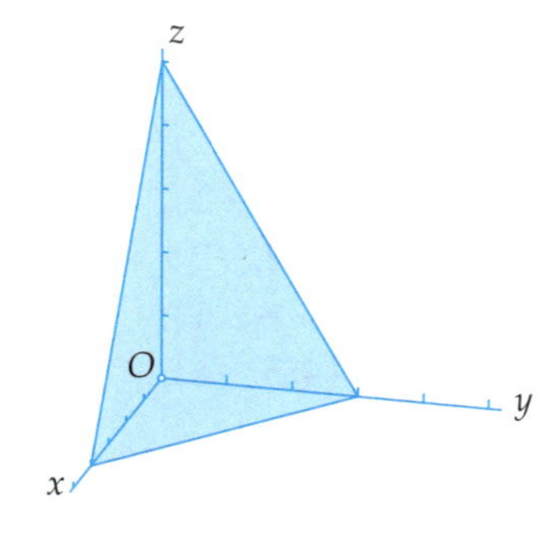


However, if the coefficient of $x$, $y$ or $z$ in the equation of the plane is zero, there is no intersection point with the corresponding axis. The plane then is parallel to that axis. For instance, the plane $2x + 3y = 4$ is parallel to the $z$-axis.

The *perpendicular bisecting plane* of two points $A = (a_1, a_2, a_3)$ and $B = (b_1, b_2, b_3)$ is the set of all points $P = (x, y, z)$ for which $d(P, A) = d(P, B)$.

As an example, we determine the perpendicular bisecting plane of the point $A = (3, 3, 2)$ and the origin $O = (0, 0, 0)$. The equation $d(P, A) = d(P, O)$ with $P = (x, y, z)$ after squaring and reordering yields (check this!)

$$3x + 3y + 2z = 11$$

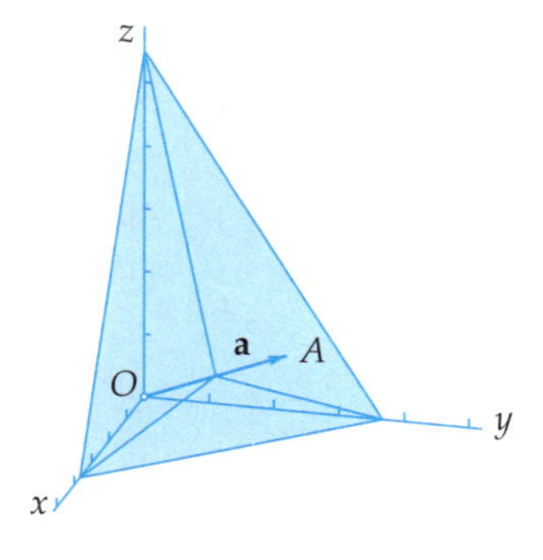


The accompanying figure shows this plane. The corresponding vector **a** from $O$ to $A$ is perpendicular to this plane. It is called a *normal vector* of the plane.

In general, the vector $\begin{pmatrix} a \\ b \\ c \end{pmatrix}$ (where at least one of $a$, $b$ and $c$ is not zero) is a normal vector to every plane $ax + by + cz = d$ (with arbitrary $d$). If $P = (x_0, y_0, z_0)$ is a point in such a plane, an equation of the plane is given by

$$ax + by + cz = ax_0 + by_0 + cz_0$$

or, written in a slightly different form,

$$a(x - x_0) + b(y - y_0) + c(z - z_0) = 0$$

15.14 The figure on the next page shows the intersection points of the planes $\alpha$ and $\beta$ with some of the edges of the given cube. Check that the given coordinates of these points are correct.

15.15 In each of the following exercises, a point $A$ and a plane $\alpha$ are given. Find an equation of the plane through $A$ parallel to $\alpha$:

a. $A = (0, 0, -4)$, $\alpha: 3x + 2y - 4z = 7$
b. $A = (1, -1, 0)$, $\alpha: 2x - 2y - 3z = 1$
c. $A = (1, 2, -1)$, $\alpha: -2x + 3y - z = 2$
d. $A = (0, 2, -2)$, $\alpha: 5x - y + 7z = 0$
e. $A = (1, -2, 1)$, $\alpha: x + 2z = 3$
f. $A = (4, 5, -6)$, $\alpha: x = 7$

15.16 Find the intersection point of the intersecting line of the planes $\alpha$ and $\beta$ with the plane $z = 1$ (if it exists):

a. $\begin{cases} x - 3y + 2z = 6 & (\alpha) \\ 2x - y - z = 2 & (\beta) \end{cases}$

b. $\begin{cases} 4x + 2y - 2z = 6 & (\alpha) \\ -2x + 3y + 5z = 1 & (\beta) \end{cases}$

c. $\begin{cases} 3x - 3y - 4z = 5 & (\alpha) \\ 4x - 2z = 6 & (\beta) \end{cases}$

d. $\begin{cases} -x + 5y - 3z = 7 & (\alpha) \\ 5x + y = 2 & (\beta) \end{cases}$

e. $\begin{cases} x + 3y + 5z = 5 & (\alpha) \\ -x + y + z = 5 & (\beta) \end{cases}$

f. $\begin{cases} x = 4 & (\alpha) \\ 3x + y + z = 9 & (\beta) \end{cases}$

g. $\begin{cases} 3x + 5y - 2z = 3 & (\alpha) \\ -x - y + z = 0 & (\beta) \end{cases}$

h. $\begin{cases} x + 2y - 2z = 2 & (\alpha) \\ 3x + 6y - z = 1 & (\beta) \end{cases}$

## Parallel planes and intersecting planes

Two distinct planes are either parallel or they intersect in a line. If they are parallel, this is immediately clear from their equations, since in that case the normal vectors are multiples of each other. If the two planes intersect, there are infinitely many points lying in both planes, and together they form the intersecting line.

We will give an example of the latter case: plane $\alpha$ with equation $x - 2y + 2z = 1$ and plane $\beta$ with equation $2x + y - z = 2$. All the intersection points $(x, y, z)$ then satisfy the system of equations

$$\begin{cases} x - 2y + 2z = 1 & (\alpha) \\ 2x + y - z = 2 & (\beta) \end{cases}$$

This is a system of two equations with three unknown variables. It has infinitely many solutions. Indeed, one of the unknown variables – in this case we may take $z$ – can be chosen freely, after which the other two variables can be solved from the system.

For instance, if we take $z = 0$, we get $x - 2y = 1$ and $2x + y = 2$, with $x = 1$ and $y = 0$ as solutions. Hence, the point $(1, 0, 0)$ is on the intersecting line. A different choice, e.g. $z = 1$, yields a different point on the intersecting line, namely $(1, 1, 1)$, which you can verify yourself. For $z = -1$, you get the point $(1, -1, -1)$. With this method you can find as many points on the intersecting line as you want.

The figure shows the planes $\alpha$ and $\beta$, as far as they fall inside the cube with faces $x = \pm 1$, $y = \pm 1$ and $z = \pm 1$. (The origin is the centre of the cube.) The intersecting line of $\alpha$ and $\beta$ is the line through $(1, -1, -1)$ and $(1, 1, 1)$. Note that in this case, the intersecting line is in the plane $x = 1$. This means that, unlike $y$ or $z$, the value of $x$ *cannot* be chosen freely if you want to find points on the intersecting line. The first coordinate of each point on this line is always equal to 1. You can verify that each point on the intersecting line has coordinates $(1, t, t)$ for some $t$.

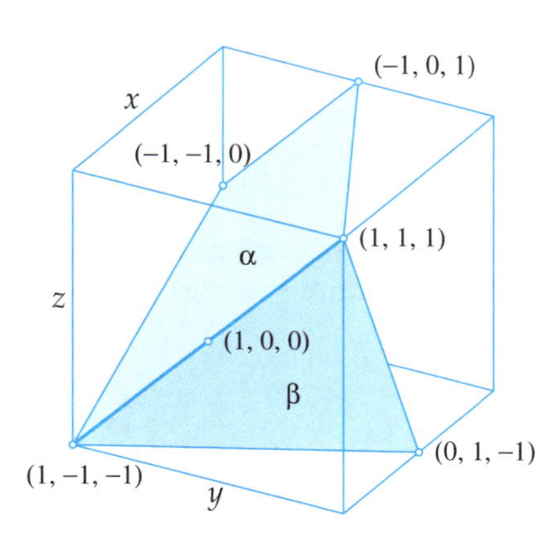

15.17 In each of the following exercises, three planes $\alpha$, $\beta$ and $\gamma$ are given. Describe their mutual positions. Find the intersection point, if they meet in one point, and two points on the intersecting line if they meet in one line:

a. $\begin{cases} x + 2y + 2z = 3 & (\alpha) \\ x - 3y + 3z = 4 & (\beta) \\ 3x - y + z = 4 & (\gamma) \end{cases}$

b. $\begin{cases} 2x - 3y - 3z = 3 & (\alpha) \\ x - 4y - 2z = 0 & (\beta) \\ 2x + y + 3z = -5 & (\gamma) \end{cases}$

c. $\begin{cases} -x - 4y + 3z = -3 & (\alpha) \\ 2x - 3y - z = -5 & (\beta) \\ 2x + 2y = 0 & (\gamma) \end{cases}$

d. $\begin{cases} x - 2y + 2z = 1 & (\alpha) \\ -2x + y + z = 5 & (\beta) \\ -3x + 6y - 6z = 4 & (\gamma) \end{cases}$

e. $\begin{cases} x - 3y + 2z = 2 & (\alpha) \\ x - 2y + 4z = 1 & (\beta) \\ - y - 2z = 1 & (\gamma) \end{cases}$

f. $\begin{cases} 4x + 4z = 8 & (\alpha) \\ x + 3y - 3z = -5 & (\beta) \\ 3x - y - z = 3 & (\gamma) \end{cases}$

g. $\begin{cases} x + 3y - 3z = 1 & (\alpha) \\ x - 3y + 2z = 4 & (\beta) \\ 2x + 6y - 6z = 2 & (\gamma) \end{cases}$

h. $\begin{cases} x - 6y - 3z = 3 & (\alpha) \\ 2x - 2y + 3z = -3 & (\beta) \\ 2x + 4y - 2z = 2 & (\gamma) \end{cases}$

15.18 Give a geometric explanation of what happened with each of the systems of equations in the exercises on page 86.

## The three planes theorem

With three distinct planes, there are several possibilities for their mutual positions. The planes are given by three linear equations, and common points (if any) can be found by solving the corresponding system of three equations.

If at least two of the planes are parallel, there are no common points. This can be seen immediately from the equations, since parallel planes have normal vectors that are multiples of each other.

If no two planes are parallel, there still are three possibilities. These are summarized in the following theorem:

> **Three planes theorem:** For three distinct planes, no two of them parallel, there are three possibilities:
> – they intersect in one point, or
> – they intersect in one line, or
> – they intersect pair-wise in three parallel lines.

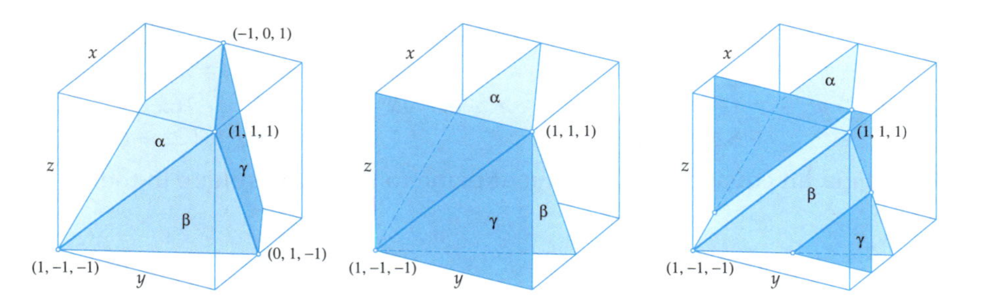


The possibilities are illustrated above for planes $\alpha$, $\beta$ and $\gamma$, where in each case we have chosen

$$\alpha: \quad x - 2y + 2z = 1 \qquad \text{and} \qquad \beta: \quad 2x + y - z = 2$$

In the first case, for $\gamma$ the plane $2x - 4y - z = -3$ through the points $(1, 1, 1)$, $(-1, 0, 1)$ and $(0, 1, -1)$ is taken. The intersection point of $\alpha$, $\beta$ and $\gamma$ then is $(1, 1, 1)$.

In the second case, for $\gamma$ the plane $x = 1$ is taken. The common intersection line here is the line through $(1, 1, 1)$ and $(1, -1, -1)$.

In the third case, $\gamma$ is the plane $x = \frac{1}{2}$.

It is clear that the intersecting line of $\alpha$ and $\beta$ in the first case intersects the plane $\gamma$, that in the second case the intersecting line lies completely in $\gamma$, and that in the third case the intersecting line is parallel to $\gamma$.

In each of the following exercises, a centre $M$ and a radius $r$ are given. Find an equation of the sphere with this centre and radius and write it as $x^2 + y^2 + z^2 + ax + by + cz + d = 0$:

15.19

a. $M = (0, 0, 1)$ and $r = 2$
b. $M = (2, 2, 0)$ and $r = 2$
c. $M = (1, 0, -3)$ and $r = 5$
d. $M = (1, 2, -2)$ and $r = 3$
e. $M = (-2, 2, 0)$ and $r = 7$

15.20

a. $M = (4, 0, 1)$ and $r = 1$
b. $M = (3, 1, -2)$ and $r = \sqrt{13}$
c. $M = (2, 0, -1)$ and $r = 5$
d. $M = (1, 7, -2)$ and $r = 7$
e. $M = (-5, 2, 1)$ and $r = 3$

Investigate whether the following equations represent spheres. If so, find their centre and radius:

15.21

a. $x^2 + y^2 + z^2 + 4x - 2y + 2z = 0$
b. $x^2 + y^2 + z^2 + x - y - 1 = 0$
c. $x^2 + y^2 + z^2 + 2x + 4z = 0$
d. $x^2 + y^2 + z^2 - 8z + 12 = 0$
e. $x^2 + y^2 + z^2 - 4x - 2y - 8z + 36 = 0$

15.22

a. $x^2 + y^2 + z^2 = 4x - 5$
b. $x^2 + y^2 + z^2 = 4z + 5$
c. $x^2 + y^2 + z^2 = 4y + 4z - 4$
d. $3x^2 + 3y^2 + 3z^2 = 2y$
e. $4x^2 + 4y^2 + 4z^2 - 16x - 8y + 12z + 60 = 0$

15.23 Find an equation of the tangent plane to the given sphere in the given point:

a. $x^2 + y^2 + z^2 = 9, \quad A = (1, 2, 2)$
b. $x^2 + y^2 + z^2 = 2, \quad A = (1, 0, -1)$
c. $x^2 + y^2 + z^2 - 2x - 4y + 3 = 0, \quad A = (1, 1, 1)$
d. $x^2 + y^2 + z^2 + 2x + 6y - 2z - 11 = 0, \quad A = (2, 0, -1)$
e. $x^2 + y^2 + z^2 + 6x - 4z - 8 = 0, \quad A = (1, -1, 0)$

15.24

a. Find equations in $Oyz$-coordinates or $Oxz$-coordinates of the intersection circles of the sphere in the figure on the next page, with the planes $x = 0$ and $y = 0$, respectively.
b. Find the centre and the radius of these circles.
c. Find the coordinates of all the intersection points of the sphere and the three coordinate axes.
d. Find an equation of the tangent plane to the sphere in each of these intersection points.

## Spheres and tangent planes

The sphere with centre $M = (m_1, m_2, m_3)$ and radius $r$ is the set of all points $P = (x, y, z)$ for which $d(P, M) = r$. Expanding this in coordinates yields the equation

$$(x - m_1)^2 + (y - m_2)^2 + (z - m_3)^2 = r^2$$

which is similar to the equation of a circle in the plane. The figure below shows the sphere with centre $M = (1, 2, 2)$ and radius $r = 4$. Its equation is

$$(x - 1)^2 + (y - 2)^2 + (z - 2)^2 = 16$$

which can be simplified to

$$x^2 + y^2 + z^2 - 2x - 4y - 4z - 7 = 0$$

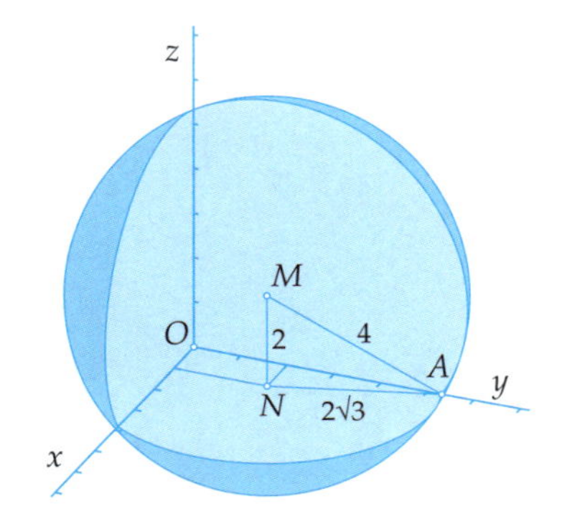


The intersecting circle with the $xy$-plane (the plane with equation $z = 0$) in $Oxy$-coordinates is given by the equation

$$x^2 + y^2 - 2x - 4y - 7 = 0$$

Its centre is $N = (1, 2, 0)$ and its radius is $2\sqrt{3}$. The intersection with the positive $y$-axis is the point $A = (0, 2 + \sqrt{11}, 0)$. The intersecting circles with the planes $x = 0$ and $y = 0$ can be found in a similar way. The figure only shows the arcs of these circles in the first octant ($x \geq 0, y \geq 0, z \geq 0$).

On page 111, we derived the equation of the tangent line to a circle in a given point on the circle. The equation of the *tangent plane* to a sphere can be found in the same way:

> *If $M = (m_1, m_2, m_3)$ is the centre of a sphere and $A = (a_1, a_2, a_3)$ is a point on the sphere, an equation of the tangent plane to the sphere in $A$ is given by*
> $$(m_1 - a_1)(x - a_1) + (m_2 - a_2)(y - a_2) + (m_3 - a_3)(z - a_3) = 0$$

A point on the sphere in the example above is $A = (0, 2 + \sqrt{11}, 0)$. Since we have $M = (1, 2, 2)$, the tangent plane in $A$ is given by the equation

$$x - \sqrt{11}(y - 2 - \sqrt{11}) + 2z = 0$$

# VI Functions

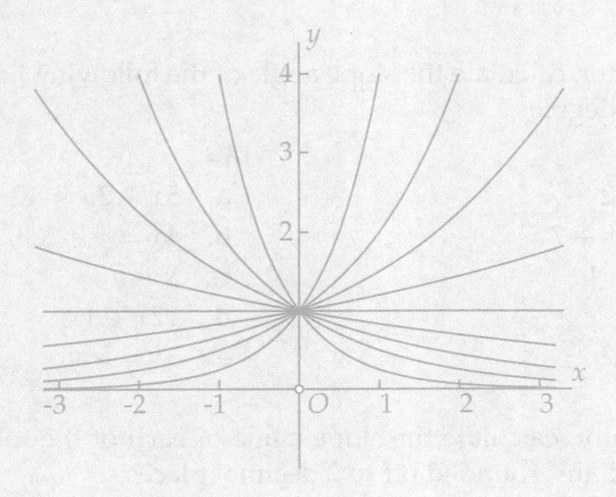


A function is a well-defined rule that assigns to each number another number: its *function value*. For instance, the root function assigns to any number $x$ its root $\sqrt{x}$. Think of the root key of a calculator, which, for any 'input number' $x$ you enter on the number block, produces the number $\sqrt{x}$ on the screen (rounded off to a fixed number of decimals).

In part VI of this book, we will look at all sorts of frequently used functions and their graphs. In the last chapter, we describe parametric curves in the plane and in space.

# 16 Functions and graphs

Calculate the slope of the following lines in the plane:

16.1

a. $3x + 5y = 4$
b. $2x = y + 7$
c. $-4x + 2y = 3$
d. $5y = 7$
e. $-x - 5y = 1$

16.2

a. $2x - 7y = -2$
b. $x = 3y - 2$
c. $-5x + 2y = -3$
d. $2x - 11y = 0$
e. $x = 2y$

Using a calculator, calculate the slope angle of the following lines in the plane, rounded off to degrees:

16.3

a. $x - 3y = 2$
b. $-3x = -y + 7$
c. $4x + 3y = 1$
d. $y = 7x$
e. $x - 4y = 2$

16.4

a. $5x - 2y = 12$
b. $4x = y + 8$
c. $x - y = 3$
d. $12x + 11y = 12$
e. $3x = -y$

Using a calculator, calculate the slope angle of each of the following lines in the plane in radians, rounded off to 2 decimal places:

16.5

a. $x - 3y = 2$
b. $-3x = -y + 7$
c. $4x + 3y = 1$
d. $y = 7x$
e. $x - 4y = 2$

16.6

a. $5x - 2y = 12$
b. $4x = y + 8$
c. $x - y = 3$
d. $12x + 11y = 12$
e. $3x = -y$

Find the linear function with the line through $P$ with slope $m$ as its graph:

16.7

a. $P = (0, 0)$, $m = 3$
b. $P = (1, -1)$, $m = -2$
c. $P = (1, 2)$, $m = 0.13$
d. $P = (-1, 1)$, $m = -1$
e. $P = (2, -3)$, $m = 4$

16.8

a. $P = (4, 0)$, $m = -4$
b. $P = (3, -4)$, $m = 2.22$
c. $P = (-1, -3)$, $m = 0$
d. $P = (1, -1)$, $m = -1.5$
e. $P = (-1, -2)$, $m = 0.4$

## Linear functions

The linear equation $4x + 3y = 12$ can also be written as

$$y = -\frac{4}{3}x + 4$$

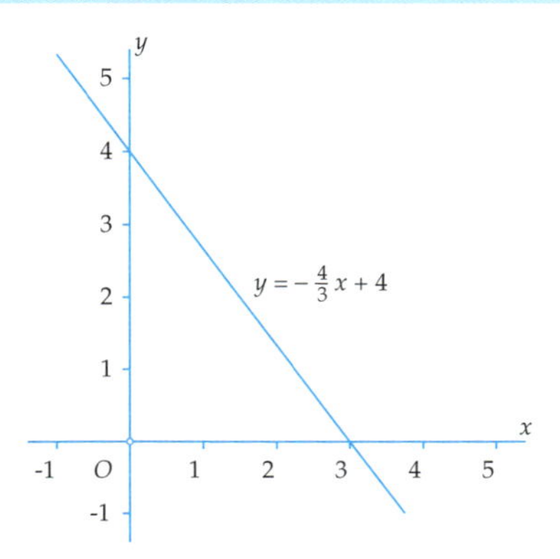


Here, $y$ is given as a function of $x$: for each $x$, the right-hand side $-\frac{4}{3}x + 4$ yields the corresponding value of $y$. The straight line in the $Oxy$-plane that is represented by the equation, is the graph of that function.

In general, each linear equation $ax + by = c$ for which $b \neq 0$ is true, can be written in the form

$$y = mx + p$$

(take $m = -a/b$ and $p = c/b$). The condition $b \neq 0$ implies that the corresponding line in the $Oxy$-plane is not vertical.

The function $f(x) = mx + p$ is called a *function of degree* 1 of $x$. Since its graph is a straight line, the function is also called a *linear* function of $x$. The expression $mx + p$ is the defining rule of the function.

The coefficient $m$ of $x$ is called the *slope*. Any increase of $h$ units in the $x$-direction along the graph corresponds to an increase of $mh$ units in the $y$-direction. A positive $m$ corresponds to an increasing graph; a negative $m$ corresponds to a decreasing graph.
If the scales on both axes are equal, $m$ is also the *tangent* of the angle $\alpha$ between the graph and the direction of the $x$-axis. This angle $\alpha$ is called the *slope angle*.

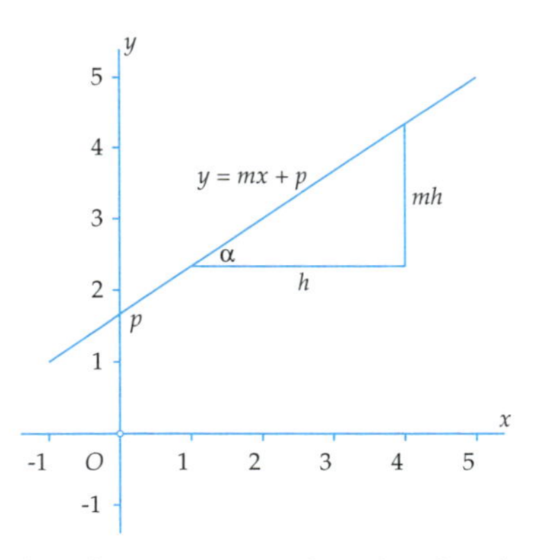


When angles are measured in degrees, $\alpha$ is taken between $-90°$ and $90°$; when measured in radians, $\alpha$ is taken between $-\frac{\pi}{2}$ and $\frac{\pi}{2}$. Note that the graph of the function $y = mx + p$ intersects the $y$-axis in the point $(0, p)$ since $x = 0$ yields $y = p$.

Find the coordinates $(x_v, y_v)$ of the vertex of the following parabolas:

16.9

a. $y = x^2 - 1$
b. $y = -3x^2 + 7$
c. $y = (x+1)^2$
d. $y = -2(x-2)^2 + 1$
e. $y = x^2 + 2x$

16.10

a. $y = (x+3)^2 + 4$
b. $y = 2x^2 - 8x$
c. $y = -3x^2 + 7x + 2$
d. $y = 2x^2 + 12x - 5$
e. $y = 5x^2 + 20x - 6$

16.11

a. $y = x^2 + 2x - 3$
b. $y = x^2 - 2x - 3$
c. $y = x^2 + 2x - 8$
d. $y = 2x^2 + x - 1$
e. $y = 3x^2 - x - 2$

16.12

a. $y = -x^2 - 2x + 3$
b. $y = -x^2 + 4x - 3$
c. $y = -x^2 - x + 2$
d. $y = 2x^2 - 3x - 2$
e. $y = 3x^2 + 2x - 1$

Find the equation $y = ax^2 + bx + c$ of the parabola with the given vertex $V$ passing through the given point $P$:

16.13

a. $V = (0,0), P = (1,2)$
b. $V = (0,0), P = (-1,-2)$
c. $V = (0,0), P = (2,1)$
d. $V = (0,0), P = (2,-2)$
e. $V = (0,0), P = (-1,-5)$

16.14

a. $V = (0,1), P = (1,2)$
b. $V = (0,-1), P = (2,-2)$
c. $V = (0,-2), P = (-1,-5)$
d. $V = (3,0), P = (-1,-2)$
e. $V = (-2,0), P = (2,1)$

16.15

a. $V = (1,2), P = (2,3)$
b. $V = (-1,2), P = (1,6)$
c. $V = (2,-1), P = (1,1)$
d. $V = (0,3), P = (1,4)$
e. $V = (-3,0), P = (-2,3)$

16.16

a. $V = (0,0), P = (3,6)$
b. $V = (\frac{1}{2}, -\frac{1}{2}), P = (1, -\frac{1}{4})$
c. $V = (\frac{1}{3}, -1), P = (\frac{2}{3}, \frac{2}{9})$
d. $V = (0, \frac{3}{2}), P = (\frac{1}{2}, 2)$
e. $V = (-\frac{3}{4}, \frac{3}{4}), P = (-\frac{1}{2}, \frac{7}{8})$

## Quadratic functions and parabolas

The function rule

$$f(x) = -x^2 + 4x + 1$$

defines a *function of degree* 2, also called a *quadratic function*.

Its graph is a *parabola*, in this case a *mountain parabola* with its *vertex* at the point $V = (2, 5)$. This becomes immediately clear when writing the defining rule after completing the square as

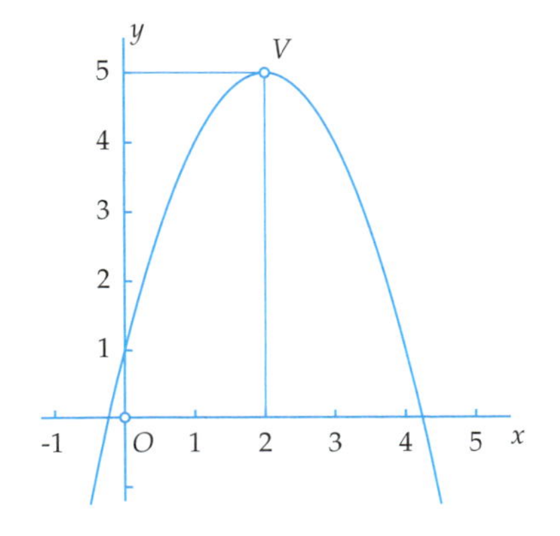


$$\begin{aligned} f(x) &= -(x^2 - 4x + 4) + 5 \\ &= -(x-2)^2 + 5 \end{aligned}$$

Indeed, the term $-(x-2)^2$ is always negative or zero, and only zero for $x = 2$. Then $f(x) = 5$ and the coordinates $(x_v, y_v)$ of the vertex $V$ are $(x_v, y_v) = (2, 5)$.

The parabola that is the graph of the function $f(x) = -x^2 + 4x + 1$ has $y = -x^2 + 4x + 1$ as its equation. All points $(x, y)$ in the $Oxy$-plane satisfying this equation are on the parabola and, conversely, the coordinates $(x, y)$ of any point on the parabola satisfy this equation.

In general, $f(x) = ax^2 + bx + c$ is the defining rule of a quadratic function, provided, of course, that $a \neq 0$. Its graph is a parabola with equation

$$y = ax^2 + bx + c$$

It is a *valley parabola* if $a > 0$, and a *mountain parabola* if $a < 0$ (as in the example above).

The highest point of the mountain, or the lowest point of the valley, is called the *vertex*. Its coordinates $(x_v, y_v)$ can be found as shown above by first calculating $x_v$ by completing the square, and then finding $y_v$ by substituting $x_v$ into the function rule $y_v = f(x_v)$.

The general equation of a parabola with vertex $(x_v, y_v)$ is

$$y = a(x - x_v)^2 + y_v$$

The constant $a$ can be found as soon as you know the coordinates of one other point $P$ on the parabola.

Plot the graphs of the functions $f$ and $g$ in one figure and find their intersection points:

16.17

a. $f(x) = x^2 + x - 2$
   $g(x) = x + 2$

b. $f(x) = -x^2 - 2x - 1$
   $g(x) = 2x + 3$

c. $f(x) = 2x^2 + x - 3$
   $g(x) = -x - 3$

d. $f(x) = -2x^2 + 5x - 2$
   $g(x) = 2x - 1$

e. $f(x) = 3x^2 + x - 4$
   $g(x) = -3x - 5$

16.18

a. $f(x) = x^2 + 1$
   $g(x) = -x^2 + 3$

b. $f(x) = x^2 + x - 2$
   $g(x) = x^2 + 2x - 3$

c. $f(x) = 2x^2 - x - 1$
   $g(x) = -x^2 + 8x - 7$

d. $f(x) = -2x^2 + 3x + 2$
   $g(x) = x^2 + x + 1$

e. $f(x) = x^2 - 2x - 3$
   $g(x) = -x^2 + 2x - 5$

For which $x$ does the inequality $f(x) \geq g(x)$ hold?

16.19

a. $f(x) = x^2 + x - 3$
   $g(x) = -1$

b. $f(x) = x^2 - 2x - 3$
   $g(x) = -2x - 2$

c. $f(x) = -x^2 + 2x - 1$
   $g(x) = x - 3$

d. $f(x) = 2x^2 - x - 1$
   $g(x) = 2x + 2$

e. $f(x) = -3x^2 + x + 2$
   $g(x) = -x + 2$

16.20

a. $f(x) = x^2 + x - 3$
   $g(x) = -x^2 + 3x - 3$

b. $f(x) = x^2 - x - 2$
   $g(x) = 2x^2 - 3x - 4$

c. $f(x) = -x^2 + 2x - 1$
   $g(x) = x^2 - 3x + 2$

d. $f(x) = 2x^2 - x - 1$
   $g(x) = -x^2 + x + 4$

e. $f(x) = -3x^2 + x + 2$
   $g(x) = 2x^2 - 5x - 6$

16.21 Find out for which real numbers $p$ the graph of $f$ doesn't intersect the $x$-axis:

a. $f(x) = x^2 + px + 1$

b. $f(x) = x^2 - x + p$

c. $f(x) = px^2 + 2x - 1$

d. $f(x) = x^2 + px + p$

e. $f(x) = -x^2 + px + p - 3$

16.22 Find out for which real numbers $p$ the graph of $f$ and the $x$-axis intersect in two distinct points:

a. $f(x) = x^2 + 2px - 1$

b. $f(x) = -x^2 + x + p + 1$

c. $f(x) = px^2 + 2x - 3$

d. $f(x) = x^2 + px + p + 3$

e. $f(x) = (p + 1)x^2 - px - 1$

## Intersection points of graphs

When given two quadratic functions

$$f(x) = 2x^2 + 3x - 2$$

and

$$g(x) = -x^2 - 3x + 7$$

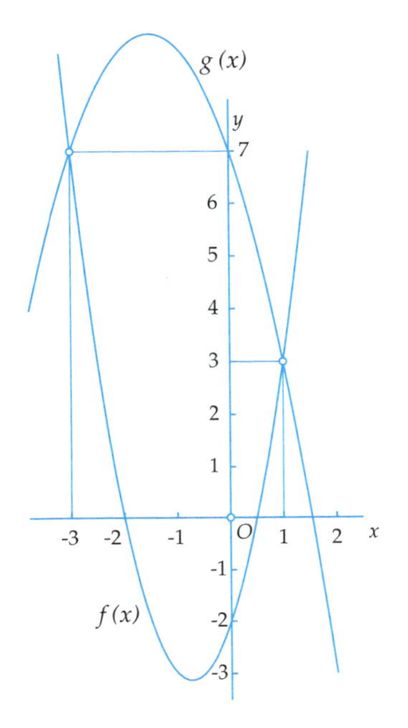


the $x$-coordinate of an intersection point of the graphs must satisfy $f(x) = g(x)$, meaning

$$2x^2 + 3x - 2 = -x^2 - 3x + 7$$

or $3x^2 + 6x - 9 = 0$. The solutions are $x = 1$ and $x = -3$. Since $f(1) = g(1) = 3$ and $f(-3) = g(-3) = 7$, the coordinates of the intersection points of the two parabolas are $(1,3)$ and $(-3,7)$.
The graphs also show that $f(x) \leq g(x)$ if $-3 \leq x \leq 1$ and that $f(x) \geq g(x)$ if $x \leq -3$ or $x \geq 1$.
This method works in general. If you are looking for the intersection points of the graphs of two functions $f$ and $g$, you first solve the equation $f(x) = g(x)$ to find the $x$-coordinates of the intersection points. Substituting such an $x$-coordinate into the defining rule of one of the functions yields the corresponding $y$-coordinate.

If both functions are quadratic, the equation that you have to solve to find the intersection points is quadratic, or possibly linear, meaning there are at most two intersection points. The same is true if one of the functions is quadratic and the other is linear.

Plot the graphs of the following functions. In particular, find the horizontal and the vertical asymptote and the intersection points with the coordinate axes (if any):

16.23

a. $f(x) = \dfrac{1}{x-1}$

b. $f(x) = \dfrac{x+1}{x}$

c. $f(x) = \dfrac{3}{2x-4}$

d. $f(x) = \dfrac{2x}{x-5}$

e. $f(x) = \dfrac{x+2}{x-2}$

16.24

a. $f(x) = \dfrac{-x}{2x-3}$

b. $f(x) = \dfrac{x+2}{3x-1}$

c. $f(x) = \dfrac{2x-4}{1-5x}$

d. $f(x) = \dfrac{x+3}{4+7x}$

e. $f(x) = \dfrac{3-2x}{4x-2}$

Using the graph of $f$, determine which numbers $x$ satisfy $-1 \leq f(x) \leq 1$:

16.25

a. $f(x) = \dfrac{1}{x+3}$

b. $f(x) = \dfrac{2x-1}{x+1}$

c. $f(x) = \dfrac{5}{x-5}$

d. $f(x) = \dfrac{2x}{3-2x}$

e. $f(x) = \dfrac{3x-1}{2x+2}$

16.26

a. $f(x) = \dfrac{2-x}{2x-1}$

b. $f(x) = \dfrac{2x-2}{3x+4}$

c. $f(x) = \dfrac{-x+7}{1-3x}$

d. $f(x) = \dfrac{2x+2}{4-5x}$

e. $f(x) = \dfrac{3-x}{2x+4}$

Find the intersection points of the graphs of the functions $f$ and $g$:

16.27

a. $f(x) = \dfrac{8}{x+3}$, $g(x) = 2x$

b. $f(x) = \dfrac{2x-4}{x+1}$, $g(x) = 2-x$

c. $f(x) = \dfrac{8}{5-x}$, $g(x) = x+4$

d. $f(x) = \dfrac{2x-1}{3-2x}$, $g(x) = 3-2x$

16.28

a. $f(x) = \dfrac{x+3}{2x+1}$, $g(x) = 3x-5$

b. $f(x) = \dfrac{2}{x+3}$, $g(x) = x+2$

c. $f(x) = \dfrac{3x+2}{x+1}$, $g(x) = 2-x$

d. $f(x) = \dfrac{2+2x}{2x-4}$, $g(x) = 3x-5$

## Fractional linear functions

A function like

$$f(x) = \frac{2x+4}{x-3}$$

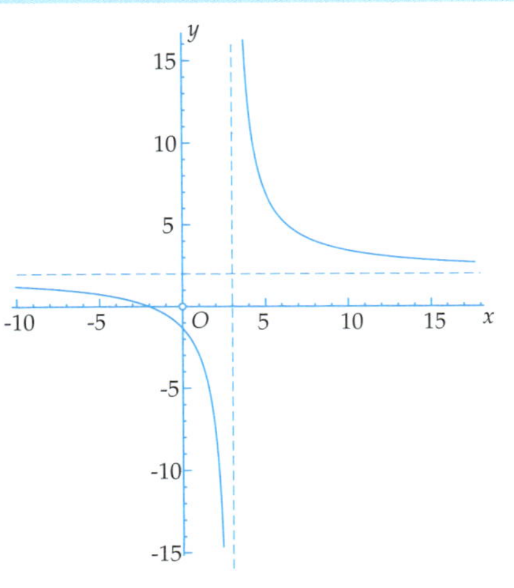


which contains a 'fraction' with linear expressions in both the numerator and the denominator, is called a *fractional linear function*. For $x = 3$, the denominator becomes zero, so $f(x)$ cannot be calculated. The graph of $f$ has the line $x = 3$ as a *vertical asymptote*. If $x$ approximates 3 from above, $f(x)$ goes to $+\infty$; if $x$ approximates 3 from below, $f(x)$ goes to $-\infty$.

Dividing the numerator and denominator by $x$, yields

$$f(x) = \frac{2+\frac{4}{x}}{1-\frac{3}{x}}$$

For very big positive or negative values of $x$, the numbers $\frac{4}{x}$ and $\frac{3}{x}$ are very small. Then $f(x)$ is almost equal to $\frac{2}{1} = 2$. Therefore, the horizontal line $y = 2$ is a *horizontal asymptote* of the graph of $f$.

The general form of a fractional linear function is

$$f(x) = \frac{ax+b}{cx+d}$$

where we assume that $c \neq 0$ (otherwise it would be an ordinary linear function). The horizontal asymptote is the line $y = \frac{a}{c}$. The function is not defined if the denominator is zero, so if $x = -\frac{d}{c}$. The vertical line $x = -\frac{d}{c}$ is the vertical asymptote of the graph, unless the numerator is also zero for $x = -\frac{d}{c}$.

The latter occurs, for instance, with the function

$$f(x) = \frac{2x-4}{-6x+12}$$

The denominator is zero if $x = 2$, but then the numerator is zero as well. For all $x \neq 2$, we have $f(x) = -\frac{1}{3}$ (divide the numerator and denominator by $2x-4$), so the graph is just the horizontal line $y = -\frac{1}{3}$ with a perforation at the point $(2, -\frac{1}{3})$.

Plot the graph of the following functions. Do not expand the brackets!

16.29

a. $f(x) = (x-1)^3$

b. $f(x) = x^3 - 1$

c. $f(x) = 1 - x^4$

d. $f(x) = 1 + (x+1)^3$

e. $f(x) = (2x-1)^3$

16.30

a. $f(x) = \sqrt{x-1}$

b. $f(x) = \sqrt[3]{x+1}$

c. $f(x) = 1 - \sqrt[4]{2-x}$

d. $f(x) = \sqrt[3]{4+7x}$

e. $f(x) = \sqrt[5]{x-2}$

16.31

a. $f(x) = \sqrt{x^3}$

b. $f(x) = \sqrt[3]{x^2}$

c. $f(x) = \sqrt{|x|}$

d. $f(x) = \sqrt{|x|^3}$

e. $f(x) = |\sqrt[3]{x}|$

16.32

a. $f(x) = |x|^3$

b. $f(x) = |x-1|^3$

c. $f(x) = |1-x^2|$

d. $f(x) = |1+x^3|$

e. $f(x) = |1-(x+1)^2|$

Solve the following equations and inequalities, always by using a figure:

16.33

a. $x^4 \leq x^3$

b. $x^4 = |x|$

c. $x^4 \geq |x|^3$

d. $x^4 \geq \sqrt{x}$

e. $x^4 \leq |\sqrt[3]{x}|$

16.34

a. $|2x+3| = 2$

b. $|2x-3| = -2$

c. $|2x+3| = 4x$

d. $|2x+3| \geq |4x|$

e. $|x^2-2x| < 1$

For the following exercises it is also recommended to first draw a (rough) figure. *Example:* Solve $\sqrt{x+2} = |x|$. The figure below shows that there are two solutions. They can be found by squaring the equation (if $x$ satisfies the original equation, it also satisfies the squared equation). This yields $x+2 = x^2$, with solutions $x = -1$ and $x = 2$.

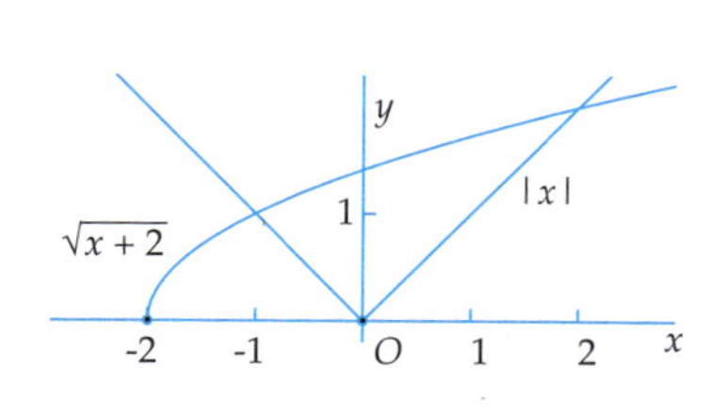


16.35 Solve:

a. $\sqrt{x} = |x|$

b. $\sqrt{x-1} = |x-2|$

c. $\sqrt{x+1} \leq x-1$

d. $\sqrt{|1-x|} \geq x$

e. $\sqrt{2x+1} \leq |x+1|$

## Power and root functions and the absolute value function

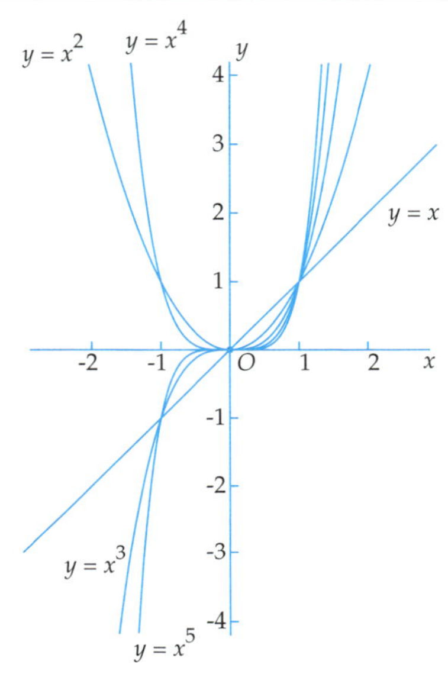


The figure to the right shows the graphs of the *power functions* $f(x) = x^n$ for $n = 1, 2, 3, 4, 5$. They all go through the origin and through the point $(1, 1)$.

For each $n > 1$, the graph of $x^n$ is tangent in the origin to the $x$-axis.
For even values of $n$, the functions there attain their minimum value.

For even values of $n$, the graphs are symmetric with respect to the $y$-axis. Then $f(-x) = f(x)$ for all $x$.

For odd values of $n$, the graphs are point symmetric with respect to the origin. Then $f(-x) = -f(x)$ for all $x$.

In general, a function $f(x)$ is called *even* if $f(-x) = f(x)$ for all $x$, and *odd* if $f(-x) = -f(x)$ for all $x$. But be careful: not every function is even or odd. Most functions are neither even nor odd. For instance, take $f(x) = x + 1$, which is neither even nor odd, as can be checked easily.

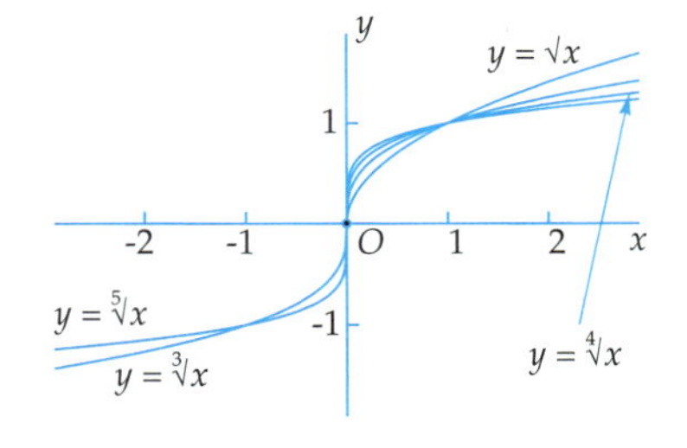


In the figure to the right, the graphs of the *root functions* $f(x) = \sqrt[n]{x}$ are drawn for $n = 2, 3, 4, 5$. They all pass through the origin and the point $(1, 1)$. For even values of $n$, they are only defined for $x \geq 0$, for odd $n$, they are defined for all $x$. All graphs are tangent to the $y$-axis in the origin.

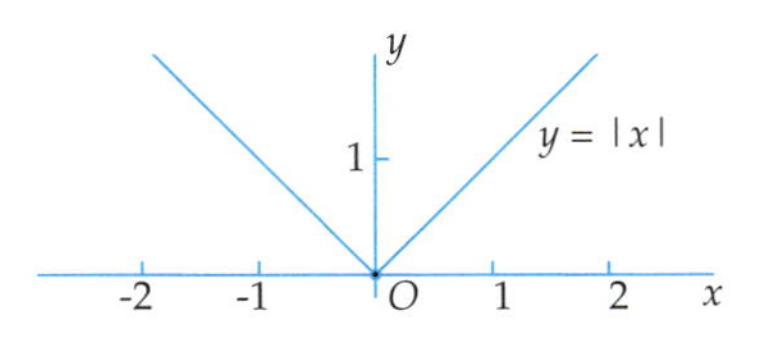


In the bottom figure, the graph of the *absolute value function* $f(x) = |x|$, defined by $|x| = x$ if $x \geq 0$ and $|x| = -x$ if $x \leq 0$, is drawn.
For all $x$ we have $|x|^2 = x^2$.

Note that $\sqrt[n]{x^n} = |x|$ if $n$ is even, and $\sqrt[n]{x^n} = x$ if $n$ is odd.

16.36 Give examples of:

a. a polynomial of degree 5 with five distinct zeroes,
b. a polynomial of degree 5 with four distinct zeroes,
c. a polynomial of degree 5 with three distinct zeroes,
d. a polynomial of degree 5 with two distinct zeroes,
e. a polynomial of degree 5 with one zero,
f. a polynomial of degree 6 with no zeroes.

According to the factor theorem on the opposite page, for every zero $x = a$ of a polynomial $f(x)$ of degree greater than or equal to 1, there is a polynomial $g(x)$ for which $f(x) = (x - a)g(x)$. For instance, take

$$f(x) = 3x^4 - 7x^3 + 3x^2 - x - 2$$

Substituting $x = 2$ shows that $f(2) = 0$, meaning $x = 2$ is a zero of $f(x)$. The long division below yields the polynomial $g(x) = 3x^3 - x^2 + x + 1$.

$$\begin{array}{rllllll}
x-2 \;/ & 3x^4 & -7x^3 & +3x^2 & -x & -2 & \backslash\; 3x^3 - x^2 + x + 1 \\
 & \underline{3x^4} & \underline{-6x^3} & & & & \\
 & & -x^3 & +3x^2 & & & \\
 & & \underline{-x^3} & \underline{+2x^2} & & & \\
 & & & x^2 & -x & & \\
 & & & \underline{x^2} & \underline{-2x} & & \\
 & & & & x & -2 & \\
 & & & & \underline{x} & \underline{-2} & \\
 & & & & & 0 &
\end{array}$$

This means that $3x^4 - 7x^3 + 3x^2 - x - 2 = (x - 2)(3x^3 - x^2 + x + 1)$ (check this!).

Verify that $a$ is a zero of the given polynomial $f(x)$. Then find the polynomial $g(x)$ for which $f(x) = (x - a)g(x)$ by long division:

16.37

a. $f(x) = x^2 - x - 2,\quad a = 2$
b. $f(x) = 2x^2 - 2,\quad a = 1$
c. $f(x) = x^3 + 1,\quad a = -1$
d. $f(x) = x^6 - 1,\quad a = 1$
e. $f(x) = 2x^3 - 4x + 8,\quad a = -2$
f. $f(x) = x^4 - 2x^2 + 1,\quad a = 1$
g. $f(x) = -x^3 - 3x^2 + 12x - 4,\quad a = 2$

16.38

a. $f(x) = 2x^4 - 2,\quad a = 1$
b. $f(x) = x^3 + x^2 + 4,\quad a = -2$
c. $f(x) = x^3 + 8,\quad a = -2$
d. $f(x) = x^4 - 16,\quad a = 2$
e. $f(x) = x^3 - 3x^2 + 2x,\quad a = 1$
f. $f(x) = 2x^3 - 4x + 8,\quad a = -2$
g. $f(x) = x^4 - 9x^3 - 6x^2 - 4,\quad a = -1$

## Polynomials

The figure to the right shows the graph of the function

$$f(x) = x^5 - 5x^3 - x^2 + 4x + 2$$

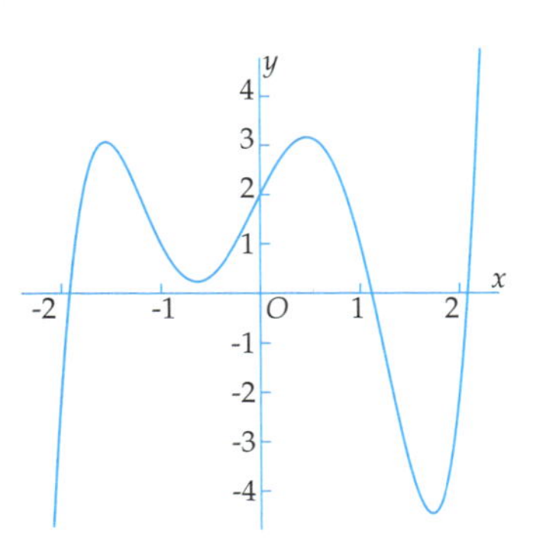


The scales on the axes are not equal.
In the figure, three *zeroes* are visible, i.e. three values of $x$ for which $f(x) = 0$. Whether $f(x)$ has more zeroes is not clear from the figure: perhaps there are more zeroes to the right or to the left. To find this out, further investigations would be necessary.

In general, each function of the form

$$f(x) = a_n x^n + a_{n-1} x^{n-1} + \cdots + a_1 x + a_0$$

is called a *polynomial function*, or *polynomial*, for short. The numbers $a_i$ are called the *coefficients*. We always assume that the highest coefficient $a_n$ is not zero (since otherwise the term $a_n x^n$ could be left out). Some, or even all of the other coefficients, however, can be zero. The number $n$ is called the *degree* of the polynomial. In the example above, we have a polynomial of degree 5.

The following *factor theorem* is frequently used:

*If $f(x)$ is a polynomial of degree $n$ with $n \geq 1$ and $a$ is a real number for which $f(a) = 0$, there is a polynomial $g(x)$ of degree $n-1$ for which $f(x) = (x-a)g(x)$.*

Take for example $f(x) = 3x^4 - 7x^3 + 3x^2 - x - 2$. Substituting $x = 2$ yields $f(2) = 3 \cdot 16 - 7 \cdot 8 + 3 \cdot 4 - 2 - 2 = 0$, which means that 2 is a zero. The corresponding polynomial $g(x)$ can be found by a *long division*, as is shown on the opposite page. The result is $g(x) = 3x^3 - x^2 + x + 1$, hence we have $f(x) = (x-2)(3x^3 - x^2 + x + 1)$.

If $g(x)$ also has a zero, again a factor of degree 1 can be divided out, and so on, each time reducing the degree by 1, until a polynomial without zeroes remains. Therefore, each polynomial $f(x)$ can be written as

$$f(x) = (x - a_1) \cdots (x - a_k) h(x)$$

where $h(x)$ is a polynomial without zeroes. Conclusion:

*Each polynomial of degree $n$ with $n \geq 1$ has at most $n$ zeroes.*

Without proof, we give the following theorem:

*Each polynomial of degree $n$ with $n$ odd has at least one zero.*

In particular, any polynomial of degree 3 has at least one zero.

16.39 For each of the following graphs, find the corresponding defining function rule. Explain your answer. Don't use a graphing calculator.

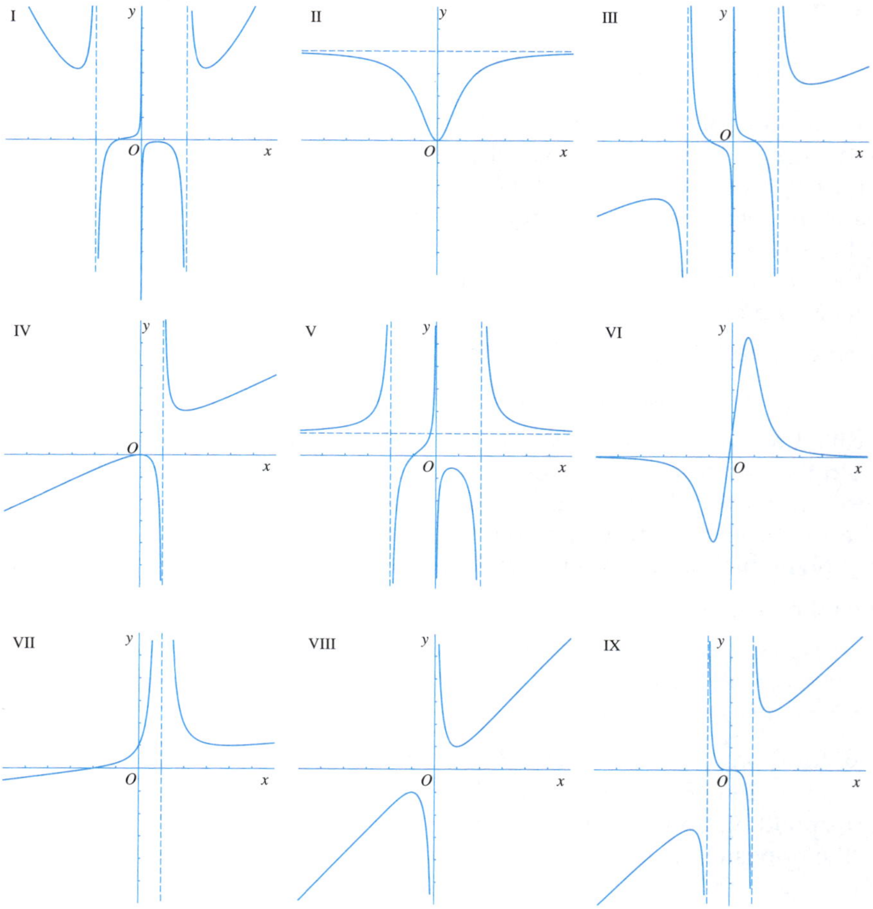

I y O x

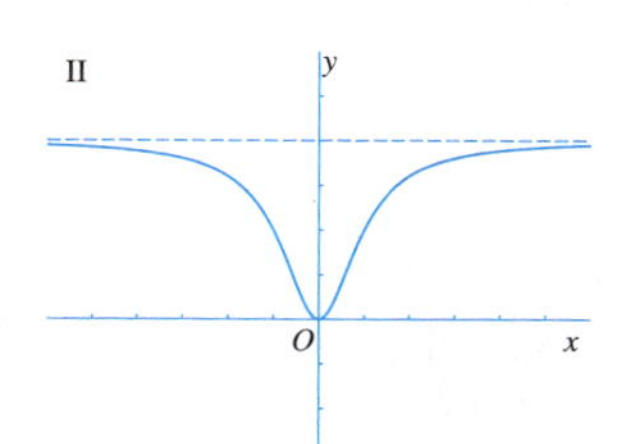


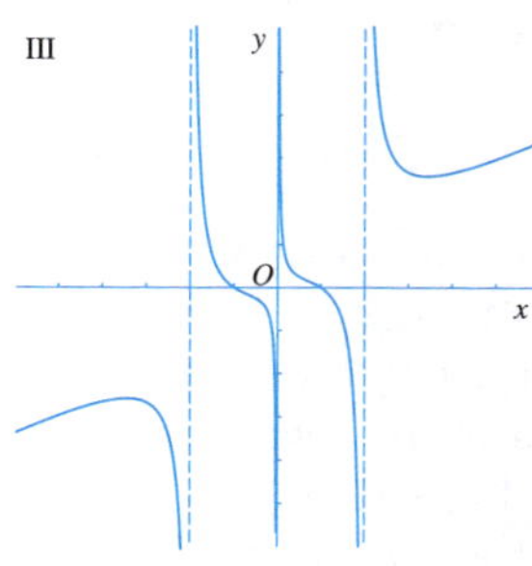


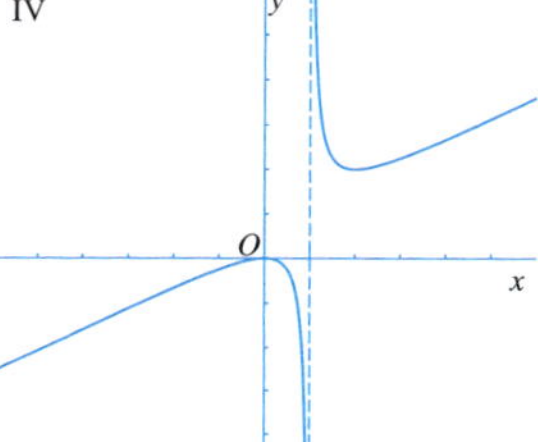


V y O x

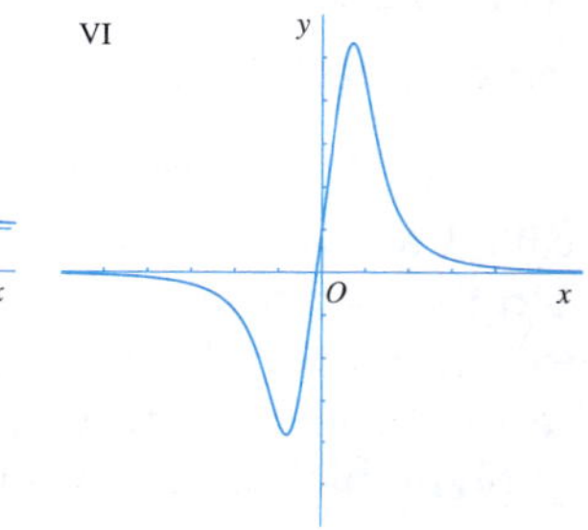


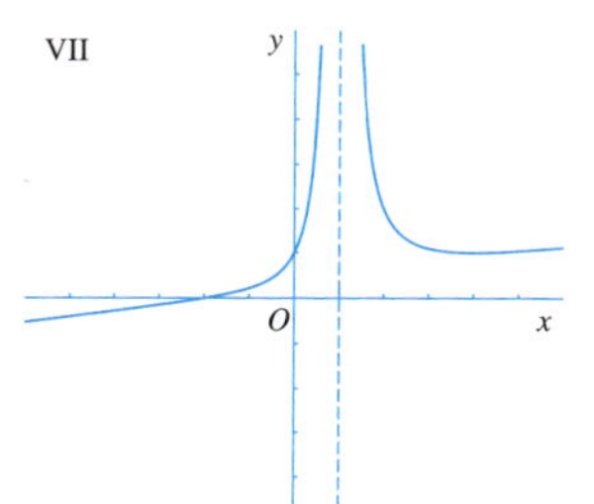


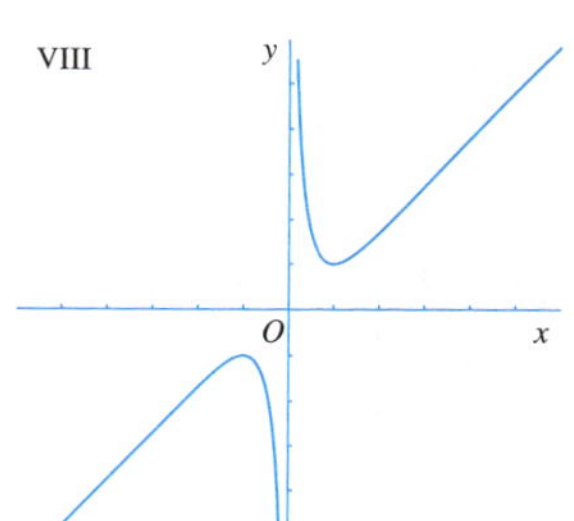


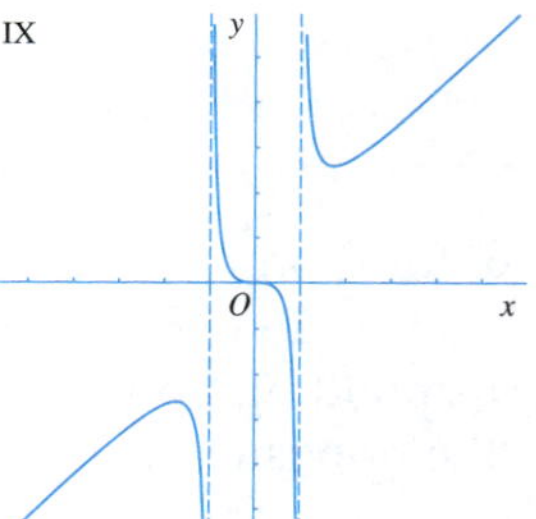


a. $\dfrac{x^4+1}{x^3+x}$

b. $\dfrac{x^3+1}{x^3-4x}$

c. $\dfrac{4x^2}{x^2+1}$

d. $\dfrac{x^4-1}{2x^3-8x}$

e. $\dfrac{x^5+1}{5x^3-20x}$

f. $\dfrac{x^2}{2x-2}$

g. $\dfrac{x^3+8}{8(x-1)^2}$

h. $\dfrac{x^3}{x^2-1}$

i. $\dfrac{8x+1}{x^4+1}$

## Rational functions

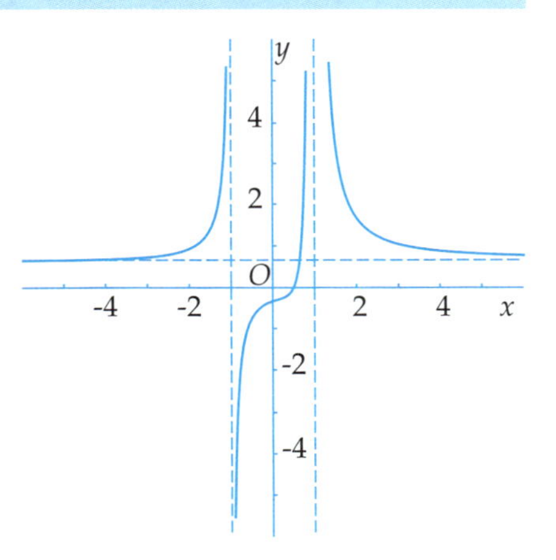


The figure to the right shows a graph of the function

$$f(x) = \frac{(x^2+1)(2x-1)}{3(x-1)^2(x+1)}$$

It exhibits two vertical asymptotes and one horizontal asymptote. The vertical asymptotes are the lines $x = 1$ and $x = -1$, exactly at the values of $x$ for which the denominator becomes zero. The horizontal asymptote is the line $y = \frac{2}{3}$. It indicates the limiting value of $f(x)$ for $x \longrightarrow +\infty$ and $x \longrightarrow -\infty$.

These limits can easily be found by writing $f(x)$ as

$$f(x) = \frac{2x^3 - x^2 + 2x - 1}{3x^3 - 3x^2 - 3x + 3} = \frac{2 - \frac{1}{x} + \frac{2}{x^2} - \frac{1}{x^3}}{3 - \frac{3}{x} - \frac{3}{x^2} + \frac{3}{x^3}}$$

by expanding the brackets and dividing the numerator and denominator by the highest power of $x$. The only zero of the function is the value of $x$ for which the numerator becomes zero, which is $x = \frac{1}{2}$.

In general, each function for which the defining rule can be written as the quotient of two polynomials, is called a *rational function*. Such a function is of the form

$$f(x) = \frac{a(x)}{b(x)} = \frac{a_n x^n + a_{n-1}x^{n-1} + \cdots + a_1 x + a_0}{b_m x^m + b_{m-1}x^{m-1} + \cdots + b_1 x + b_0}$$

We can assume that the polynomial $a(x)$ in the numerator and the polynomial $b(x)$ in the denominator don't have common zeroes, since otherwise they could be canceled out by the factor theorem. After this has been done, the zeroes of the numerator $a(x)$ are also the zeroes of $f(x)$. The zeroes of the denominator $b(x)$ are called the *poles* of $f(x)$. Each pole corresponds with a vertical asymptote.

A horizontal asymptote only occurs if $n \leq m$. If $n = m$, it is the line $y = \dfrac{a_n}{b_n}$. This happened in the example above. There we had $a_3 = 2$ and $b_3 = 3$.

If $n < m$, the $x$-axis is the horizontal asymptote. Then the limit of $f(x)$ for $x \longrightarrow \pm\infty$ is equal to zero.

# 17 Trigonometry

Convert the following angles, given in degrees, to radians. Give exact answers!

17.1

a. $30^\circ$
b. $45^\circ$
c. $60^\circ$
d. $70^\circ$
e. $15^\circ$

17.2

a. $20^\circ$
b. $50^\circ$
c. $80^\circ$
d. $100^\circ$
e. $150^\circ$

17.3

a. $130^\circ$
b. $135^\circ$
c. $200^\circ$
d. $240^\circ$
e. $330^\circ$

Convert the following angles, given in radians, to degrees. Give exact answers!

17.4

a. $\frac{1}{6}\pi$
b. $\frac{7}{6}\pi$
c. $\frac{1}{3}\pi$
d. $\frac{2}{3}\pi$
e. $\frac{1}{4}\pi$

17.5

a. $\frac{5}{4}\pi$
b. $\frac{5}{12}\pi$
c. $\frac{11}{24}\pi$
d. $\frac{15}{8}\pi$
e. $\frac{23}{12}\pi$

17.6

a. $\frac{71}{72}\pi$
b. $\frac{41}{24}\pi$
c. $\frac{25}{18}\pi$
d. $\frac{13}{24}\pi$
e. $\frac{31}{36}\pi$

For the given angles of rotation, calculate the angle $\alpha$ with $0 \leq \alpha < 360^\circ$ which yields the same rotational result:

17.7

a. $-30^\circ$
b. $445^\circ$
c. $-160^\circ$
d. $700^\circ$
e. $515^\circ$

17.8

a. $-220^\circ$
b. $-650^\circ$
c. $830^\circ$
d. $1000^\circ$
e. $1550^\circ$

17.9

a. $-430^\circ$
b. $935^\circ$
c. $1200^\circ$
d. $-240^\circ$
e. $730^\circ$

17.10

a. Find the area of the circle sector with radius 3 and a central angle equal to 1 radian.
b. Find the area of a half circle with a *diameter* equal to 1.
c. Find the circumference of a circle with an area equal to 1.

## Angle measurement

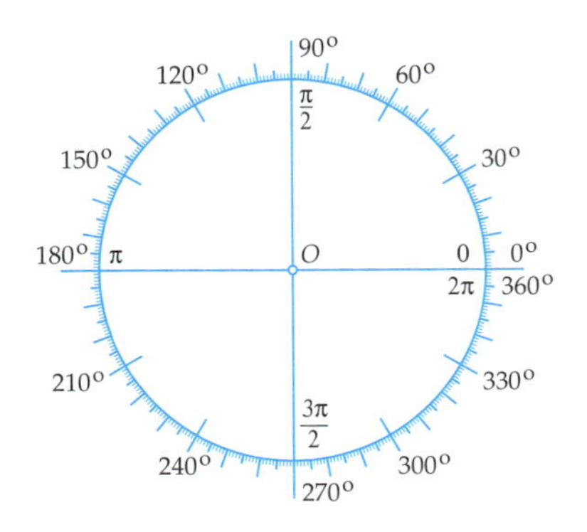


Angles are measured in *degrees* or in *radians*. The figure shows the unit circle in the plane (the circle with radius 1 and the origin $O$ as its centre) on which both scales are marked. A full turn measures 360 degrees, or $2\pi$ radians. Angles of rotation can also be measured in degrees or in radians. The *direction of rotation* matters: by convention, anticlockwise rotations are given a positive sign, clockwise rotations a negative sign.

For rotations, the angle of rotation can, of course, also be greater than $360°$. However, for the final result, it is immaterial whether integer multiples of $360°$ (or $2\pi$ radians) are added or subtracted.

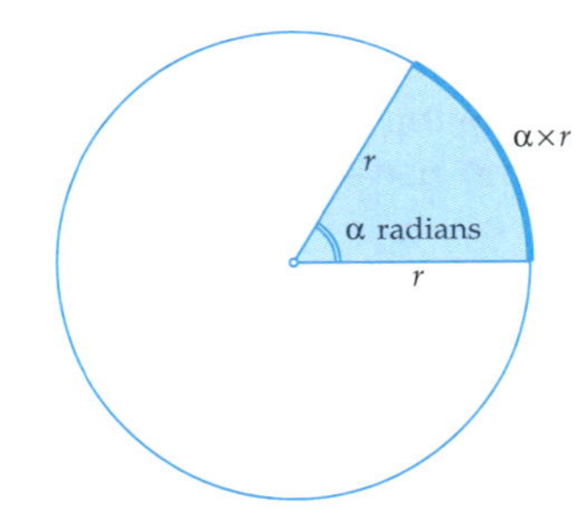


The word *radian* is related to *radius*. If you mark an arc with a central angle of $\alpha$ radians on a circle with radius $r$, its length is exactly equal to $\alpha \times r$. Therefore, the angle measurement in radians yields the ratio between the arc length and the radius, hence the name radian. For a circle with radius 1, the arc length is exactly *equal* to the central angle $\alpha$ in radians.

The circumference of a circle with radius $r$ is $2\pi r$. Thus, with a full turn around a circle corresponds a rotational angle of $2\pi$ radians. An angle of 1 radian is slightly less than 60 degrees, namely, rounded off to eight decimal places, 57.29577950 degrees. The exact value is $360/(2\pi)$.

The area of a circle with radius $r$ is $\pi r^2$. It follows that the area of a sector with a central angle of $\alpha$ radians is $\dfrac{\alpha}{2\pi} \times \pi r^2 = \dfrac{1}{2}\alpha r^2$.

Later in this chapter, the *trigonometric functions* $\sin x$, $\cos x$ and $\tan x$ will be introduced. These functions are defined for $x$ given in *radians*. This is why, from now on, we will always measure angles in radians, unless stated otherwise.

17.11

a. Using Pythagoras' theorem (see page 143), show that the length of the sides of a square with diagonals of length 1 is equal to $\frac{1}{2}\sqrt{2}$.

b. Using Pythagoras' theorem, show that the length of each median in an equilateral triangle with sides of length 1 is equal to $\frac{1}{2}\sqrt{3}$ (a median in a triangle is a line joining a vertex to the midpoint of the opposite side).

c. Use the results above to explain the values for $\alpha = \frac{1}{3}\pi$, $\alpha = \frac{1}{4}\pi$ and $\alpha = \frac{1}{6}\pi$ in the table on the opposite page (also see the figures below).

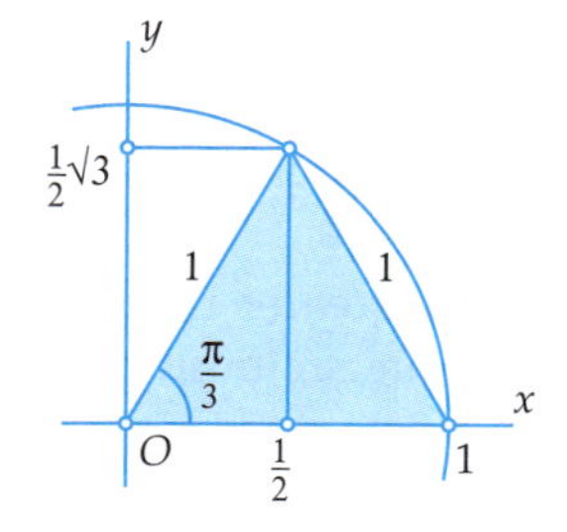


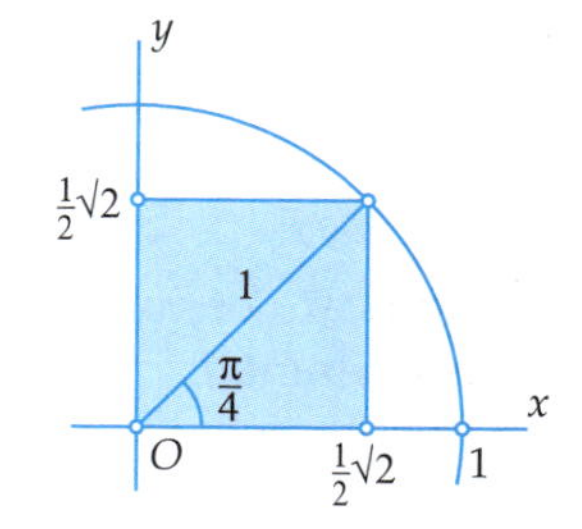


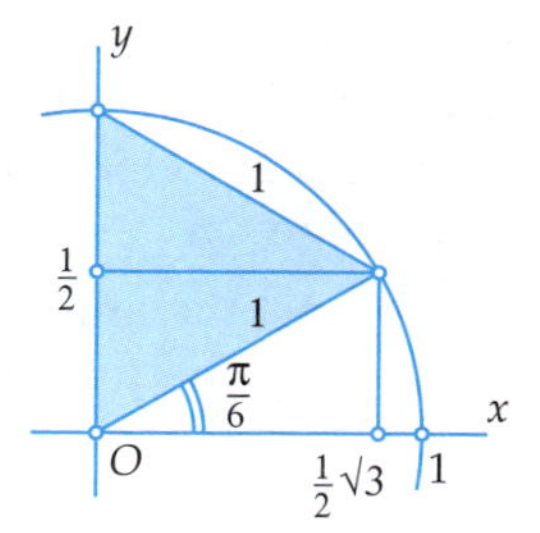


Use the table on the opposite page and a drawing of the unit circle to calculate the following sines, cosines and tangents. Clearly mark each angle on the unit circle and use symmetry if necessary. Give exact answers!

17.12

a. $\sin \frac{2}{3}\pi$

b. $\cos \frac{3}{4}\pi$

c. $\cos \frac{11}{6}\pi$

d. $\tan \frac{5}{4}\pi$

e. $\sin \frac{5}{6}\pi$

17.13

a. $\sin 3\pi$

b. $\tan 7\pi$

c. $\cos -5\pi$

d. $\tan 12\pi$

e. $\sin -5\pi$

17.14

a. $\sin -\frac{2}{3}\pi$

b. $\tan \frac{7}{4}\pi$

c. $\cos -\frac{7}{6}\pi$

d. $\tan -\frac{5}{3}\pi$

e. $\sin \frac{13}{4}\pi$

17.15

a. $\tan \frac{4}{3}\pi$

b. $\sin -\frac{3}{4}\pi$

c. $\cos \frac{11}{3}\pi$

d. $\tan -\frac{15}{4}\pi$

e. $\cos -\frac{23}{6}\pi$

17.16

a. $\cos 13\pi$

b. $\tan 17\pi$

c. $\sin -7\pi$

d. $\tan 11\pi$

e. $\cos -8\pi$

17.17

a. $\sin \frac{23}{6}\pi$

b. $\tan -\frac{17}{4}\pi$

c. $\sin -\frac{7}{3}\pi$

d. $\tan -\frac{25}{6}\pi$

e. $\sin \frac{23}{4}\pi$

## The sine, the cosine and the tangent

Any given angle of rotation $\alpha$ corresponds to a rotation in the plane around the origin over this angle. Positive angles correspond to anticlockwise rotations and negative angles correspond to clockwise rotations. Such a rotation can be visualized by means of an arc of the unit circle with central angle $\alpha$ starting at the point $(1, 0)$. The coordinates $(x, y)$ of the end point then are the *cosine* and the *sine* of $\alpha$, respectively. So

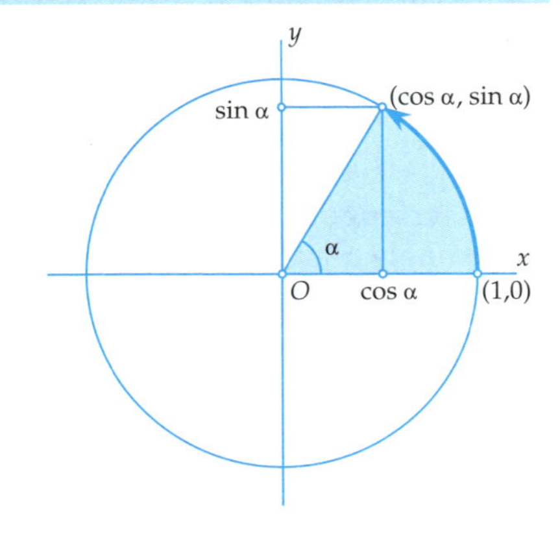


$$x = \cos\alpha \qquad \text{and} \qquad y = \sin\alpha$$

Since $(x, y)$ is on the unit circle, we have $x^2 + y^2 = 1$, so

$$\cos^2\alpha + \sin^2\alpha = 1$$

N.B.: $\cos^2\alpha$ means $(\cos\alpha)^2$ and $\sin^2\alpha$ means $(\sin\alpha)^2$. These are the normal notations. However, $\cos\alpha^2$ means $\cos(\alpha^2)$. Use brackets in cases of possible ambiguities!

By definition, the *tangent* of $\alpha$ is the quotient of the sine and the cosine, in formula:

$$\tan\alpha = \frac{\sin\alpha}{\cos\alpha}$$

From these definitions of the sine and the cosine by means of the unit circle the following *symmetry properties* are evident:

$$\sin(-\alpha) = -\sin\alpha, \qquad \cos(-\alpha) = \cos\alpha, \qquad \tan(-\alpha) = -\tan\alpha$$

There are angles $\alpha$ with special values for the sine, the cosine and the tangent. For $0 \leq \alpha \leq \frac{\pi}{2}$ (in radians) these are listed in the following table. You should know these values *by heart*.

| $\alpha$ | $0$ | $\frac{1}{6}\pi$ | $\frac{1}{4}\pi$ | $\frac{1}{3}\pi$ | $\frac{1}{2}\pi$ |
|---|---|---|---|---|---|
| $\sin\alpha$ | $0$ | $\frac{1}{2}$ | $\frac{1}{2}\sqrt{2}$ | $\frac{1}{2}\sqrt{3}$ | $1$ |
| $\cos\alpha$ | $1$ | $\frac{1}{2}\sqrt{3}$ | $\frac{1}{2}\sqrt{2}$ | $\frac{1}{2}$ | $0$ |
| $\tan\alpha$ | $0$ | $\frac{1}{3}\sqrt{3}$ | $1$ | $\sqrt{3}$ | $-$ |

17.18 For each angle $\alpha$ the inequalities $-1 \leq \cos\alpha \leq 1$ and $-1 \leq \sin\alpha \leq 1$ hold. Are there similar bounds for the tangent? Using the 'tangent line figure', investigate what happens with the tangent if a point rotates around the unit circle.

In a rectangular triangle $ABC$, the acute angle $\alpha$ and one of the sides are given. Using a calculator, calculate the other two sides, rounded off to four decimal places:

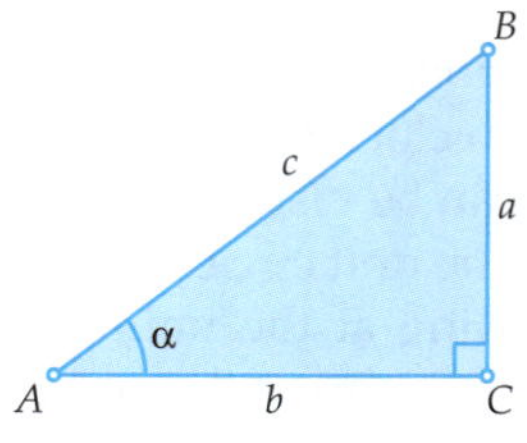


17.19
- a. $\alpha = 32°, c = 3$
- b. $\alpha = 63°, c = 2$
- c. $\alpha = 46°, a = 2$
- d. $\alpha = 85°, c = 7$
- e. $\alpha = 12°, b = 3$

17.20
- a. $\alpha = 23°, a = 3$
- b. $\alpha = 49°, b = 2$
- c. $\alpha = 76°, c = 8$
- d. $\alpha = 21°, b = 2$
- e. $\alpha = 17°, b = 4$

17.21
- a. $\alpha = 1.1$ rad, $c = 3$
- b. $\alpha = 0.5$ rad, $c = 4$
- c. $\alpha = 0.2$ rad, $a = 2$
- d. $\alpha = 1.2$ rad, $b = 7$
- e. $\alpha = 0.7$ rad, $a = 3$

In the following exercises, the sine or the cosine of an angle $\alpha$ with $0 < \alpha < \frac{1}{2}\pi$ is given. Calculate the other trigonometric function (cosine or sine) of $\alpha$, and also the tangent. Give exact answers!

*Example:* $\sin\alpha = \frac{5}{7}$.
Draw a rectangular triangle $ABC$ as shown with $a = 5$ and $c = 7$. Then, indeed, $\sin\alpha = \frac{5}{7}$ and Pythagoras' theorem yields $b = \sqrt{49-25} = \sqrt{24} = 2\sqrt{6}$, meaning $\cos\alpha = \frac{b}{c} = \frac{2}{7}\sqrt{6}$ and $\tan\alpha = \frac{a}{b} = \frac{5}{12}\sqrt{6}$.

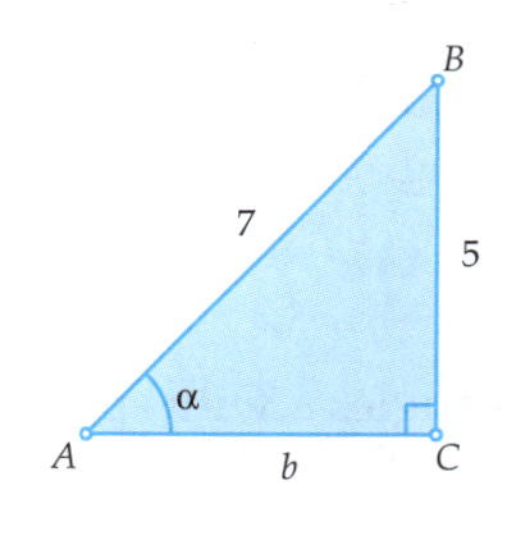


17.22
- a. $\sin\alpha = \frac{1}{5}$
- b. $\cos\alpha = \frac{2}{7}$
- c. $\sin\alpha = \frac{3}{8}$
- d. $\cos\alpha = \frac{2}{5}$
- e. $\cos\alpha = \frac{5}{7}$

17.23
- a. $\sin\alpha = \frac{3}{4}$
- b. $\cos\alpha = \frac{1}{6}$
- c. $\sin\alpha = \frac{1}{8}$
- d. $\cos\alpha = \frac{5}{8}$
- e. $\cos\alpha = \frac{5}{13}$

17.24
- a. $\sin\alpha = \frac{6}{31}$
- b. $\cos\alpha = \frac{4}{23}$
- c. $\sin\alpha = \frac{1}{3}\sqrt{5}$
- d. $\cos\alpha = \frac{1}{3}\sqrt{7}$
- e. $\cos\alpha = \frac{1}{4}\sqrt{10}$

## The tangent on the vertical tangent line

For any point on the unit circle with central angle $\alpha$ you can find $\cos\alpha$ and $\sin\alpha$ as projections on the $x$-axis and the $y$-axis, respectively. But the tangent of $\alpha$ can also be readily visualized. To do this, draw the vertical tangent line to the circle in the point $(1,0)$ and intersect this vertical line with the (elongated) radius line as shown below. The $y$-coordinate of the intersection point then is exactly equal to $\tan\alpha$.

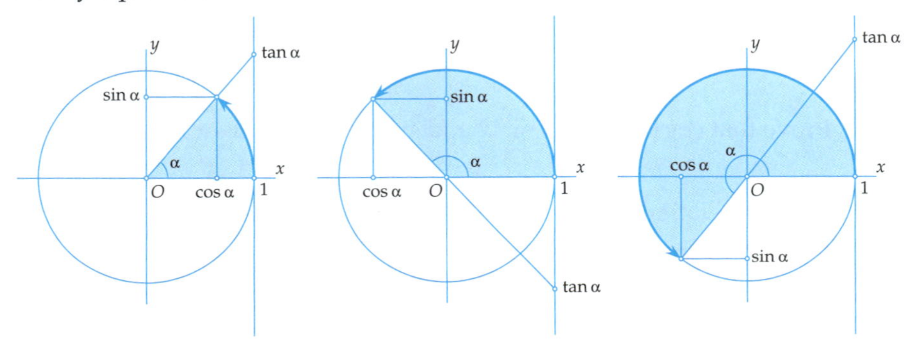


Above, the situation is given for various values of $\alpha$ in, respectively, the first, the second and the third quadrant. In the second quadrant, the sine is positive and the cosine negative, so $\tan\alpha = \frac{\sin\alpha}{\cos\alpha}$ is negative. In the third quadrant, the cosine and the sine are both negative, so then the tangent is positive. In the fourth quadrant (not drawn), the cosine is positive, while the sine is negative, so again the tangent is negative.

The figure clearly shows that for any angle $\alpha$ we have

$$\tan(\alpha + \pi) = \tan\alpha$$

## The rectangular triangle

If $ABC$ is a rectangular triangle with a right angle at $C$, an acute angle $\alpha$ at $A$ and sides of length $a$, $b$ and $c$ opposite to $A$, $B$ and $C$, respectively, then the following relations hold. You should learn them by heart!

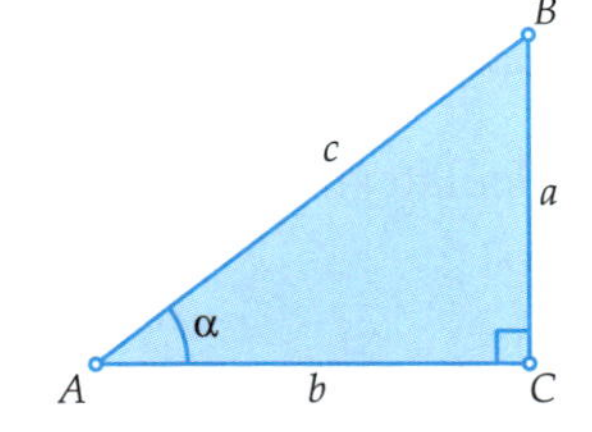


$$\sin\alpha = \frac{a}{c}, \qquad \cos\alpha = \frac{b}{c}, \qquad \tan\alpha = \frac{a}{b}$$

$$a^2 + b^2 = c^2 \qquad \textit{(Pythagoras' theorem)}$$

17.25 Derive the following formulas. You can use the formulas on page 141 and the addition formulas on the opposite page:

a. $\cos 2\alpha = 2\cos^2\alpha - 1 = 1 - 2\sin^2\alpha$

b. $1 + \tan^2\alpha = \dfrac{1}{\cos^2\alpha}$

c. $\tan(\alpha + \beta) = \dfrac{\tan\alpha + \tan\beta}{1 - \tan\alpha\tan\beta}$

d. $\tan 2\alpha = \dfrac{2\tan\alpha}{1 - \tan^2\alpha}$

17.26 Prove that $\cos\left(\frac{\pi}{2} - x\right) = \sin x$ and $\sin\left(\frac{\pi}{2} - x\right) = \cos x$ for all $x$.

17.27 Use the relations $\alpha = \frac{\alpha+\beta}{2} + \frac{\alpha-\beta}{2}$ and $\beta = \frac{\alpha+\beta}{2} - \frac{\alpha-\beta}{2}$ to derive the following formulas:

a. $\sin\alpha + \sin\beta = 2\sin\frac{\alpha+\beta}{2}\cos\frac{\alpha-\beta}{2}$

b. $\sin\alpha - \sin\beta = 2\sin\frac{\alpha-\beta}{2}\cos\frac{\alpha+\beta}{2}$

c. $\cos\alpha + \cos\beta = 2\cos\frac{\alpha+\beta}{2}\cos\frac{\alpha-\beta}{2}$

d. $\cos\alpha - \cos\beta = -2\sin\frac{\alpha+\beta}{2}\sin\frac{\alpha-\beta}{2}$

The double angle formulas $\cos 2\alpha = 2\cos^2\alpha - 1 = 1 - 2\sin^2\alpha$ (see above) can also be used to calculate $\sin\alpha$ and $\cos\alpha$ if $\cos 2\alpha$ is known:

$$\sin\alpha = \pm\sqrt{\tfrac{1}{2}(1 - \cos 2\alpha)} \qquad \text{and} \qquad \cos\alpha = \pm\sqrt{\tfrac{1}{2}(1 + \cos 2\alpha)}$$

Use this, and if necessary the addition formulas, in working out the exercises. Always give exact answers.

*Example:* $\cos\frac{5}{12}\pi = \pm\sqrt{\frac{1}{2}(1 + \cos\frac{5}{6}\pi)} = \pm\sqrt{\frac{1}{2}(1 - \frac{1}{2}\sqrt{3})} = \frac{1}{2}\sqrt{2 - \sqrt{3}}$ since $\cos\frac{5\pi}{6} = -\frac{1}{2}\sqrt{3}$ and the angle $\frac{5}{12}\pi$ is in the first quadrant, meaning that its cosine is positive.

17.28 Calculate:

a. $\sin\frac{1}{8}\pi$

b. $\cos\frac{1}{8}\pi$

c. $\tan\frac{1}{8}\pi$

d. $\sin\frac{1}{12}\pi$

e. $\cos\frac{1}{12}\pi$

17.29 Calculate:

a. $\sin\frac{3}{8}\pi$

b. $\cos\frac{7}{8}\pi$

c. $\tan\frac{5}{8}\pi$

d. $\sin\frac{7}{12}\pi$

e. $\tan\frac{7}{12}\pi$

17.30 Calculate:

a. $\sin\frac{11}{8}\pi$

b. $\cos\frac{17}{12}\pi$

c. $\tan\frac{15}{8}\pi$

d. $\tan\frac{13}{12}\pi$

e. $\cos\frac{13}{8}\pi$

## Addition formulas and double angle formulas

Apart from the fundamental formula $\sin^2\alpha + \cos^2\alpha = 1$ and the symmetry formulas on page 141, there are two more basic trigonometric formulas:

$$\begin{aligned}\cos(\alpha+\beta) &= \cos\alpha\cos\beta - \sin\alpha\sin\beta\\ \sin(\alpha+\beta) &= \sin\alpha\cos\beta + \cos\alpha\sin\beta\end{aligned}$$

The validity of these formulas is illustrated in the figure below. But first, we will derive some more formulas from the above addition formulas. In the exercises on the opposite page, we ask you to derive even more trigonometric formulas. *A complete formula overview can be found on page 313.*

Replacing $\beta$ by $-\beta$ in the addition formulas yields

$$\begin{aligned}\cos(\alpha-\beta) &= \cos\alpha\cos\beta + \sin\alpha\sin\beta\\ \sin(\alpha-\beta) &= \sin\alpha\cos\beta - \cos\alpha\sin\beta\end{aligned}$$

Taking $\alpha = \beta$ yields the *double angle formulas*:

$$\begin{aligned}\cos 2\alpha &= \cos^2\alpha - \sin^2\alpha\\ \sin 2\alpha &= 2\sin\alpha\cos\alpha\end{aligned}$$

In exercise 17.25a you are asked to extend the formula for $\cos 2\alpha$ to

$$\cos 2\alpha = \cos^2\alpha - \sin^2\alpha = 2\cos^2\alpha - 1 = 1 - 2\sin^2\alpha$$

**A geometric illustration of the addition formulas**

The dark coloured rectangular triangles in the figure both have an acute angle $\alpha$. The hypotenuse $OQ$ of the bottom triangle has length $\cos\beta$, as $OQ$ is also a side of the rectangular triangle $OQP$ with acute angle $\beta$ and hypotenuse $OP = 1$. Similarly, $PQ = \sin\beta$ holds.

Since the hypotenuse $OQ$ of the dark triangle at the bottom has length $\cos\beta$, the lengths of the rectangular sides are $\cos\alpha\cos\beta$ (horizontal side) and $\sin\alpha\cos\beta$ (vertical side). Similarly, the rectangular sides of the top dark triangle have lengths $\cos\alpha\sin\beta$ (vertical) and $\sin\alpha\sin\beta$ (horizontal).

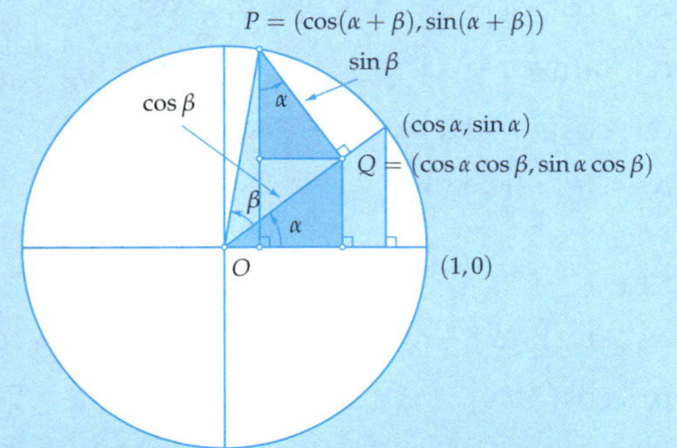


The $x$-coordinate of point $P$ is equal to $\cos(\alpha+\beta)$, as $P$ corresponds to a rotation angle $\alpha+\beta$, and, on the other hand, it is equal to $\cos\alpha\cos\beta - \sin\alpha\sin\beta$ (the difference between the horizontal rectangular sides of the dark triangles). This yields the first addition formula. The second can be derived in a similar way from the fact that the $y$-coordinate of $P$ is equal to the sum of the lengths of the vertical rectangular sides of the dark triangles.

From the drawing of the unit circle and the table on page 141, it follows that all solutions of the equation $\sin x = \frac{1}{2}$ can be written in the form $x = \frac{1}{6}\pi + 2k\pi$ or $x = \frac{5}{6}\pi + 2k\pi$ for an integer value of $k$. In a similar way, find all the solutions of the next equations. Make sure you always use the unit circle and the table on page 141:

17.31

a. $\sin x = -\frac{1}{2}$

b. $\cos x = \frac{1}{2}$

c. $\tan x = -1$

17.32

a. $\sin x = \frac{1}{2}\sqrt{2}$

b. $\cos x = -\frac{1}{2}\sqrt{3}$

c. $\tan x = -\sqrt{3}$

17.33

a. $\tan x = \frac{1}{3}\sqrt{3}$

b. $\cos x = -\frac{1}{2}\sqrt{2}$

c. $\cos x = 0$

Plot the graph of each of the following functions. A rough sketch suffices, but always give the period length and the intersection points of the graph and the $x$-axis:

17.34

a. $\sin(-x)$

b. $\cos(-x)$

c. $\tan(-x)$

d. $\cos(x + \frac{1}{2}\pi)$

e. $\sin(x - \frac{1}{2}\pi)$

17.35

a. $\tan(x + \frac{1}{2}\pi)$

b. $\cos(x - \frac{1}{6}\pi)$

c. $\sin(x + \frac{2}{3}\pi)$

d. $\cos(x - \frac{5}{4}\pi)$

e. $\tan(x - \frac{1}{3}\pi)$

17.36

a. $\tan(x + \frac{1}{6}\pi)$

b. $\cos(x - 3\pi)$

c. $\sin(x + \frac{20}{3}\pi)$

d. $\cos(x - \frac{15}{4}\pi)$

e. $\tan(x - \frac{17}{6}\pi)$

17.37

a. $\sin(\frac{3}{2}\pi - x)$

b. $\cos(\frac{1}{4}\pi + x)$

c. $\tan(\frac{2}{3}\pi - x)$

d. $\cos(\frac{5}{3}\pi + x)$

e. $\sin(\frac{1}{3}\pi - x)$

17.38

a. $\tan(2x)$

b. $\cos(3x)$

c. $\sin(\frac{2}{3}x)$

d. $\cos(\frac{5}{4}x)$

e. $\tan(8x)$

17.39

a. $\tan(3x + \frac{1}{6}\pi)$

b. $\cos(2x - 3\pi)$

c. $\sin(2x + \frac{20}{3}\pi)$

d. $\cos(\frac{1}{2}x - \frac{15}{4}\pi)$

e. $\tan(\frac{1}{3}x - \frac{17}{6}\pi)$

17.40

a. $\sin(2\pi x)$

b. $\cos(3\pi x)$

c. $\tan(\pi x)$

d. $\cos(2\pi x + \frac{1}{2}\pi)$

e. $\sin(2\pi x - \frac{1}{3}\pi)$

17.41

a. $\tan(6\pi x + \pi)$

b. $\cos(4\pi x - 7\pi)$

c. $\sin(2\pi x + \frac{2}{3}\pi)$

d. $\cos(\frac{5}{4}\pi x - \pi)$

e. $\tan(\frac{1}{3}\pi x + 2\pi)$

17.42

a. $\tan(\pi x + \frac{1}{6}\pi)$

b. $\cos(\frac{1}{2}\pi x - 3\pi)$

c. $\sin(7\pi x + \frac{2}{3}\pi)$

d. $\cos(5\pi x - \frac{5}{4}\pi)$

e. $\tan(\frac{3}{4}\pi x - \frac{7}{6}\pi)$

## Graphs of trigonometric functions

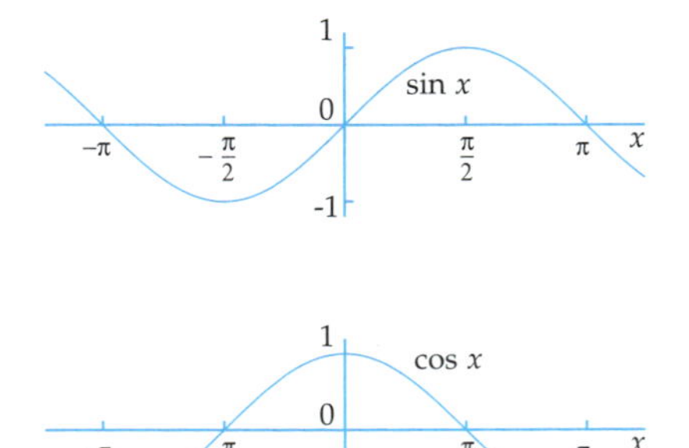


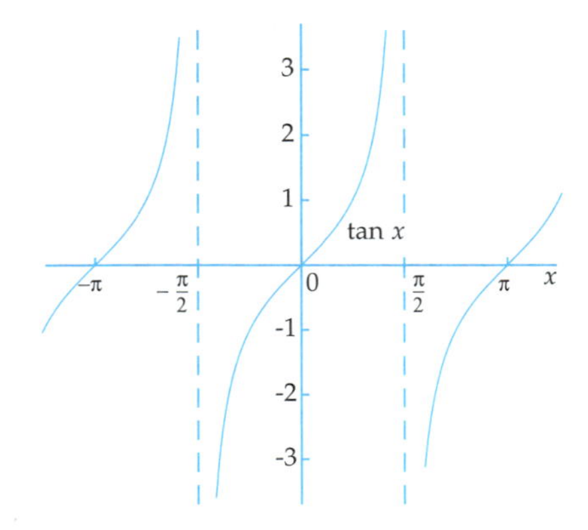


The figure above shows graphs of the functions $\sin x$, $\cos x$ and $\tan x$, with $x$ in radians. These functions are *periodic*: the sine and the cosine with period $2\pi$, the tangent with period $\pi$. The tangent has vertical asymptotes at $x = \frac{\pi}{2} + k\pi$ with integer $k$, as the cosine is zero for these values of $x$, so then $\tan x = (\sin x)/(\cos x)$ is not defined. For instance, take the asymptote $x = \frac{\pi}{2}$. If $x$ approximates $\frac{\pi}{2}$ *from below*, $\tan x$ goes to $+\infty$, but if $x$ approximates $\frac{\pi}{2}$ *from above*, $\tan x$ goes to $-\infty$. Notation:

$$\lim_{x \uparrow \frac{\pi}{2}} \tan x = +\infty \qquad \text{and} \qquad \lim_{x \downarrow \frac{\pi}{2}} \tan x = -\infty$$

The symmetry properties of the sine, the cosine and the tangent (see page 141) can also be seen in the graphs. Remember that

$$\sin(-x) = -\sin x, \qquad \cos(-x) = \cos x, \qquad \tan(-x) = -\tan x$$

The figure to the right shows examples of two frequently used transformations. The top drawing shows the graph of the function $\sin x$ (in black) and also the graph of $\sin 2x$ (in blue). In the latter, the sine graph is compressed horizontally by a factor 2. The period is halved: $\pi$ instead of $2\pi$.
The bottom drawing shows the graph of $\cos x$ (in black) but also the graph of the function $\cos(x - \frac{\pi}{3})$ (in blue). In the latter, the cosine graph has been translated horizontally over a distance $\frac{\pi}{3}$ to the right.

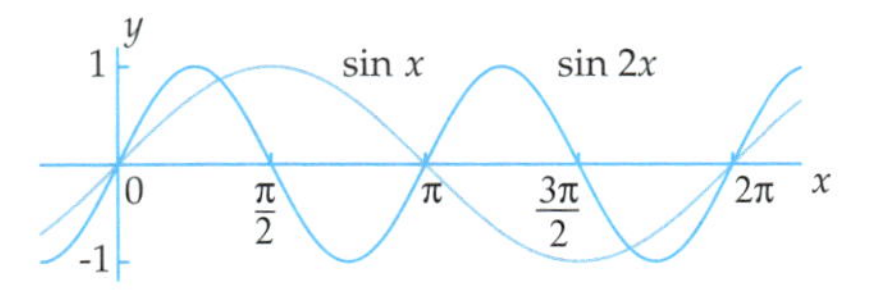


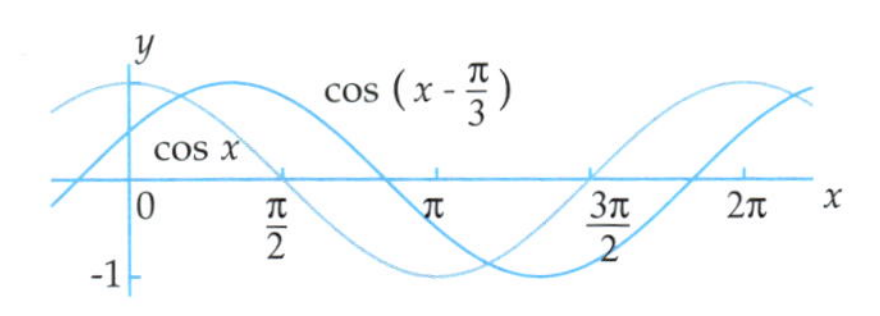

Calculate:

17.43

a. $\arcsin -1$

b. $\arccos 0$

c. $\arctan -1$

d. $\arcsin \frac{1}{2}\sqrt{2}$

e. $\arccos -\frac{1}{2}\sqrt{3}$

17.44

a. $\arctan -\frac{1}{3}\sqrt{3}$

b. $\arccos -\frac{1}{2}\sqrt{2}$

c. $\arcsin \frac{1}{2}\sqrt{3}$

d. $\arctan -\sqrt{3}$

e. $\arccos -1$

17.45

a. $\arctan(\tan \pi)$

b. $\arccos(\cos -\pi)$

c. $\arcsin(\sin 3\pi)$

d. $\arcsin(\sin \frac{2}{3}\pi)$

e. $\arctan(\tan \frac{5}{4}\pi)$

17.46 Assume $\alpha = \arcsin \frac{1}{3}$. Calculate:

a. $\cos \alpha$

b. $\tan \alpha$

c. $\sin 2\alpha$

d. $\cos\left(\alpha + \frac{\pi}{4}\right)$

e. $\cos \frac{1}{2}\alpha$

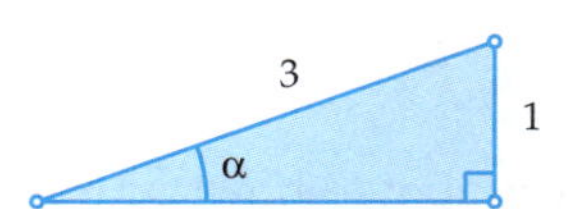


Calculate the following values. Always draw a suitably chosen rectangular triangle.

17.47

a. $\sin(\arcsin -\frac{5}{7})$

b. $\sin(\arccos 0)$

c. $\tan(\arctan \frac{3}{4})$

d. $\cos(\arctan 1)$

e. $\sin(\arctan -1)$

17.48

a. $\cos(\arcsin \frac{3}{5})$

b. $\sin(\arccos \frac{2}{3})$

c. $\cos(-\arctan \frac{3}{4})$

d. $\tan(\arcsin \frac{5}{7})$

e. $\sin(\arctan -4)$

17.49 Calculate:

a. $\arcsin(\cos \frac{1}{5}\pi)$ *Hint:* $\cos \alpha = \sin(\frac{\pi}{2} - \alpha)$

b. $\arccos(\sin \frac{3}{7}\pi)$

c. $\arcsin(\cos \frac{2}{3}\pi)$

d. $\arcsin(\cos \frac{7}{5}\pi)$

e. $\arctan(\tan \frac{9}{5}\pi)$

## The inverse trigonometric functions

If an angle $x$ (in radians) satisfies $\sin x = \frac{1}{2}$, there are infinitely many possibilities for $x$. Indeed, the sine is a periodic function and, moreover, during one period every value (except 1 and $-1$) is taken twice. Therefore, $\sin x = \frac{1}{2}$ if $x = \frac{1}{6}\pi$ and if $x = \frac{5}{6}\pi$, and an arbitrary integer multiple of $2\pi$ can be added to each of these values.

These choices, however, disappear if *by convention* we restrict $x$ to the interval $[-\frac{\pi}{2}, \frac{\pi}{2}]$, i.e. $-\frac{\pi}{2} \leq x \leq \frac{\pi}{2}$. In the figure below, the corresponding part of the graph is marked by a thick curve.

For the cosine and the tangent, for which similar problems occur, one also has chosen intervals of preference: $[0, \pi]$ for the cosine and $\langle -\frac{\pi}{2}, \frac{\pi}{2} \rangle$ for the tangent.

Now, if a value $y_0$ is given, there is always a *unique* $x_0$ in the preferred interval for which $\sin x_0 = y_0$, $\cos x_0 = y_0$ or $\tan x_0 = y_0$, respectively, holds. In the figure below we have indicated how $x_0$ corresponding with $y_0$ is found.

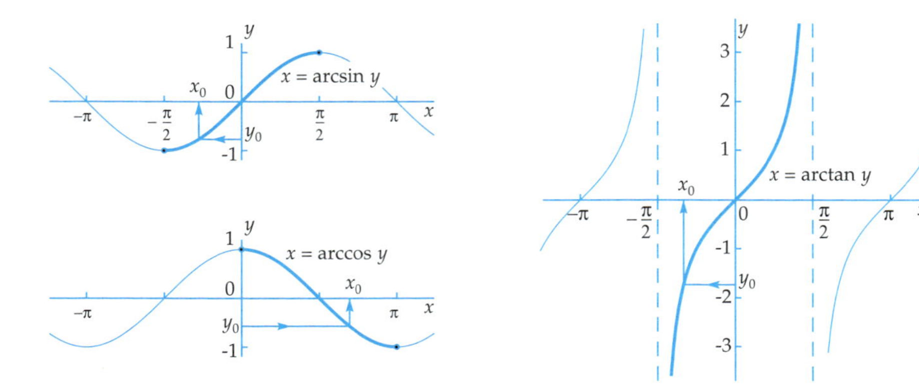


The corresponding functions are called, respectively, arcsin, arccos and arctan. They are called *inverse trigonometric functions*. These functions can also be found on most calculators, sometimes as $\sin^{-1}$, $\cos^{-1}$ and $\tan^{-1}$.

Thus we have:

$$
\begin{array}{lclcl}
x = \arcsin y & \Longleftrightarrow & y = \sin x & \text{and} & -\frac{1}{2}\pi \leq x \leq \frac{1}{2}\pi \\
x = \arccos y & \Longleftrightarrow & y = \cos x & \text{and} & 0 \leq x \leq \pi \\
x = \arctan y & \Longleftrightarrow & y = \tan x & \text{and} & -\frac{1}{2}\pi < x < \frac{1}{2}\pi
\end{array}
$$

Plot a graph of each of the following functions. A rough sketch suffices. First, investigate for which values of $x$ the function is defined and which values the function can take. Also, find the points of intersection with the coordinate axes and any asymptotes that might occur:

17.50

a. $\arcsin 2x$

b. $\arccos 2x$

c. $\arctan -x$

d. $\arcsin -2x$

e. $\arccos -\frac{1}{3}x$

17.51

a. $\arcsin \frac{1}{3}x$

b. $\arccos -\frac{1}{2}x$

c. $\arctan -3x$

d. $\arcsin(1-x)$

e. $\arccos(1+x)$

17.52 Calculate the following limits:

a. $\lim_{x\to\infty} \arctan 2x$

b. $\lim_{x\to\infty} \arctan -\frac{1}{5}x$

c. $\lim_{x\to-\infty} \arctan(x+3)$

d. $\lim_{x\to\infty} \arctan(2-5x)$

e. $\lim_{x\to-\infty} \arctan(x^2)$

17.53 Plot a graph of the following functions:

a. $\arctan(-3x+1)$

b. $\arcsin(1-2x)$

c. $\frac{1}{2}\pi + \arcsin x$

d. $\arctan(1-x^2)$

e. $\arctan \frac{1}{x}$

## The graphs of the inverse trigonometric functions

The functions $\arcsin x$ and $\arccos x$ both have the closed interval $-1 \leq x \leq 1$ as their domain. Their graphs can be derived from the graphs op page 149 by reflecting the thick curves in the line $y = x$. The results are shown below.

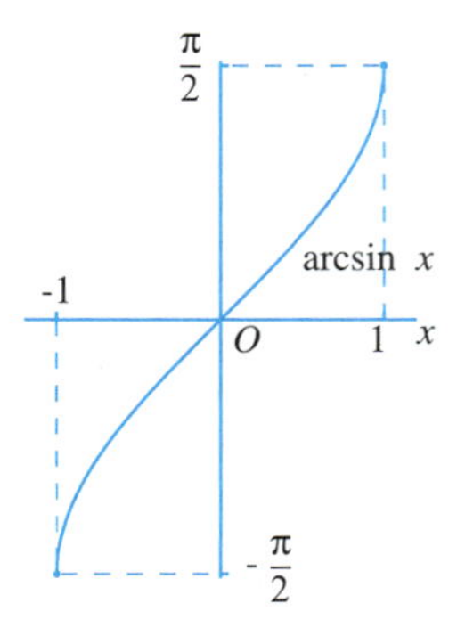


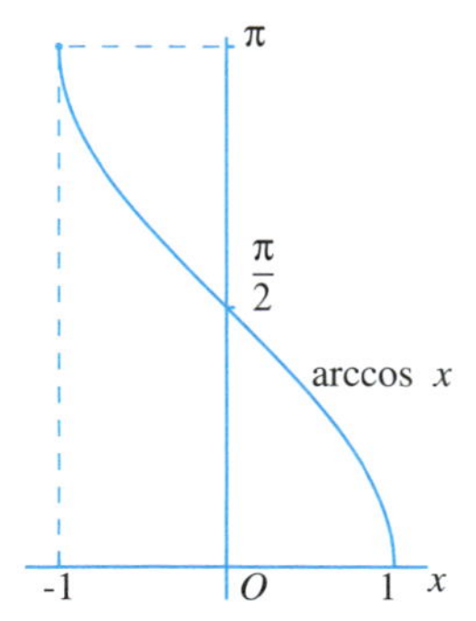


Similarly, the graph of the function $\arctan x$ can be obtained. This function is defined for *all* values of $x$. The graph has two horizontal asymptotes: the lines $y = -\frac{1}{2}\pi$ and $y = \frac{1}{2}\pi$.

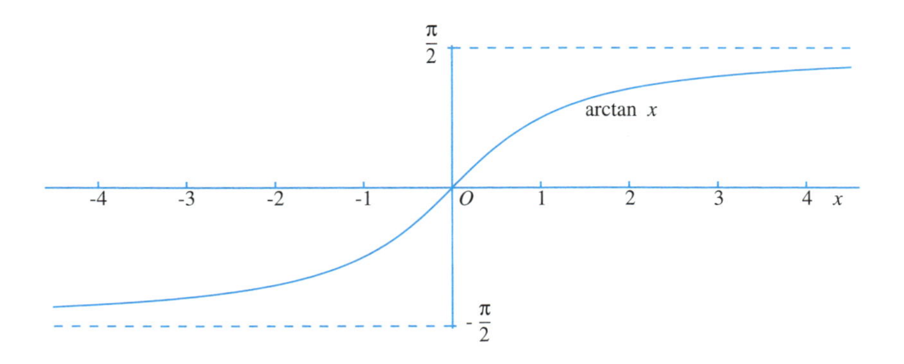


The horizontal asymptotes express the following limits: for $x \to \infty$, the function $\arctan x$ gradually approximates the *limiting value* $\frac{1}{2}\pi$.
In formula: $\lim_{x\to\infty} \arctan x = \frac{1}{2}\pi$. For $x \to -\infty$, the function $\arctan x$ gradually approximates the limiting value $-\frac{1}{2}\pi$. In formula: $\lim_{x\to-\infty} \arctan x = -\frac{1}{2}\pi$.

Calculate the following limits using the standard limit on the opposite page. Look at the following two examples:

*Example 1:*

$$\lim_{x\to 0}\frac{\sin 3x}{x}=\lim_{x\to 0}\frac{\sin 3x}{3x}\times 3=3\times\lim_{y\to 0}\frac{\sin y}{y}=3\times 1=3$$

At the second equality sign, we took $y=3x$. Note that $x\to 0$ implies $y\to 0$ and conversely.

*Example 2:*

$$\lim_{x\to 0}\frac{\sin 5x}{\sin 4x}=\lim_{x\to 0}\frac{\sin 5x}{5x}\times\frac{4x}{\sin 4x}\times\frac{5}{4}=\frac{5}{4}\times\frac{\lim\limits_{x\to 0}\dfrac{\sin 5x}{5x}}{\lim\limits_{x\to 0}\dfrac{\sin 4x}{4x}}=\frac{5}{4}\times\frac{1}{1}=\frac{5}{4}$$

17.54

a. $\lim\limits_{x\to 0}\dfrac{x}{\sin 2x}$

b. $\lim\limits_{x\to 0}\dfrac{x}{\tan x}$

c. $\lim\limits_{x\to 0}\dfrac{\sin 7x}{\sin 3x}$

d. $\lim\limits_{x\to 0}\dfrac{\sin 4x}{x\cos x}$

e. $\lim\limits_{x\to 0}\dfrac{\tan x}{\sin 3x}$

17.55

a. $\lim\limits_{x\to 0}\dfrac{\sin^2 x}{x^2}$

b. $\lim\limits_{x\to 0}\dfrac{\tan^2 4x}{x\sin x}$

c. $\lim\limits_{x\to 0}\dfrac{1-\cos 2x}{3x^2}$

d. $\lim\limits_{x\to 0}\dfrac{1-\cos x}{x}$

e. $\lim\limits_{x\to 0}\dfrac{x^2}{\sin x\tan 3x}$

Calculate the following limits using a suitably chosen substitution. For the first three limits, you might take $y=x-\pi$, $y=x-\frac{\pi}{2}$ and $y=\arcsin x$, respectively, to write them in a form in which the standard limit can be applied.

17.56

a. $\lim\limits_{x\to \pi}\dfrac{\sin x}{x-\pi}$

b. $\lim\limits_{x\to \frac{\pi}{2}}\dfrac{\cos x}{2x-\pi}$

c. $\lim\limits_{x\to 0}\dfrac{\arcsin x}{3x}$

d. $\lim\limits_{x\to 0}\dfrac{\frac{\pi}{2}-\arccos x}{x}$

e. $\lim\limits_{x\to \frac{\pi}{4}}\dfrac{\sin x-\cos x}{x-\frac{\pi}{4}}$

17.57

a. $\lim\limits_{x\to 0}\dfrac{\arcsin x}{\arctan x}$

b. $\lim\limits_{x\to -1}\dfrac{\sin 2\pi x}{\tan 3\pi x}$

c. $\lim\limits_{x\to \infty}x\sin\dfrac{1}{x}$

d. $\lim\limits_{x\to 1}\dfrac{\arctan(x-1)}{x-1}$

e. $\lim\limits_{x\to -\infty}\cos\dfrac{1}{x}$

## A standard limit

An important limit in many applications of trigonometry is

$$\lim_{x \to 0} \frac{\sin x}{x} = 1$$

It means that for small values of $x$, the quotient $\dfrac{\sin x}{x}$ is nearly 1; the closer $x$ is to 0, the closer the quotient is to 1. A different way to express this, is saying that $\sin x$ is nearly equal to $x$ for small angles $x$ (measured in radians), or $\sin x \approx x$ if $x$ is small. This also can be seen in the graph of the sine on page 147. For small values of $x$, it almost coincides with the line $y = x$. A geometric proof of this limit is given below.

A frequently used application of this limit is

$$\lim_{x \to 0} \frac{\tan x}{x} = 1$$

It is proven as follows:

$$\lim_{x \to 0} \frac{\tan x}{x} = \lim_{x \to 0} \frac{\frac{\sin x}{\cos x}}{x} = \lim_{x \to 0} \frac{\sin x}{x \cos x} = \lim_{x \to 0} \frac{\sin x}{x} \times \frac{1}{\cos x} = 1 \times 1 = 1$$

as $\lim\limits_{x \to 0} \cos x = \cos 0 = 1$.

### A geometric proof of the standard limit

Since $\frac{\sin(-x)}{(-x)} = \frac{\sin x}{x}$, it is sufficient to consider only positive values of $x$.

The figure shows a sector $OQP$ of the unit circle with central angle $2x$ where the chord $PQ$ and the tangent lines $PR$ and $QR$ are also drawn. Arc $PQ$ has length $2x$. $PS = \sin x$ and $PR = \tan x$. Because the connecting straight line $PQ$ is shorter than the circular arc $PQ$, which in turn is shorter than the detour $PR + RQ$, we have $2 \sin x < 2x < 2 \tan x$. From the first inequality, it follows that $\frac{\sin x}{x} < 1$ and from the second one it follows that $\frac{\sin x}{x} > \cos x$, as $\tan x = \frac{\sin x}{\cos x}$. Combining both inequalities yields

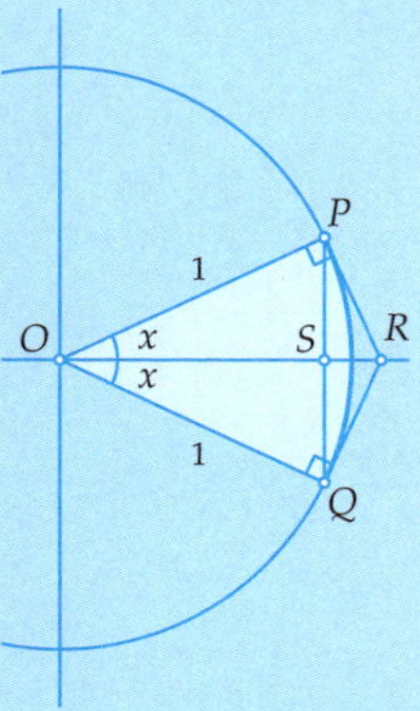


$$\cos x < \frac{\sin x}{x} < 1$$

When $x$ goes to zero, $\cos x$ goes to 1, and for the quotient $\frac{\sin x}{x}$, located between $\cos x$ and 1, the same must hold. This completes the proof.

17.58 In each of the following exercises, some sides and angles of a triangle are given. Using a calculator, calculate the requested angle or side and also the area $O$ of the triangle. Round off your answers to four decimal places (with angles in radians).

a. $\alpha = 43°$, $\beta = 82°$, $a = 3$. Calculate $b$ and $O$.
b. $\alpha = 113°$, $\beta = 43°$, $b = 2$. Calculate $a$ and $O$.
c. $\alpha = 26°$, $\beta = 93°$, $a = 4$. Calculate $c$ and $O$.
d. $\alpha = 76°$, $a = 5$, $c = 3$. Calculate $\gamma$ and $O$.
e. $\beta = 36°$, $a = 2$, $b = 4$. Calculate $\alpha$ and $O$.
f. $\beta = 1.7$ rad, $a = 3$, $b = 4$. Calculate $\gamma$ and $O$.
g. $a = 4$, $b = 5$, $c = 6$. Calculate $\gamma$ and $O$.
h. $a = 5$, $b = 5$, $c = 6$. Calculate $\alpha$ and $O$.
i. $\beta = 0.75$ rad, $b = 6$, $c = 5$. Calculate $a$ and $O$.
j. $\alpha = 71°$, $b = 2$, $c = 3$. Calculate $a$ and $O$.
k. $\alpha = 58°$, $a = 6$, $b = 5$. Calculate $c$ and $O$.

## Sides, angles and area of a triangle

For any triangle $ABC$ with angles $\alpha$, $\beta$ and $\gamma$ at $A$, $B$ and $C$, sides of length $a$, $b$ and $c$ opposite to $A$, $B$ and $C$, and area $O$, the following relations hold:

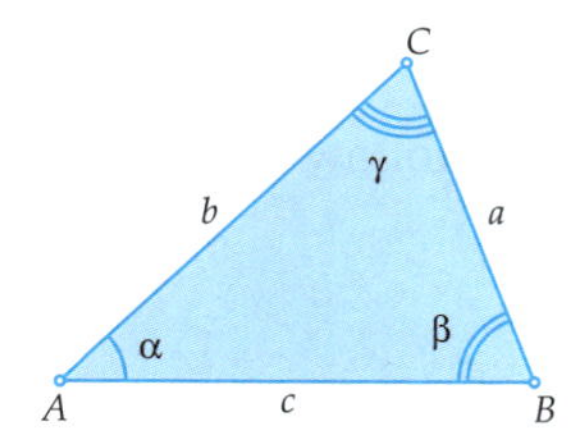


$$\frac{a}{\sin\alpha} = \frac{b}{\sin\beta} = \frac{c}{\sin\gamma} \qquad \textit{(rule of sines)}$$

$$a^2 = b^2 + c^2 - 2bc\cos\alpha \qquad \textit{(cosine rule)}$$

$$O = \tfrac{1}{2}bc\sin\alpha = \tfrac{1}{2}ca\sin\beta = \tfrac{1}{2}ab\sin\gamma \qquad \textit{(area formula)}$$

*Proof of the rule of sines:*

$h = b\sin\alpha = a\sin\beta$ so $\dfrac{a}{\sin\alpha} = \dfrac{b}{\sin\beta}$

and similarly $\dfrac{a}{\sin\alpha} = \dfrac{c}{\sin\gamma}$.

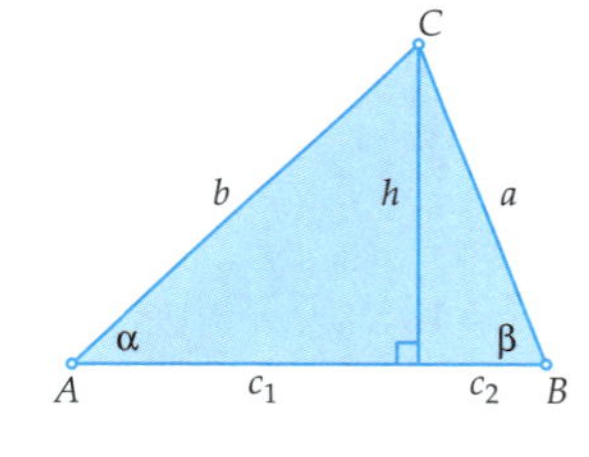


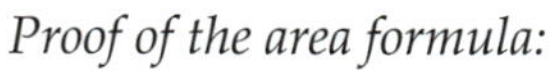

*Proof of the area formula:*

$O = \frac{1}{2}hc = \frac{1}{2}bc\sin\alpha$, et cetera.

*Proof of the cosine rule:*

$$a^2 = h^2 + c_2^2 = (b^2 - c_1^2) + (c - c_1)^2 = b^2 + c^2 - 2c_1c = b^2 + c^2 - 2bc\ \cos\alpha$$

If angle $\alpha$ is obtuse, the proofs of the rule of sines and the area formula are exactly the same. For the proof of the cosine rule, a small adaptation is required (see the figure to the right).

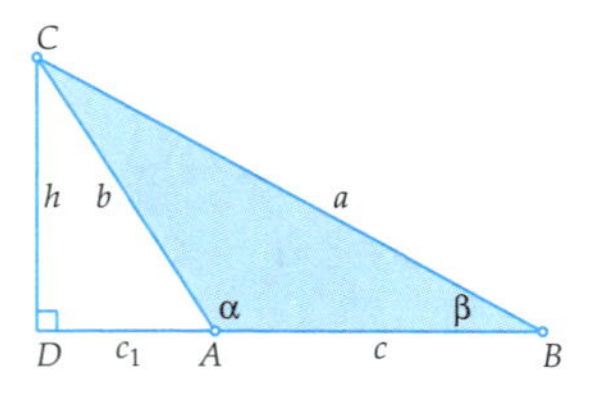


If $AD = c_1$ and $BD = c_2$, we have $c = AB = c_2 - c_1$ and $c_1 = -b\cos\alpha$ since $\frac{\pi}{2} < \alpha < \pi$ which means $\cos\alpha$ is negative. Hence:

$$a^2 = h^2 + c_2^2 = (b^2 - c_1^2) + (c + c_1)^2 = b^2 + c^2 + 2c_1c = b^2 + c^2 - 2bc\ \cos\alpha$$

which proves the cosine rule in this case as well.

If $\alpha$ is a right angle, the cosine rule equals Pythagoras' theorem, this time in the form $a^2 = b^2 + c^2$.

# 18 Exponentials and logarithms

Simplify the following expressions as much as possible:

18.1

a. $2^{(3^5)}$

b. $(2^3)^5\,(2^5)^3$

c. $2^{(3^5)} : 2^{(5^3)}$

d. $2^{1+x}\,3^x$

e. $4^{2x} : 2^x$

18.2

a. $\left(2^{3-2x}\right)^4$

b. $\sqrt{9^{x-1}}$

c. $\sqrt[5]{10^{20x+10}}$

d. $2^x \times 4^{1-x} : 8^x$

e. $\left(10^{-2x}\right)^2$

Plot the graph of the following functions. A rough sketch suffices, but always find the intersection points (if any) with the coordinate axes and the horizontal asymptote. The scales on both axes do not need to be the same:

18.3

a. $f(x) = 2^{x-1}$

b. $f(x) = 2^{1-x}$

c. $f(x) = 10^{2x}$

d. $f(x) = (11/10)^{x+2}$

e. $f(x) = 3^{3x}$

18.4

a. $f(x) = 3^{2x-1}$

b. $f(x) = (1/3)^{3-x}$

c. $f(x) = (1/10)^{x+1}$

d. $f(x) = (9/10)^{x-2}$

e. $f(x) = (2/3)^{2x+2}$

18.5

a. $f(x) = 2^{x-1} - \frac{1}{2}$

b. $f(x) = 2^{1-x} - 8$

c. $f(x) = 10^{2x} - 100$

d. $f(x) = (11/10)^{x+2} + 1$

e. $f(x) = 3^{3x} - 9$

18.6

a. $f(x) = 4^{(x-1)/2} - 4$

b. $f(x) = 7^{1-2x} - 49$

c. $f(x) = (1/3)^{2+x} + 2$

d. $f(x) = (4/3)^{(x+2)/2} + 1$

e. $f(x) = 13^{3x} - 13$

18.7 Plot the graphs of the following functions in one figure:

$$f(x) = 2^x, \quad g(x) = 2^{-x} \quad \text{and} \quad h(x) = 2^x + 2^{-x}$$

18.8 Plot the graphs of the following functions in one figure:

$$f(x) = 3^x, \quad g(x) = -3^{-x} \quad \text{and} \quad h(x) = 3^x - 3^{-x}$$

## Exponential functions

Functions of the form $f(x) = a^x$ with $a > 0$ are called *exponential functions*. Below, the graph of $a^x$ is drawn for various values of $a$.

All graphs go through the point $(0, 1)$, since $a^0 = 1$ for every value of $a$. The graph is increasing if $a > 1$ and decreasing if $0 < a < 1$.
For $a = 1$, the graph is just the horizontal line $y = 1$, since $1^x = 1$ for every $x$.
The graphs of $a^x$ and $(1/a)^x$ are mirror images in the $y$-axis since $(1/a)^x = \left(a^{-1}\right)^x = a^{-x}$.

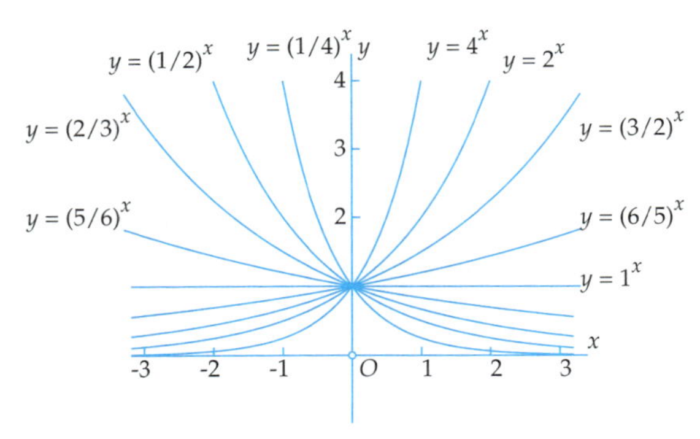


For each $a \neq 1$, the $x$-axis is a horizontal asymptote of the graph. Indeed, $\lim_{x \to -\infty} a^x = 0$ for $a > 1$ and $\lim_{x \to +\infty} a^x = 0$ for $0 < a < 1$.

All graphs are in the upper half plane, since $a^x > 0$ for all $x$ and each $a > 0$.

The rules for powers from chapter 3 (see page 27) that were given for rational exponents, also generally hold for arbitrary real numbers as exponents. Here, we present them in a slightly different form. They are valid for all positive $a$ and $b$ and all real numbers $x$ and $y$.

$$
\begin{aligned}
a^x \times a^y &= a^{x+y} \\
a^x : a^y &= a^{x-y} \\
(a^x)^y &= a^{x \times y} \\
(a \times b)^x &= a^x \times b^x \\
\left(\frac{a}{b}\right)^x &= \frac{a^x}{b^x}
\end{aligned}
$$

Simplify the following expressions as much as possible:

18.9

a. $\log_2 \frac{1}{4}$

b. $\log_4 8$

c. $\log_2 5 + \log_2 3$

d. $\log_3 \frac{2}{9} - \log_3 \frac{8}{27}$

e. $\log_{10}(2^6) - \log_{10} \frac{1}{2}$

18.10

a. $\log_5 8 + \log_5 4$

b. $\log_5 8 - \log_5 4$

c. $\log_5 8 \times \log_5 4$

d. $\log_5 8 : \log_5 4$

e. $\log_{1/2} 5 + \log_2 5$

Plot the graph of the following functions. A rough sketch suffices, but indicate the intersection points (if any) with the coordinate axes and the vertical asymptote. The scales on both axes do not need to be the same:

18.11

a. $f(x) = \log_2(x-1)$

b. $f(x) = \log_2(1-x)$

c. $f(x) = \log_{10}(2x)$

d. $f(x) = \log_3(x+2)$

e. $f(x) = \log_{3/2}(3x)$

18.12

a. $f(x) = \log_{1/2}(x-2)$

b. $f(x) = \log_{2/3}(4x)$

c. $f(x) = \log_{1/10}(3x+10)$

d. $f(x) = \log_{3/4}(3x-2)$

e. $f(x) = \log_{1/2}(32x)$

18.13

a. $f(x) = \log_2 |x|$

b. $f(x) = \log_2 |4x|$

c. $f(x) = \log_{10} |x-1|$

d. $f(x) = \log_3 \sqrt{x}$

e. $f(x) = \log_{3/2}(x^2)$

18.14

a. $f(x) = \log_{1/2}(2/x)$

b. $f(x) = \log_{2/3}(\frac{2}{3}x^2)$

c. $f(x) = \log_{1/10} |10-3x|$

d. $f(x) = \log_4(64x^5)$

e. $f(x) = \log_{10}(1000/x^2)$

18.15 Plot the graphs of the following functions in one figure:

$$f(x) = \log_{10} x, \quad g(x) = \log_{10} 10x \quad \text{and} \quad h(x) = \log_{10} 100x$$

18.16 Plot the graphs of the following functions in one figure:

$$f(x) = \log_2(1-x), \quad g(x) = \log_2(1+x) \quad \text{and} \quad h(x) = \log_2(1-x^2)$$

Solve the following equations:

18.17

a. $2^x = 5$

b. $5^{x+5} = 2$

c. $10^{1-x} = \frac{1}{100}$

d. $10^{2x} = 25$

e. $(49\sqrt{7})^x = 49$

18.18

a. $\log_{10} x = 2$

b. $\log_{10} x = \frac{1}{4}$

c. $\log_5(2x-1) = 2$

d. $\log_2 x^2 = 3$

e. $\log_{1/3} x = -4$

## Logarithmic functions

For $a > 0$, $a \neq 1$, the *logarithm* $\log_a x$ to the base $a$ of the real number $x > 0$ is defined as the number $y$ for which $a^y = x$ holds, so

$$\log_a x = y \qquad \Longleftrightarrow \qquad a^y = x$$

Functions of the form $f(x) = \log_a x$ are called *logarithmic functions*. The logarithmic function to the base $a$ and the exponential function with base $a$ are mutually inverse functions, which means that the effect of the one is nullified by the other. Hence we have

$$a^{\log_a x} = x \quad \text{for all } x > 0$$

and

$$\log_a(a^y) = y \quad \text{for all } y$$

in which $a > 0$, $a \neq 1$ always is assumed, since otherwise the logarithm doesn't exist.

The graph of a logarithmic function is obtained by reflecting the graph of the corresponding exponential function in the line $y = x$.

This way, the figure to the right can be obtained from the figure on page 157. All graphs go through $(1, 0)$, since for each $a$ we have $\log_a 1 = 0$. The graph is increasing if $a > 1$ and decreasing if $0 < a < 1$.
The graphs of $\log_a x$ and $\log_{1/a} x$ are mirror images in the $x$-axis.

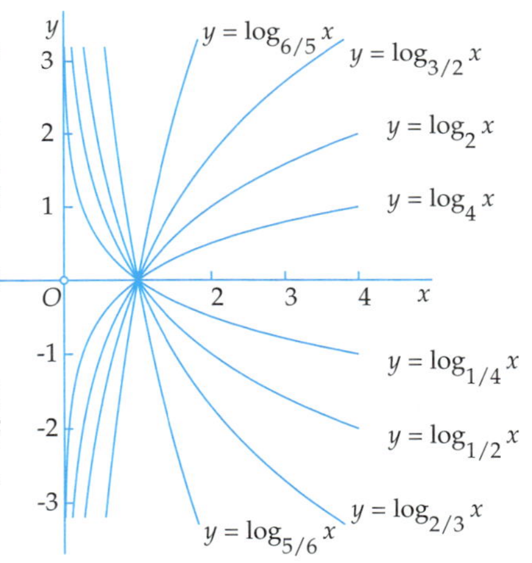


From the properties of exponential functions on page 157 properties of logarithmic functions can be derived. The most important are:

$$\begin{aligned}
\log_a(xy) &= \log_a x + \log_a y && (x, y > 0)\\
\log_a(x/y) &= \log_a x - \log_a y && (x, y > 0)\\
\log_a(x^p) &= p \log_a x && (x > 0)\\
\log_a x &= \frac{\log_b x}{\log_b a} && (x > 0)
\end{aligned}$$

Using the last formula, logarithms to a given base $a$ can be written as logarithms to a different base $b$.

Plot the graph of the following functions. A rough sketch suffices. The scales on the axes do not need to be the same:

18.19

a. $f(x) = \mathrm{e}^{1-x}$

b. $f(x) = \mathrm{e}^{-|x|}$

c. $f(x) = \mathrm{e}^{-x^2}$

d. $f(x) = \mathrm{e}^{-|x-1|}$

e. $f(x) = \mathrm{e}^{1-x^2}$

18.20

a. $f(x) = \ln(1+x)$

b. $f(x) = \ln(1-x)$

c. $f(x) = \ln|x|$

d. $f(x) = \ln|1-x|$

e. $f(x) = \ln|1-x^2|$

18.21 The *hyperbolic cosine*, the *hyperbolic sine* and the *hyperbolic tangent*, abbreviated to *cosh*, *sinh* and *tanh*, respectively, are defined by

$$\cosh x = \frac{\mathrm{e}^{x} + \mathrm{e}^{-x}}{2}, \qquad \sinh x = \frac{\mathrm{e}^{x} - \mathrm{e}^{-x}}{2}, \qquad \tanh x = \frac{\sinh x}{\cosh x}$$

Plot the graph of these functions and prove the following formulas:

a. $\cosh^2 x - \sinh^2 x = 1$

b. $\tanh^2 x + 1/\cosh^2 x = 1$

c. $\sinh(x+y) = \sinh x \cosh y + \cosh x \sinh y$

d. $\cosh(x+y) = \cosh x \cosh y + \sinh x \sinh y$

It can be proven that the graph of the hyperbolic cosine exhibits the form of a hanging chain. This graph is also called a *catenary* (from the Latin word *catena*, meaning chain).

18.22 Calculate:

a. $\lim\limits_{x\to 0} \dfrac{\mathrm{e}^{-x} - 1}{x}$

b. $\lim\limits_{x\to 0} \dfrac{\mathrm{e}^{2x} - 1}{x}$

c. $\lim\limits_{x\to 0} \dfrac{\mathrm{e}^{-3x} - 1}{x}$

d. $\lim\limits_{x\to 0} \dfrac{\mathrm{e}^{ax} - 1}{x}$

e. $\lim\limits_{x\to 0} \dfrac{\mathrm{e}^{x} - 1}{\sqrt{x}}$

18.23 Calculate:

a. $\lim\limits_{x\to 1} \dfrac{\mathrm{e}^{x} - \mathrm{e}}{x-1}$

b. $\lim\limits_{x\to 2} \dfrac{\mathrm{e}^{x} - \mathrm{e}^{2}}{x-2}$

c. $\lim\limits_{x\to a} \dfrac{\mathrm{e}^{x} - \mathrm{e}^{a}}{x-a}$

d. $\lim\limits_{x\to 0} \dfrac{\sinh x}{x}$

e. $\lim\limits_{x\to 0} \dfrac{\tanh x}{x}$

## The function $e^x$ and the natural logarithm

The graphs of the exponential functions $f(x) = a^x$ with $a > 0$ all intersect the $y$-axis in the point $(0, 1)$. All tangent lines in the point $(0, 1)$ are different, and all have an equation of the form $y = 1 + mx$ for some $m$.

There is exactly one value of $a$ for which $m = 1$ is true, which means that the line $y = 1 + x$ is the tangent line to the graph of $f(x) = a^x$ in $(0, 1)$. That number is called e and the corresponding function $f(x) = e^x$ is of paramount importance in the differential and integral calculus. The graph can be seen to the right. It can be proven that the number e, like the number $\pi$ or the number $\sqrt{2}$, is *irrational*. In decimal form:
$e = 2.718281828459\ldots$

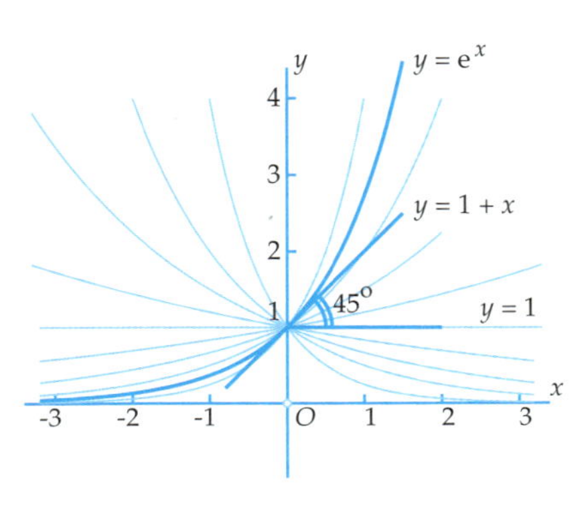


For small values of $x$, the graph of $f(x) = e^x$ and the tangent line $y = 1 + x$ almost coincide, so for $x$ small we have $e^x \approx 1 + x$. It is even true that $\dfrac{e^x - 1}{x} \approx 1$ for $x \approx 0$, or, more precisely, formulated by means of a limit:

$$\lim_{x \to 0} \frac{e^x - 1}{x} = 1$$

The logarithm to the base e is called the *natural logarithm*. Instead of $\log_e x$, we usually write $\ln x$.

Since the exponential function with base e and the natural logarithm are mutually inverse functions, for any $x > 0$, we have $x = e^{\ln x}$. Applying this to $a^x$ instead of $x$, yields $a^x = e^{\ln a^x}$ and since $\ln a^x = x \ln a$, the exponential function $a^x$ can be written in the following way as an exponential function with base e, which, as will be shown later, has many useful applications.

$$a^x = e^{x \ln a}$$

As a first application, we derive

$$\lim_{x \to 0} \frac{a^x - 1}{x} = \ln a$$

Proof: $\dfrac{a^x - 1}{x} = \dfrac{e^{x \ln a} - 1}{x} = \dfrac{e^{x \ln a} - 1}{x \ln a} \ln a \to \ln a$ for $x \to 0$.

Plot the graph of the following functions. A rough sketch suffices.
*Hint*: if necessary, rewrite the function rule in a more manageable form first, using the calculation rules of the logarithm. First, find the domain of the function, i.e. the values of $x$ for which the function rule can be applied.

18.24

a. $f(x) = \ln(x-4)$
b. $f(x) = \ln 4x$
c. $f(x) = \ln(4-x)$
d. $f(x) = \ln(4x-4)$
e. $f(x) = \ln(2x-3)$
f. $f(x) = \ln(2-3x)$
g. $f(x) = \ln|x-3|$
h. $f(x) = \ln \dfrac{1}{x}$
i. $f(x) = \ln \dfrac{1}{x-1}$
j. $f(x) = \ln \dfrac{2}{x-2}$

18.25

a. $f(x) = \ln \dfrac{1}{x^2}$
b. $f(x) = \ln \dfrac{2}{1-2x}$
c. $f(x) = \ln \dfrac{1}{\sqrt{x}}$
d. $f(x) = \ln \dfrac{1}{|x|}$
e. $f(x) = \ln \dfrac{2}{|x-2|}$
f. $f(x) = \ln \dfrac{3}{x^3}$
g. $f(x) = \ln \left| \dfrac{x-1}{x+1} \right|$

18.26 Prove that $\log_a b = \dfrac{1}{\log_b a}$.

Calculate the following limits.

18.27

a. $\lim\limits_{x \to 0} \dfrac{\ln(1-x)}{x}$
b. $\lim\limits_{x \to 0} \dfrac{\ln(1+2x)}{x}$
c. $\lim\limits_{x \to 0} \dfrac{\ln(1-3x)}{2x}$
d. $\lim\limits_{x \to 0} \dfrac{\ln(1+x^2)}{x}$
e. $\lim\limits_{x \to 0} \dfrac{\ln(1+x)}{\ln(1-x)}$

18.28

a. $\lim\limits_{x \to 1} \dfrac{{}^2\log x}{x-1}$
b. $\lim\limits_{x \to 1} \dfrac{{}^3\log x}{1-x}$
c. $\lim\limits_{x \to 2} \dfrac{\ln \frac{x}{2}}{x-2}$ (*Hint*: take $y = \frac{x}{2}$)
d. $\lim\limits_{x \to 3} \dfrac{\ln x - \ln 3}{x-3}$
e. $\lim\limits_{x \to a} \dfrac{\ln x - \ln a}{x-a}$

## More on the natural logarithm function

The figure below shows the graph of the natural logarithm function amidst a number of graphs of other logarithmic functions.

Its characteristic property is that the line $y = x - 1$, intersecting the $x$-axis at an angle of 45°, is the tangent line to the graph in the point $(1, 0)$. For $x$ close to 1, the graph and the tangent line almost coincide. Then, not only $\ln x \approx x - 1$ as $x \approx 1$, but it is even true that

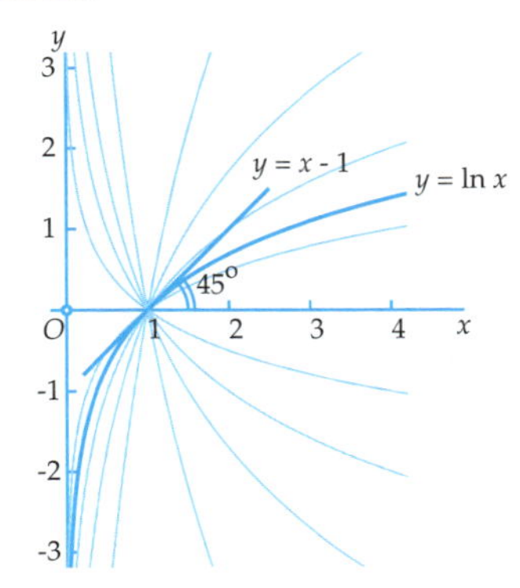


$$\lim_{x \to 1} \frac{\ln x}{x - 1} = 1$$

Substituting $x = 1 + u$, we get a limit for $u \to 0$ that is frequently used in applications:

$$\lim_{u \to 0} \frac{\ln(1 + u)}{u} = 1$$

Furthermore, if, in the formula

$$\log_a x = \frac{\log_b x}{\log_b a}$$

(see page 159), we substitue for the number $b$ the base e of the natural logarithm, we get

$$\log_a x = \frac{1}{\ln a} \ln x$$

which clearly shows that every logarithmic function, up to a constant factor, is equal to the *natural* logarithm function $\ln x$. This can also be seen in the figure above. In the differential and integral calculus this property is frequently used.

Calculate the following limits:

18.29

a. $\lim_{x\to\infty} \frac{x^2}{2^x}$

b. $\lim_{x\to\infty} \frac{x^2}{3^{x-1}}$

c. $\lim_{x\to\infty} \frac{x^{20}}{(3/2)^x}$

d. $\lim_{x\to\infty} \frac{x^{30}}{4^{2x}}$

e. $\lim_{x\to\infty} \frac{x^{70}}{2^{x-5}}$

18.30

a. $\lim_{x\to\infty} \frac{(x+1)^2}{2^x}$

b. $\lim_{x\to\infty} \frac{x^2+5x}{3^{x-1}}$

c. $\lim_{x\to-\infty} \frac{x^{20}}{(4/3)^{-x}}$

d. $\lim_{x\to-\infty} \frac{x^{30}}{4^{1-2x}}$

e. $\lim_{x\to-\infty} \frac{x^{70}}{2^{2-2x}}$

18.31

a. $\lim_{x\to-\infty} \frac{2^x}{x^{200}}$

b. $\lim_{x\to\infty} \frac{2^{-x}}{x^{-1}}$

c. $\lim_{x\to-\infty} \frac{3^{-x}}{x^5}$

d. $\lim_{x\to-\infty} \frac{3^{2x}}{2^{3x}}$

e. $\lim_{x\to-\infty} \frac{x^7}{7^{x-5}}$

Calculate the following limits:

18.32

a. $\lim_{x\to\infty} \frac{\log_{10} x}{x+1}$

b. $\lim_{x\to\infty} \frac{\log_{10}(x+1)}{x}$

c. $\lim_{x\to\infty} \frac{\log_{10}(x^2)}{x^2+1}$

d. $\lim_{x\to\infty} \frac{\log_{10}(x^{100})}{\sqrt[100]{x}}$

e. $\lim_{x\to\infty} \frac{\log_{1/10}(1000x)}{1000x+1}$

18.33

a. $\lim_{x\downarrow 0} x\left(\log_2 x\right)$

b. $\lim_{x\downarrow 0} \sqrt{x}\left(\log_{1/2} x\right)$

c. $\lim_{x\downarrow 0} x\left(\log_{10}(100x)\right)$

d. $\lim_{x\downarrow 0} x^3\left(\log_{1/3} x\right)$

e. $\lim_{x\to 0} \sqrt[3]{x}\left(\log_3 |x|\right)$

*Hint for these limits:* write $y = \frac{1}{x}$

## Standard limits

For $a > 1$, the function $f(x) = a^x$ increases to infinity as $x$ goes to infinity. It increases faster than any function of the form $g(x) = x^p$, even if $p$ is very large. Indeed, for every $p$ it is true that

$$\lim_{x \to +\infty} \frac{x^p}{a^x} = 0 \quad \text{if} \quad a > 1$$

For example, take $p = 1\,000\,000$ ($= 10^6$) and $a = 10$. For $x > 1$, we have that, initially, $x^{1000000}$ is much greater than $10^x$. But for, e.g., $x = 10^{100}$, the number $10^x$ counts $10^{100} + 1$ digits, while $x^{1000000}$ counts no more than $10^8 + 1$ digits!

For $a > 1$, the function $f(x) = \log_a x$ is increasing to infinity as $x$ goes to infinity, but only very slowly. In fact, such a logarithmic function increases slower than any function of the form $g(x) = x^q$ with $q > 0$, even if $q$ is very small (but positive). For example, take $a = 10$ and $q = 1/1\,000\,000$. Although $f(x)$ and $g(x)$ both go to infinity for $x \to +\infty$, the quotient $f(x)/g(x)$ goes to zero. The general formula is:

$$\lim_{x \to +\infty} \frac{\log_a x}{x^q} = 0 \quad \text{if} \quad q > 0$$

Apart from the *standard limits* just mentioned, we have come across two other important limits in this chapter. These are also called standard limits. We look at them again. The first one is about the exponential function with base e:

$$\lim_{x \to 0} \frac{\mathrm{e}^x - 1}{x} = 1$$

This limit implies that the graph of $\mathrm{e}^x$ intersects the $y$-axis in the point $(0, 1)$ at an angle of 45°.

The second limit is about the natural logarithm function $\ln x$:

$$\lim_{x \to 1} \frac{\ln x}{x - 1} = 1$$

This limit implies that the graph of the function $\ln x$ intersects the $x$-axis in the point $(1, 0)$ at an angle of 45°. An equivalent, and also frequently used form, is obtained by the substitution $x = 1 + u$:

$$\lim_{u \to 0} \frac{\ln(1 + u)}{u} = 1$$

# 19 Parametric curves

19.1 Show that $x = 3\sin t, y = 2\cos t$ is another parametrization of the ellipse on the opposite page. What is the sense of tracing (clockwise or anticlockwise) and what is $P_0$?

19.2 Find a parametrization of the ellipse on the opposite page with clockwise tracing and $P_0 = (-3, 0)$.

19.3 Find a parametrization of the ellipse $\dfrac{x^2}{16} + \dfrac{y^2}{25} = 1$.

19.4 Find a parametrization of:

a. the circle with centre $(0, 0)$ and radius 2,
b. the circle with centre $(-1, 3)$ and radius 3,
c. the circle with centre $(2, -3)$ and radius 5,
d. the parabola $x = y^2$,
e. the hyperbola $xy = 1$.

19.5 Show that $(x, y) = (t + \frac{1}{t}, t - \frac{1}{t})$ is a parametrization of the hyperbola $x^2 - y^2 = 4$ and draw the curve. Which values of $t$ correspond to the left-hand branch and the right-hand branch? How can you find the asymptotes?

19.6 For each of the following curves, find the corresponding parametrization. Explain your answers. Don't use a graphing calculator.

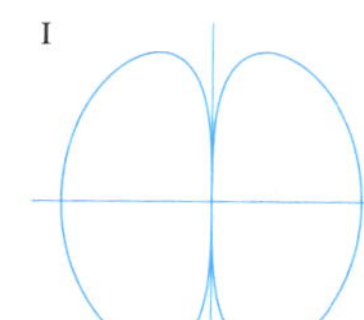


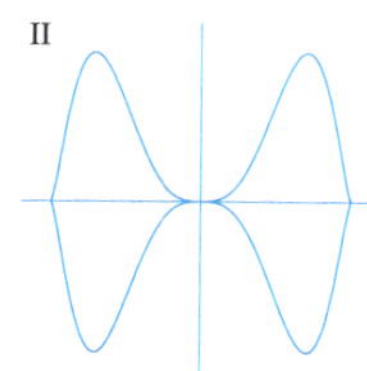


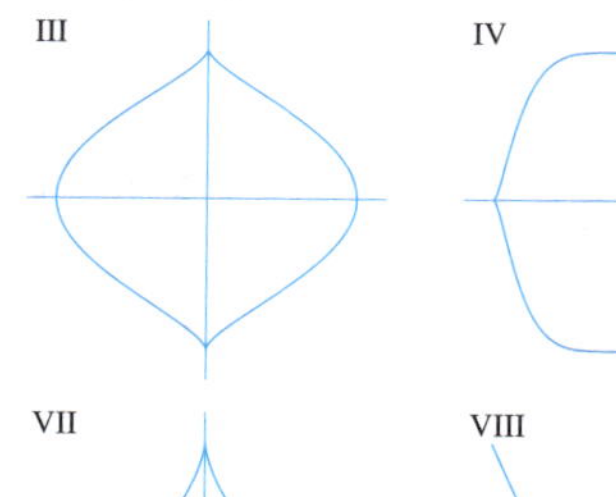


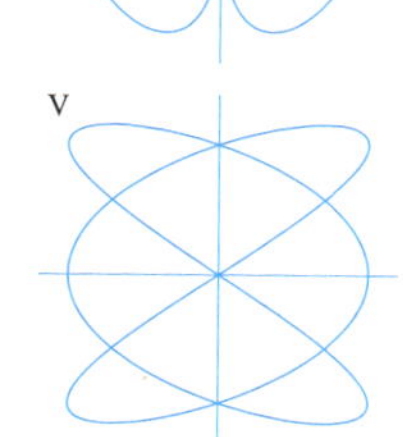


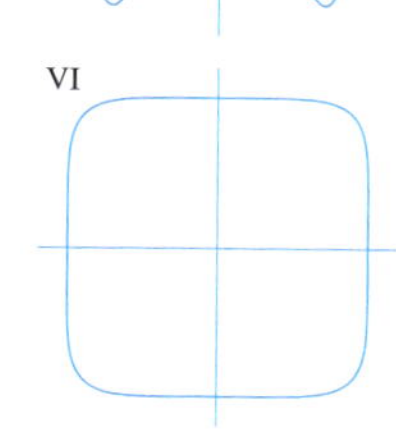


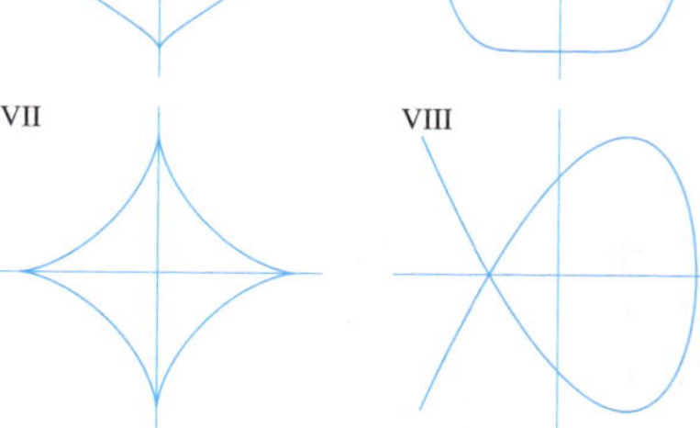


a. $(\cos 3t, \sin 2t)$
b. $(\cos 2t, \sin 3t)$
c. $(\cos^3 t, \sin^3 t)$
d. $(\cos^3 t, \sin t)$
e. $(\cos^3 t, \sin 2t)$
f. $(\cos \frac{1}{2}t, \sin^3 t)$
g. $(\sqrt[3]{\cos t}, \sqrt[3]{\sin t})$
h. $(\sqrt[3]{\cos t}, \sin^3 t)$

## Curves in the plane

The ellipse to the right consists of all points $P = (x, y)$ for which $x = 3\cos t$ and $y = 2\sin t$. So the coordinates $x$ and $y$ are functions of a variable $t$. If $t$ runs from 0 to $2\pi$, the point $P$ traces the ellipse in the direction of the arrow. For $t = 0$ we get $P = (3, 0)$, for $t = \frac{\pi}{2}$ $P = (0, 2)$, for $t = \pi$ $P = (-3, 0)$, for $t = \frac{3\pi}{2}$ $P = (0, -2)$ and for $t = 2\pi$ $P$ is back again in $(3, 0)$.

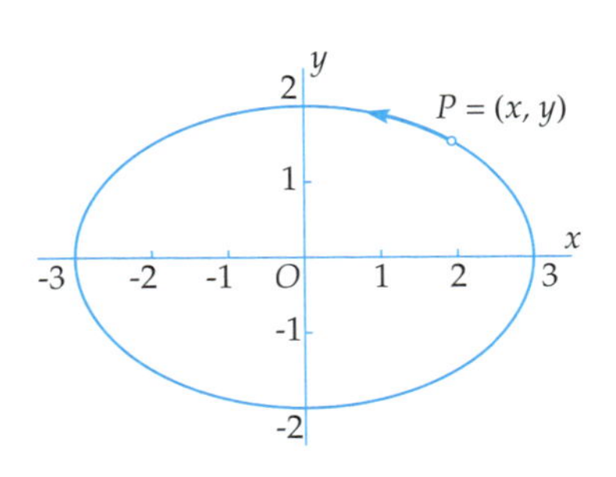


In general, by means of two functions $x = x(t)$ and $y = y(t)$ a *curve in the plane* can be described. Such a description is called a *parametrization* or *parametric representation* of the curve. The variable $t$ is called the *parameter*. We always assume that the functions $x(t)$ and $y(t)$ are *continuous* on their $t$-domain, which means that they don't have jumps. In that case, the curve $(x(t), y(t))$ also doesn't have jumps. It can be a *closed* curve, which means that it returns to a previously visited point for a certain $t$-value, as is the case in the ellipse above, but this is not necessary.

In many applications, the variable $t$ that yields the position of $P$ on the curve, represents *time*, for instance measured in seconds. In that case $(x(t), y(t))$ is the position of $P$ at time $t$. Sometimes, we write $P_t$ for the position of $P$ at time $t$.

Of course, the graph of a function $y = f(x)$ is also a curve in the plane. It is easy to find a parametrization for it, namely $x = t$, $y = f(t)$. But, conversely, not every parametric curve is the graph of a function, as is shown by the ellipse above.

Sometimes, it is possible to eliminate the parameter $t$ from the parametrization of a curve, which results in an *equation* of the curve. In the case of the ellipse above, this can be done using the well-known relation $\cos^2 t + \sin^2 t = 1$. Since $x/3 = \cos t$ and $y/2 = \sin t$, we have

$$\frac{x^2}{9} + \frac{y^2}{4} = 1$$

This is the equation of the ellipse. Compare this to the equation $x^2 + y^2 = 1$ of the unit circle.

19.7 In the plane, draw the set of all points satisfying:

a. $r < 3$

b. $0 \leq \varphi \leq \frac{\pi}{5}$

c. $1 \leq r \leq 2, \quad -\frac{\pi}{3} \leq \varphi \leq \frac{\pi}{3}$

d. $r > 2, \quad |\varphi| < \frac{\pi}{2}$

e. $r = \varphi, \quad 0 \leq \varphi \leq \pi$

f. $0 \leq r \leq \varphi, \quad 0 \leq \varphi \leq \pi$

g. $r = \dfrac{1}{\cos\varphi}, \quad 0 \leq \varphi < \frac{\pi}{2}$

19.8 Derive the *cosine rule*

$$d^2 = r_1^2 + r_2^2 - 2r_1r_2\cos\varphi$$

by calculating in the triangle with vertices $O = (0,0)$, $P = (r_1\cos\varphi, r_1\sin\varphi)$ and $Q = (r_2, 0)$ the square of the distance $d = d(P,Q)$.

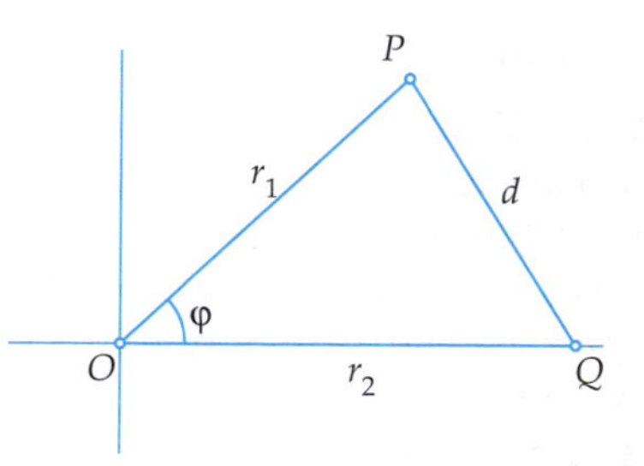


19.9 For each of the given equations in $r$ and $\varphi$, find the corresponding curve. Explain your answers. Don't use a graphing calculator.

I

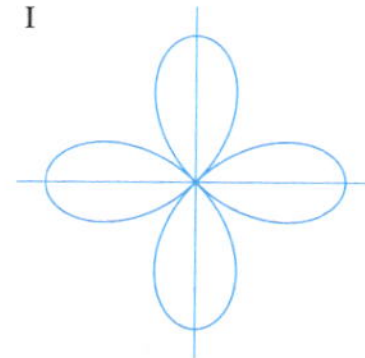

II

III

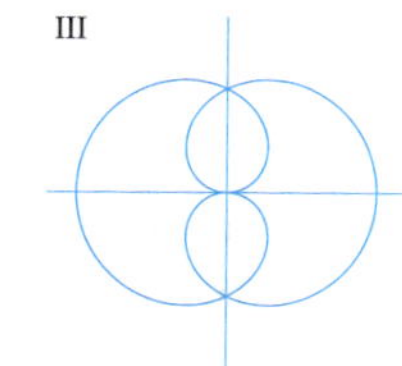

IV

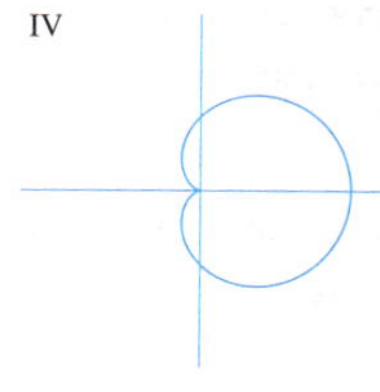

V

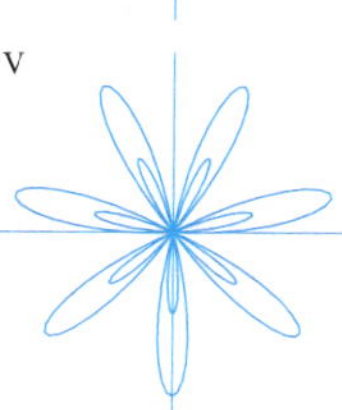

VI

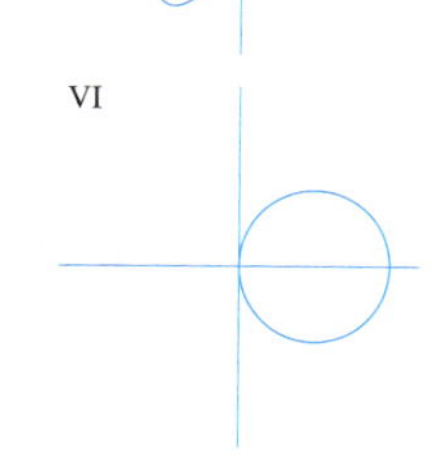

VII

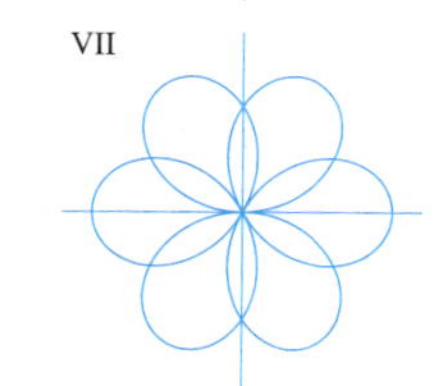

VIII

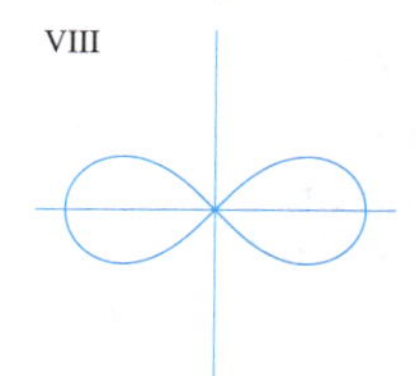

a. $r = \cos\varphi$

b. $r = \cos 2\varphi$

c. $r = \cos 3\varphi$

d. $r = \sin\frac{1}{2}\varphi$

e. $r = \cos\frac{3}{2}\varphi$

f. $r^2 = \cos 2\varphi$

g. $r = 1 + \cos\varphi$

h. $r = 1 + 3\cos 7\varphi$

## Polar coordinates

If an orthonormal coordinate system $Oxy$ in the plane is given, the position of a point $P$ with coordinates $(x, y)$ can also be given by two different 'coordinates': the distance $r = d(O, P)$ of $P$ to the origin $O$, and the angle $\varphi$ between the connecting segment $OP$ and the positive $x$-axis.

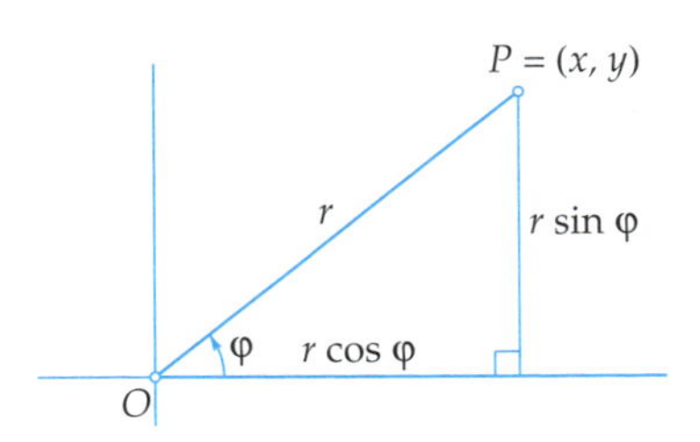


As usual, we measure this angle in radians and anticlockwise. Of course, $\varphi$ is determined up to integer multiples of $2\pi$. The numbers $r$ and $\varphi$ are called the *polar coordinates* of $P$. The origin $O$ is called the *pole*, and the positive $x$-axis is called the *polar axis*. From the polar coordinates of $P$, the ordinary coordinates can readily be calculated:

$$x = r\cos\varphi \quad \text{and} \quad y = r\sin\varphi$$

Furthermore, by Pythagoras' theorem and the definition of the tangent

$$r = \sqrt{x^2 + y^2} \quad \text{and} \quad \tan\varphi = \frac{y}{x} \quad (\text{provided } x \neq 0)$$

For $P = (0, 0)$ the angle $\varphi$ is not defined.

Some curves are easily given by describing a relation between $r$ and $\varphi$. For instance, a *logarithmic spiral* is given by an equation of the form

$$\ln r = c\varphi$$

for some constant $c$. In the drawing to the right we have chosen $c = 0.1$.

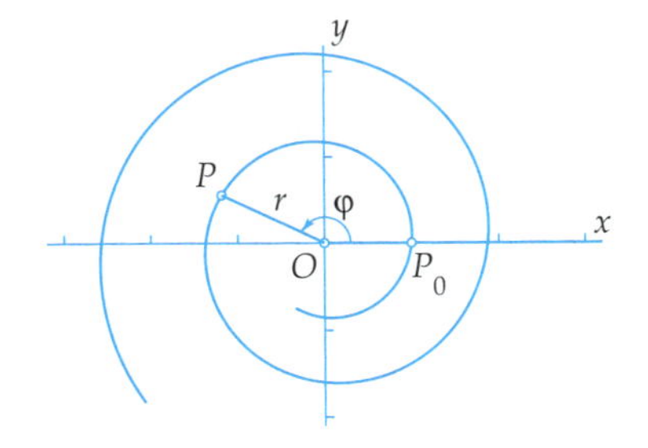


The equation can also be written as $r = \mathrm{e}^{c\varphi}$, which immediately yields the ordinary coordinates $(x, y)$ of a point $P$ on the spiral:

$$P = (x, y) = (r\cos\varphi, r\sin\varphi) = (\mathrm{e}^{c\varphi}\cos\varphi, \mathrm{e}^{c\varphi}\sin\varphi)$$

This is a parametrization of the logarithmic spiral with parameter $\varphi$.

In general, each continuous function $r = r(\varphi)$ defines a parametric curve $(x, y) = (r(\varphi)\cos\varphi, r(\varphi)\sin\varphi)$ with parameter $\varphi$. Also, in case $r(\varphi) < 0$ for certain values of $\varphi$, this definition is used, although, strictly speaking, the polar coordinates of such a point are not $r$ and $\varphi$, but $-r = |r|$ and $\varphi + \pi$.

19.10 Find a parametrization of the helix on the opposite page that traces the curve top-down, so with $P_{-1} = (1, 0, 1)$ and $P_1 = (1, 0, -1)$.

19.11 Sketch the following space curves. To enhance the spatial impression, plot them inside an enclosing cube:

a. $(t, t^2, t^3)$, $-1 \le t \le 1$
b. $(\cos 2\pi t, \sin 2\pi t, t)$, $0 \le t \le 1$
c. $(t, \sin 2\pi t, \cos 2\pi t)$, $0 \le t \le 1$
d. $(\cos t, \sin t, \cos t)$, $0 \le t \le 2\pi$

19.12 For each of the following space curves, find the corresponding parametrization. Explain your answers. Don't use a graphing calculator. Each curve is plotted inside the cube with vertices $(\pm 1, \pm 1, \pm 1)$:

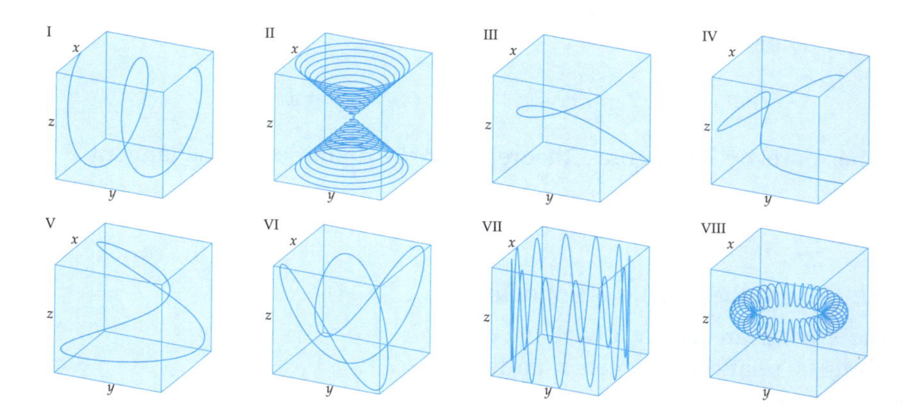


a. $(t, 2t^2 - 1, t^3)$
b. $(\sin t, \sin 2t, \cos t)$
c. $(\sin t, \sin 2t, \cos 3t)$
d. $(\sin 2\pi t, t, \cos 2\pi t)$
e. $(\sin 2\pi t, t^2 - 1, t^3)$
f. $\frac{1}{5}((4 + \sin 40t)\cos t, (4 + \sin 40t)\sin t, \cos 40t)$
g. $(\cos t, \sin t, \cos 12t)$
h. $(t\cos 24\pi t, t\sin 24\pi t, t)$

## Space curves

When in space an orthonormal coordinate system $Oxyz$ is given, a parametric space curve is defined by *three* continuous functions $x = x(t)$, $y = y(t)$ and $z = z(t)$ of a parameter $t$. Then, $(x(t), y(t), z(t))$ is the point on the curve corresponding to the parameter value $t$. If $t$ represents time and $P$ is a point that traces the curve according to this parametrization, $(x(t), y(t), z(t))$ is the position of $P$ at time $t$. This position is also denoted by $P_t$.

The figure to the right shows the space curve with parametrization

$$\begin{cases} x(t) &= \cos 8\pi t \\ y(t) &= \sin 8\pi t \\ z(t) &= t \end{cases}$$

for $-1 \leq t \leq 1$. The curve is a helix that starts at $P_{-1} = (1, 0, -1)$ and ends at $P_1 = (1, 0, 1)$ after eight complete turns. To enhance the spatial impression, it is drawn inside the cube with vertices $(\pm 1, \pm 1, \pm 1)$.

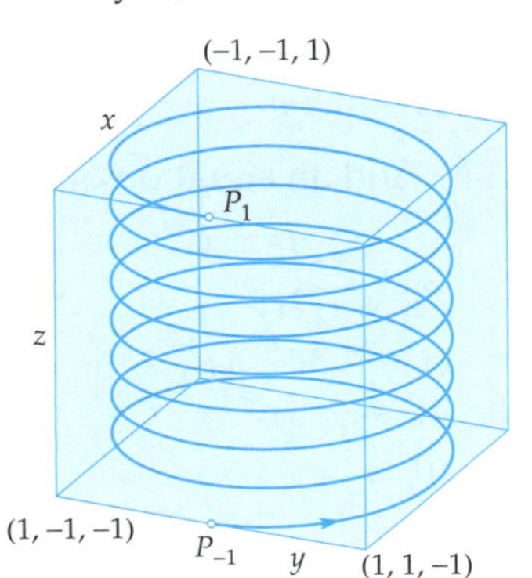


You may have already noticed that the parametrization of a planar or spatial curve is not unique in the sense that different parametrizations may represent the *same* curve (as a geometric object). In viewing the parameter $t$ as time, you may also say that the point $P = P_t$ can trace the curve in many different ways (in fact, in infinitely many ways). This means that a parametrization not only defines a curve, but also the way it is traced.

For instance, the helix above can also be parametrized as follows:

$$\begin{cases} x(t) &= \cos 8\pi t^3 \\ y(t) &= \sin 8\pi t^3 \\ z(t) &= t^3 \end{cases}$$

with, again, $-1 \leq t \leq 1$.

19.13 Find a parametrization of:

a. The line through the points $(-1,1)$ and $(1,-2)$.
b. The line through the points $(1,0)$ and $(0,2)$.
c. The line through the points $(-1,2)$ and $(1,2)$.
d. The line $x+y=1$.
e. The line $3x-4y=2$.
f. The line $5x+7y=-2$.
g. The line $x=1$.
h. The line $y=-3$.

19.14 Find an equation of each of the following lines in parametric form:

a. $(3t+2, 2t+3)$
b. $(2t-1, 2t)$
c. $(t+7, 3t-1)$
d. $(4t+2, 3)$
e. $(0, t)$
f. $(4t-2, -2t+1)$

19.15 Find a parametrization of:

a. The line through $(0,1,1)$ and $(-1,1,2)$.
b. The line through $(1,-1,1)$ and $(2,0,0)$.
c. The line through $(3,0,1)$ and $(-1,-1,0)$.
d. The line through $(1,0,-1)$ and $(-2,4,1)$.
e. The line through $(2,-1,-1)$ and $(0,0,-2)$.
f. The intersection line of the planes $x-y+2z=0$ and $2x+y-z=1$.
g. The intersection line of the planes $-x+3y+z=1$ and $2x+2y-z=-2$.
h. The intersection line of the planes $3x-y=5$ and $x-2y-3z=0$.
i. The intersection line of the planes $2x-y+2z=0$ and $2x+y-z=1$.
j. The intersection line of the planes $-x+3z=2$ and $x+y-z=3$.

19.16 Find a parametrization of:

a. The $x$-axis, the $y$-axis and the $z$-axis.
b. The intersection line of the planes $x=1$ and $z=-1$.
c. The intersection line of the planes $x=y$ and $y=z$.

## Straight lines in parametric form

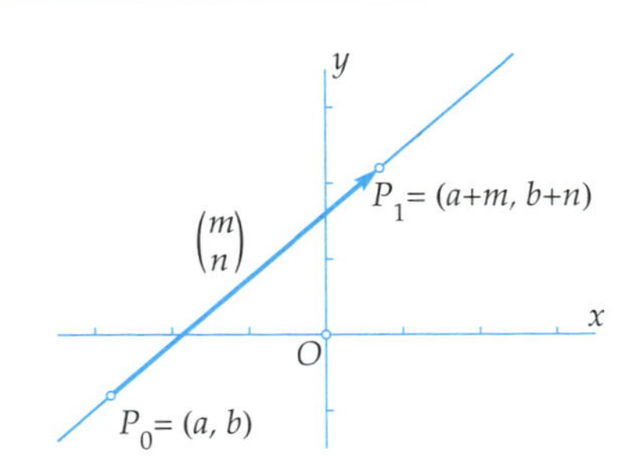


A simple situation in the plane is the case where $x$ and $y$ both are *linear* functions of $t$, so $x = a + mt$ and $y = b + nt$ for certain $a$, $b$, $m$ and $n$. Then, the 'curve' that is traced by the point $P = (x, y)$ is the straight line through $P_0 = (a, b)$ and $P_1 = (a + m, b + n)$. The vector from $P_0$ to $P_1$ has coordinates $\begin{pmatrix} m \\ n \end{pmatrix}$. It is called a *direction vector* of the line.

All the lines in the plane, also the vertical lines, can be described in this way. The only condition is that $m$ and $n$ cannot both be zero. If $m = 0$ and $n \neq 0$, the line is vertical, if $m \neq 0$ and $n = 0$, the line is horizontal.

An equation of the line is obtained by eliminating the parameter $t$. If $x = a + mt$ and $y = b + nt$, then $nx - my = na - mb$ is an equation of the line.

A linear parametrization in space also yields a straight line. If $x(t) = a + mt$, $y(t) = b + nt$, $z(t) = c + pt$, the 'curve' $(x(t), y(t), z(t)) = (a + mt, b + nt, c + pt)$ is a straight line. The *direction vector* in this case is the vector with coordinates $\begin{pmatrix} m \\ n \\ p \end{pmatrix}$.

A parametrization of the intersection line of two planes can be found by eliminating one variable from both equations and putting $t$ equal to one of the remaining variables. As an example, we take the planes

$$\begin{array}{rcrcrcrl} x & - & 2y & + & 2z & = & 1 & (\alpha) \\ 2x & + & y & - & z & = & 2 & (\beta) \end{array}$$

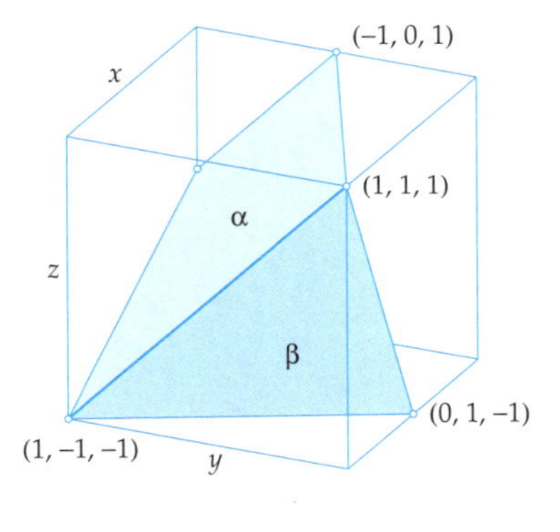


Elimination of $x$ yields $-5y + 5z = 0$. Taking, e.g., $z = t$ yields $y = t$ as well, and substituting this into the equation of $\alpha$ gives $x - 2t + 2t = 1$, so $x = 1$.

Hence, a parametrization of the intersection line is $(x, y, z) = (1, t, t)$.

# VII Calculus

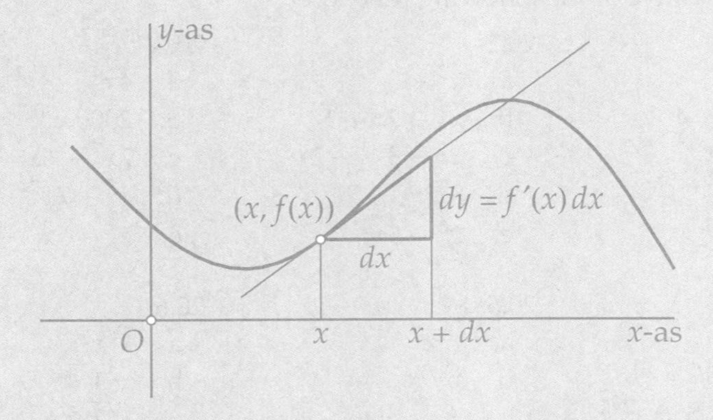


Differential and integral calculus are among the most successful branches of mathematics. Applications extend from astronomy to nanotechnology, from civil engineering to quantum mechanics, from physics and chemistry to economics and business management, from probability and statistics to population dynamics.
In this book, we focus on mathematical techniques. However, applications play an important role: our presentation aims at establishing a firm foundation for mathematics as a tool in all kinds of disciplines. This is one of the reasons why differentials are treated in full detail: in most applications they form the starting point for mathematical modeling.

# 20 Derivatives

On page 179, five differentiation rules are listed. For the exercises on this page, you only need the first two:

$$\begin{aligned}(c\,f(x))' &= c\,f'(x) \quad \text{for any constant } c\\ (f(x)+g(x))' &= f'(x)+g'(x)\end{aligned}$$

Furthermore, you should know that $(x^p)' = p\,x^{p-1}$ for any real number $p$.

Find the derivative of the following functions:

20.1
- a. $2x-3$
- b. $2$
- c. $4x^2+1$
- d. $10\,x^7$
- e. $4x+x^3$

20.2
- a. $x^3-3$
- b. $x^2-2x+1$
- c. $x^4-3x^3+2$
- d. $8x^8$
- e. $x^6-6x^4$

20.3
- a. $4x^4-3x^2+2$
- b. $2000\,x^{2000}$
- c. $7x^7-6x^6$
- d. $x^3+7x^7-12$
- e. $x^2-5x^3+x$

20.4
- a. $\sqrt{x}$
- b. $x\sqrt{x}$
  *Hint:* $x\sqrt{x} = x^{3/2}$
- c. $\sqrt{x^3}$
- d. $x^2\sqrt{x}$
- e. $\sqrt{2x}$
  *Hint:* $\sqrt{2x} = \sqrt{2}\sqrt{x}$

20.5
- a. $\sqrt[3]{x}$
- b. $x^{2/3}$
- c. $\sqrt[4]{x}$
- d. $x\sqrt[4]{x}$
- e. $x^2\sqrt[5]{x^2}$

20.6
- a. $\sqrt[7]{x^2}$
- b. $\sqrt{3x^3}$
- c. $\sqrt[3]{2x^5}$
- d. $\sqrt[4]{x^5}$
- e. $\sqrt{x^7}$

20.7
- a. $x^{-1}$
- b. $2x^{-2}$
- c. $3x^{-3}$
- d. $x^{-1/2}$
- e. $x^{-2/3}$

20.8
- a. $x^{2.2}$
- b. $x^{4.7}$
- c. $x^{-1.6}$
- d. $x^{0.333}$
- e. $x^{-0.123}$

20.9
- a. $\dfrac{1}{x}$
- b. $\dfrac{3}{2x}$
- c. $\dfrac{5}{x^5}$
- d. $\dfrac{\sqrt{x}}{x}$
- e. $\dfrac{1}{x\sqrt[3]{x}}$

## Tangent line and derivative

The graphs of many functions are smooth in all or almost all points: the more you zoom in on a smooth point of the graph, the more it resembles a straight line. That line is the tangent line to the graph at that point.

The figure to the right shows the graph of such a function $f(x)$, together with the tangent line at the point $(a, f(a))$. Indeed, close to that point the graph and the tangent line almost coincide. As an illustration, we marked a small rectangle around the point $(a, f(a))$ and enlarged it in the lower figure.

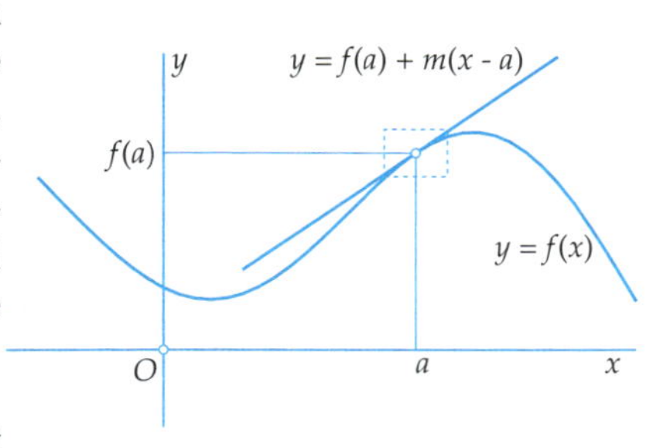


If the tangent line to the graph is not vertical, its equation can be written as $y = f(a) + m(x - a)$ for some $m$, the slope of the tangent line. For $x$ close to $a$ then clearly

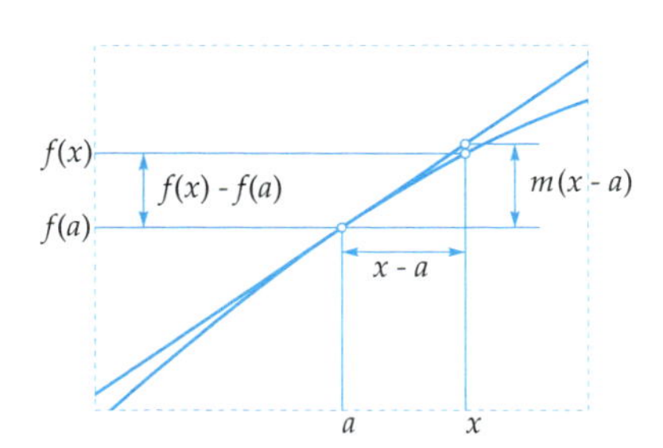


$$\frac{f(x) - f(a)}{x - a} \approx \frac{m(x - a)}{x - a} = m$$

and the approximation gets better as $x$ gets closer to $a$.

Formulated more precisely, using a limit:

$$m = \lim_{x \to a} \frac{f(x) - f(a)}{x - a}$$

The number $m$ is called the *derivative* of $f(x)$ in $a$, notation $f'(a)$.

If the graph of a function $f(x)$ has a non-vertical tangent line in all points of a certain interval, then for any $x$ in that interval the derivative $f'(x)$ exists. In this way, the derivative itself has become a function, the *derived function*, which often is simply called *the derivative*. Frequently used notations for the derivative of a function $f(x)$ are, apart from $f'(x)$, also $\frac{df}{dx}(x)$ and $\frac{d}{dx}f(x)$. The determination of the derivative of a given function is called *differentiation*.

If $f(x)$ is a *linear* function, so $f(x) = mx + c$, the graph is a straight line with slope $m$ and the graph coincides with the tangent line at each point. Hence, for each $x$, we have $f'(x) = m$. In particular, the derivative of each constant function is zero.

Using the chain rule, find the derivative of the following functions. *Examples:*

1. $\left((x^3-1)^5\right)' = 5(x^3-1)^4 \cdot 3x^2 = 15x^2(x^3-1)^4$
2. $\left(\sin(x^2+1)\right)' = \left(\cos(x^2+1)\right) \cdot 2x = 2x\cos(x^2+1)$

*N.B.:* when applying the chain rule to a composed function $f(g(x))$, it is *not* a good idea to write down the functions $f$ and $g$ separately. Simply, while differentiating, peel away the composed function $f(g(x))$ *starting with the outmost function*, just as in the examples above. At the end, in most cases the answer can be simplified:

20.10

a. $(2+3x)^3$
b. $(3-5x)^7$
c. $(1-3x^2)^{-1}$
d. $(1-\sqrt{x})^4$
e. $(x-x^4)^{-2}$

20.11

a. $(2x-3)^5$
b. $(x^2+5)^{-1}$
c. $\sqrt{3x-4}$
d. $\sqrt{x^2+x}$
e. $(x+4x^3)^{-3}$

20.12

a. $\sqrt{1+x+x^2}$
b. $\sqrt[3]{1+x+x^2}$
c. $(x^2-1)^4$
d. $\sqrt{x^3+1}$
e. $(x^2+x)^{3/2}$

Using the product rule (and the chain rule, if necessary), find the derivative of:

20.13

a. $x\sin x$
b. $x\cos 2x$
c. $x^2\ln x$
d. $(x+1)\tan x$
e. $(2x+1)\ln x$

20.14

a. $\sqrt{x+1}\ln x$
b. $(\sin x)(\ln x^2)$
c. $x\ln\sqrt[3]{x}$
d. $x\ln(\sin x)$
e. $\sqrt{x}\ln(1-x^2)$

20.15

a. $x\,(\log_2 x)$
b. $\sqrt{x}\,(\log_5 x^3)$
c. $(x-1)(\log_2 x)$
d. $x\mathrm{e}^{-x}$
e. $x^2\mathrm{e}^{-x^2}$

Using the quotient rule, find the derivative of:

20.16

a. $\dfrac{x}{x+1}$
b. $\dfrac{x-1}{x+1}$
c. $\dfrac{x^2}{x+1}$
d. $\dfrac{x}{x^2+1}$
e. $\dfrac{x-1}{x^2+x}$

20.17

a. $\dfrac{\sqrt{x}}{x-1}$
b. $\dfrac{x^2-1}{x+2}$
c. $\dfrac{x^2-1}{x^2+1}$
d. $\dfrac{2x-3}{4x+1}$
e. $\dfrac{1-x}{2-x}$

20.18

a. $\dfrac{\sin x}{1+\cos x}$
b. $\dfrac{\cos x}{x+1}$
c. $\dfrac{\arcsin x}{x+1}$
d. $\dfrac{\ln x}{\sin x}$
e. $\dfrac{\mathrm{e}^x}{1+\mathrm{e}^x}$

## Rules of differentiation and standard derivatives

*Rules of differentiation:*

$$\begin{aligned}
(c\,f(x))' &= c\,f'(x) \quad \text{for any constant } c \\
(f(x)+g(x))' &= f'(x)+g'(x) \\
(f(g(x)))' &= f'(g(x))g'(x) \quad \text{(chain rule)} \\
(f(x)g(x))' &= f'(x)g(x)+f(x)g'(x) \quad \text{(product rule)} \\
\left(\frac{f(x)}{g(x)}\right)' &= \frac{f'(x)g(x)-f(x)g'(x)}{(g(x))^2} \quad \text{(quotient rule)}
\end{aligned}$$

*Standard functions and their derivatives:*

| $f(x)$ | $f'(x)$ | |
|---|---|---|
| $x^p$ | $p\,x^{p-1}$ | for each $p$ |
| $a^x$ | $a^x \ln a$ | for each $a > 0$ |
| $\mathrm{e}^x$ | $\mathrm{e}^x$ | |
| $\log_a x$ | $\dfrac{1}{x \ln a}$ | for each $a > 0,\ a \neq 1$ |
| $\ln x$ | $\dfrac{1}{x}$ | |
| $\sin x$ | $\cos x$ | |
| $\cos x$ | $-\sin x$ | |
| $\tan x$ | $\dfrac{1}{\cos^2 x}$ | |
| $\arcsin x$ | $\dfrac{1}{\sqrt{1-x^2}}$ | |
| $\arccos x$ | $-\dfrac{1}{\sqrt{1-x^2}}$ | |
| $\arctan x$ | $\dfrac{1}{1+x^2}$ | |

Find the derivative of the following functions:

20.19

a. $\sin(x-3)$
b. $\cos(2x+5)$
c. $\sin(3x-4)$
d. $\cos(x^2)$
e. $\sin\sqrt{x}$

20.20

a. $\tan(x+2)$
b. $\tan(2x-4)$
c. $\sin(x^2-1)$
d. $\cos(1/x)$
e. $\tan\sqrt[3]{x}$

20.21

a. $\arcsin 2x$
b. $\arcsin(x+2)$
c. $\arccos(x^2)$
d. $\arctan\sqrt{x}$
e. $\ln(\cos x)$

20.22

a. $\mathrm{e}^{2x+1}$
b. $\mathrm{e}^{1-x}$
c. $2\mathrm{e}^{-x}$
d. $3\mathrm{e}^{1-x}$
e. $\mathrm{e}^{x^2}$

20.23

a. $\mathrm{e}^{x^2-x+1}$
b. $\mathrm{e}^{1-x^2}$
c. $3\mathrm{e}^{3-x}$
d. $2\mathrm{e}^{\sqrt{x}}$
e. $\mathrm{e}^{1+\sqrt{x}}$

20.24

a. $2^{x+2}$
b. $3^{1-x}$
c. $2^{2-3x}$
d. $5^{x^2}$
e. $3^{\sqrt[3]{x}}$

20.25

a. $\ln(1-2x)$
b. $\ln(3x^2-8)$
c. $\ln(3x-4x^2)$
d. $\ln(x^3+x^6)$
e. $\ln(x^2+1)$

20.26

a. $\ln\sqrt{x+1}$
b. $\ln x^2$
c. $\ln\sqrt[3]{x}$
d. $\ln\sqrt[3]{1-x}$
e. $\ln(4-x)^2$

20.27

a. $\log_2 x$
b. $\log_3 x^3$
c. $\log_{10}(x+1)$
d. $\log_{10}\sqrt{x+1}$
e. $\log_2(x^2+x+1)$

For the following functions, investigate for which values of $x$ they are defined but not differentiable. A rough sketch of the graph may be helpful!

20.28

a. $f(x)=|x-1|$
b. $f(x)=|x^2-1|$
c. $f(x)=\sqrt{|x|}$
d. $f(x)=|\ln(x-1)|$
e. $f(x)=\mathrm{e}^{|x|}$

20.29

a. $f(x)=\sin|x|$
b. $f(x)=\cos|x|$
c. $f(x)=|\sin x|$
d. $f(x)=\sin\sqrt[3]{x}$
e. $f(x)=\ln(1+\sqrt{x})$

Find the equation of the tangent line to the graph of $f(x)$ at the point $(a, f(a))$ in the following cases (see page 177):

20.30

a. $f(x)=2x^2-3,\quad a=1$
b. $f(x)=x^5-3x^2+3,\ a=-1$
c. $f(x)=4x^3+2x-3,\ a=0$
d. $f(x)=8x^4-x^7,\quad a=2$
e. $f(x)=4x-2x^2+x^3, a=-1$

20.31

a. $f(x)=x^2-3x^{-1},\quad a=1$
b. $f(x)=x^2-3\sqrt{x}-3, a=4$
c. $f(x)=x^3+x-3,\quad a=0$
d. $f(x)=x^{-4}-2,\quad a=1$
e. $f(x)=8x-2x^2,\quad a=-3$

## Differentiability

In the previous section, $f'(a)$ has been defined as

$$f'(a) = \lim_{x \to a} \frac{f(x) - f(a)}{x - a}$$

The number $f'(a)$ is the slope of the tangent line to the graph of $f(x)$ at the point $(a, f(a))$. This limit must exist and, moreover, it must be finite, since we assumed that the tangent line is not vertical. If both conditions are satisfied, $f(x)$ is called *differentiable* in $a$.

Not every function is differentiable in each point. For instance, the function $f(x) = |x|$ is not differentiable in 0, since $\dfrac{f(x) - f(0)}{x - 0}$ is equal to 1 if $x > 0$ and equal to $-1$ if $x < 0$. The limit for $x \to 0$ therefore doesn't exist. This can also be seen in the graph: it has a non-smooth corner at the origin that doesn't disappear on zooming in.

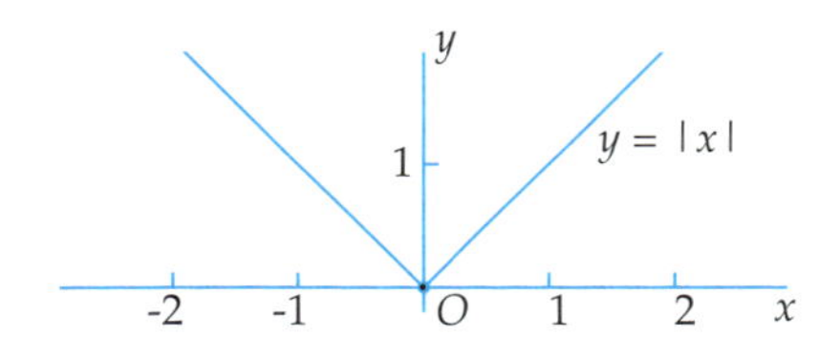


But even if a tangent line does exist, the function is not necessarily differentiable, since the tangent line may be vertical. In that case, the limit that defines the derivative is plus or minus infinity.

For instance, the tangent line to the graph of the function $f(x) = \sqrt[3]{x}$ is vertical at the origin. Indeed, we have

$$\lim_{x \to 0} \frac{\sqrt[3]{x} - \sqrt[3]{0}}{x - 0} = \lim_{x \to 0} \frac{1}{\sqrt[3]{x^2}} = +\infty$$

Therefore, the function $f(x) = \sqrt[3]{x}$ is not differentiable for $x = 0$.

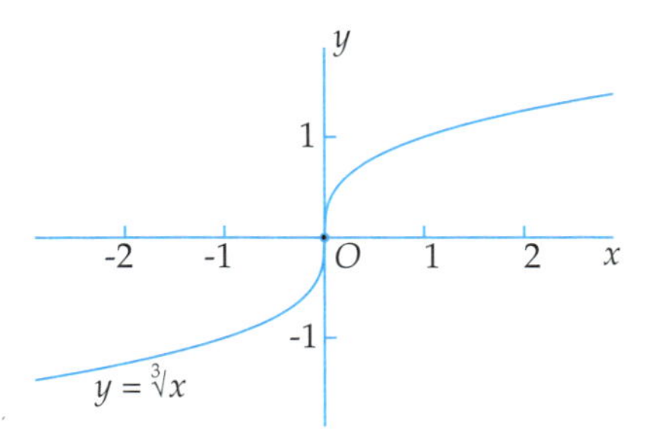

Find the second derivative of the following functions:

20.32

a. $\sqrt{x+1}$

b. $\dfrac{x-1}{x+1}$

c. $\ln(x^2+1)$

d. $x \ln x$

e. $x \sin x$

f. $x^2 \cos 2x$

20.33

a. $\sin(\sqrt{x})$

b. $\tan x$

c. $\arctan x$

d. $x\sqrt{x-1}$

e. $\dfrac{\sin x}{x}$

f. $\sin^2 x$

Find the tenth derivative of the following functions. First, try to discover a pattern by finding the first, second, third, ... derivatives.

20.34

a. $x^9$

b. $x^{10}$

c. $x^{11}$

d. $\mathrm{e}^{-x}$

e. $\mathrm{e}^{2x}$

f. $\mathrm{e}^{x+1}$

20.35

a. $\dfrac{1}{x+1}$

b. $\ln x$

c. $\sin 2x$

d. $\sin\left(x+\frac{\pi}{4}\right)$

e. $x\mathrm{e}^{x}$

f. $x\mathrm{e}^{-x}$

## Higher derivatives

If a function $f(x)$ is differentiable in all points of an interval, the derivative may again be a differentiable function. The derivative of the derivative then is called the *second derivative*. Usual notations are $f''(x)$, $\frac{d^2f}{dx^2}(x)$ and $\frac{d^2}{dx^2}f(x)$. (In the last two notations, mind the distinct places of the 'exponent' 2 in the 'numerator' and the 'denominator'!)

We can go on, and define the $n$-th derivative of a function as the derivative of the $(n-1)$-th derivative, provided the latter is a differentiable function. The $n$-th derivative with $n > 2$ is usually denoted as $f^{(n)}(x)$, $\frac{d^nf}{dx^n}(x)$ or $\frac{d^n}{dx^n}f(x)$.

Some functions can be differentiated as many times as you want: for each $n$ the $n$-th derivative exists. Such functions are called *infinitely often differentiable*. We will give a few examples:

a. $f(x) = x^n$, for any positive integer $n$.
Then $f'(x) = nx^{n-1}$, $f''(x) = n(n-1)x^{n-2}$ and so on. The exponent decreases by 1 in each step, and the $n$-th derivative is constant, namely $n!$ ($n$-factorial, see page 57). All higher derivatives are zero.

b. $f(x) = \mathrm{e}^x$. Then $f^{(n)}(x) = \mathrm{e}^x$ for each $n$.

c. $f(x) = \sin x$. Then $f'(x) = \cos x$, $f''(x) = -\sin x$, $f^{(3)}(x) = -\cos x$, $f^{(4)}(x) = \sin x$ and so on.

d. $f(x) = \cos x$. Then $f'(x) = -\sin x$, $f''(x) = -\cos x$, $f^{(3)}(x) = \sin x$, $f^{(4)}(x) = \cos x$ and so on.

e. $f(x) = \frac{1}{x}$. Since we can write $f(x) = x^{-1}$, the higher derivatives are easy to find:
$f'(x) = (-1)x^{-2} = -x^{-2}$,
$f''(x) = (-1)(-2)x^{-3} = 2!x^{-3}$,
$f^{(3)}(x) = (-1)(-2)(-3)x^{-4} = -3!x^{-4}$ and so on.

f. $f(x) = \sqrt{x} = x^{\frac{1}{2}}$. Then $f'(x) = \frac{1}{2}x^{-\frac{1}{2}}$,
$f''(x) = (\frac{1}{2})(-\frac{1}{2})x^{-\frac{3}{2}} = -\frac{1}{4}x^{-\frac{3}{2}}$,
$f^{(3)}(x) = (\frac{1}{2})(-\frac{1}{2})(-\frac{3}{2})x^{-\frac{5}{2}} = \frac{3}{8}x^{-\frac{5}{2}}$ and so on.

For each of the following functions, find the zeroes (if any) of the derivative and the intervals on which the function is monotonically increasing or decreasing:

20.36

a. $x^3+1$

b. $x^4-4x^3+4$

c. $\dfrac{x^2-1}{x^2+1}$

20.37

a. $x^3+x$

b. $x^6-6x+3$

c. $\dfrac{1}{x^2}$

20.38

a. $\dfrac{x^2+1}{x+1}$

b. $x^3-2x^2+3x-1$

c. $\arctan x^2$

20.39 Find out whether the following statements are true or false. Explain your answer; in case the statement is false, give a *counter example*, i.e. an example of a function $f(x)$ on an interval $I$ for which the statement is false:

a. If $f(x)$ is monotonically increasing on $I$, then $f(x)$ is also monotonically non-decreasing on $I$.

b. If $f(x)$ is monotonically non-increasing on $I$, then $f(x)$ is also monotonically decreasing on $I$.

c. A function cannot be monotonically increasing and decreasing on $I$ at the same time.

d. A function cannot be monotonically non-increasing and non-decreasing on $I$ at the same time.

e. If $f(x)$ is monotonically increasing and differentiable on $I$, then $f'(x) > 0$ for all $x$ in $I$.

20.40 Find out whether the following statements are true or false. Explain your answer; in case the statement is false, give a *counter example*, i.e. an example of a function $f(x)$ on an interval $I$ for which the statement is false:

a. If $f(x)$ is monotonically increasing on $I$, then $g(x) = (f(x))^2$ is also monotonically increasing on $I$.

b. If $f(x)$ is monotonically increasing on $I$, then $g(x) = (f(x))^3$ is also monotonically increasing on $I$.

c. If $f(x)$ is monotonically decreasing on $I$, then $g(x) = \mathrm{e}^{-f(x)}$ is monotonically increasing on $I$.

20.41 Find out whether the following statements are true or false. Explain your answer; in case the statement is false, give a *counter example*, i.e. an example of functions $f(x)$ and $g(x)$ on an interval $I$ for which the statement is false.

a. If $f(x)$ and $g(x)$ are monotonically increasing on $I$, then $f(x)+g(x)$ is also monotonically increasing on $I$.

b. If $f(x)$ and $g(x)$ are monotonically increasing on $I$, then $f(x)\times g(x)$ is also monotonically increasing on $I$.

## Increasing and decreasing functions and the derivative

Let a function $f(x)$ be defined on an interval $I$.

The function $f(x)$ is called *monotonically increasing on I* if $f(x_1) < f(x_2)$ for all $x_1$ and $x_2$ in $I$ with $x_1 < x_2$.
The function $f(x)$ is called *monotonically non-decreasing on I* if $f(x_1) \leq f(x_2)$ for all $x_1$ and $x_2$ in $I$ with $x_1 < x_2$.
The function $f(x)$ is called *monotonically decreasing on I* if $f(x_1) > f(x_2)$ for all $x_1$ and $x_2$ in $I$ with $x_1 < x_2$.
The function $f(x)$ is called *monotonically non-increasing on I* if $f(x_1) \geq f(x_2)$ for all $x_1$ and $x_2$ in $I$ with $x_1 < x_2$.

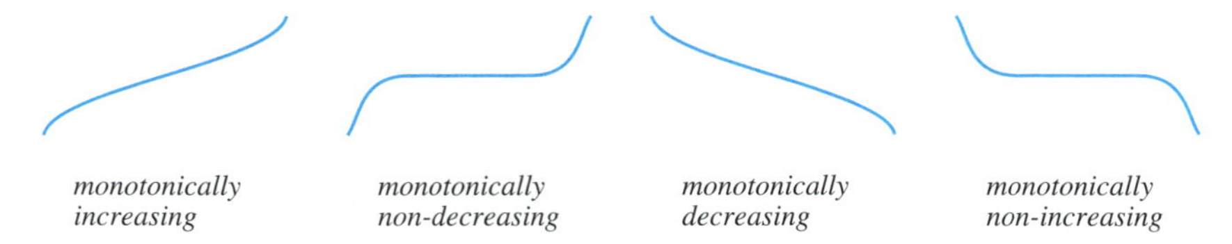

*monotonically increasing* | *monotonically non-decreasing* | *monotonically decreasing* | *monotonically non-increasing*

In this definitions, differentiability of $f(x)$ doesn't play any role. For differentiable functions, the following theorem holds:

**Theorem:** Let $f(x)$ be differentiable in all points of an interval $I$. Then:

a. if the function $f(x)$ is monotonically non-decreasing on the interval $I$, then $f'(x) \geq 0$ for all $x$ in $I$,
b. if the function $f(x)$ is monotonically non-increasing on the interval $I$, then $f'(x) \leq 0$ for all $x$ in $I$,
c. if $f'(x) > 0$ for all $x$ in $I$, then $f(x)$ is monotonically increasing on $I$,
d. if $f'(x) \geq 0$ for all $x$ in $I$, then $f(x)$ is monotonically non-decreasing on $I$,
e. if $f'(x) < 0$ for all $x$ in $I$, then $f(x)$ is monotonically decreasing on $I$,
f. if $f'(x) \leq 0$ for all $x$ in $I$, then $f(x)$ is monotonically non-increasing on $I$,
g. if $f'(x) = 0$ for all $x$ in $I$, then $f(x)$ is constant on $I$.

The proofs of items (a) and (b) are not difficult, those of the other items, however, are rather subtle. Here we omit all proofs.

For the following functions, find the $x$-value of all local and global maximum and minimum values and in each case explain what kind of extremal value it is. Always use a (rough) sketch of the graph of the function, and, if needed, also use the derivative:

20.42

a. $x^3 - x$
b. $x^4 - 2x^2$
c. $x^4 - 6x^2 + 5$
d. $|x - 1|$
e. $|x^2 - 1|$

20.43

a. $\sin x$
b. $\sin x^2$
c. $\sin \sqrt{x}$
d. $\sin |x|$
e. $|\sin x|$

20.44

a. $x \ln x$
b. $(\ln x)^2$
c. $\arcsin x$
d. $\ln \cos x$
e. $\ln |\cos x|$

20.45

a. $x\mathrm{e}^{x}$
b. $\mathrm{e}^{-x^2}$
c. $x\mathrm{e}^{-x^2}$
d. $\mathrm{e}^{\sin x}$
e. $\mathrm{e}^{-|x|}$

20.46 The figure below shows the graph of the function $f(x) = \sin \dfrac{\pi}{x}$.

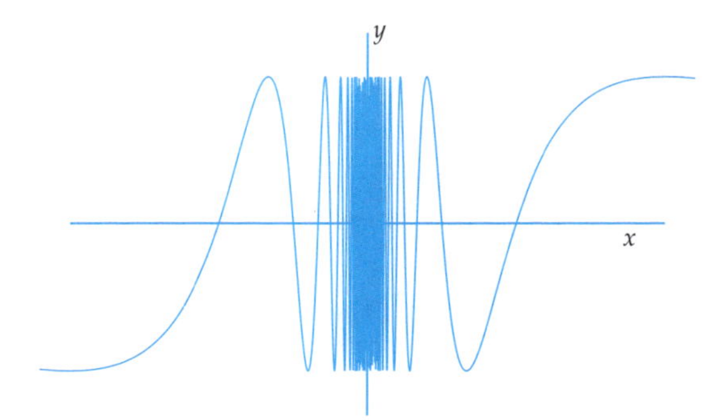


a. Find all zeroes.
b. Find the $x$-value of all maximum and minimum values.
c. Calculate $\lim_{x\to\infty} f(x)$ and $\lim_{x\to-\infty} f(x)$.
d. Does $\lim_{x\to 0} f(x)$ exist? Explain your answer.

## Extremal values

This section is about maximum and minimum values of functions. First we explain these terms.

> If $f(x) \leq f(c)$ for all $x$ in the domain of $f(x)$, then $f(c)$ is called the *global maximum value* of the function. If $f(x) \geq f(c)$ for all $x$ in the domain, then $f(c)$ is called the *global minimum value*.
>
> The value $f(c)$ is called *a local maximum value* or *local minimum value* of $f(x)$ if a number $r > 0$ exists such that $f(x) \leq f(c)$ or $f(x) \geq f(c)$, respectively, for all $x$ in the domain of $f(x)$ with $|x - c| < r$.

The general term for minimum or maximum value is *extremal value* or *extremum*. A global maximum or minimum is also always a local maximum or minimum, but the converse needs not be true. The figure to the right shows the graph of a fourth degree polynomial with three extremal values: a local minimum, a local maximum and a global minimum that, of course, is also a local minimum.There is no global maximum.

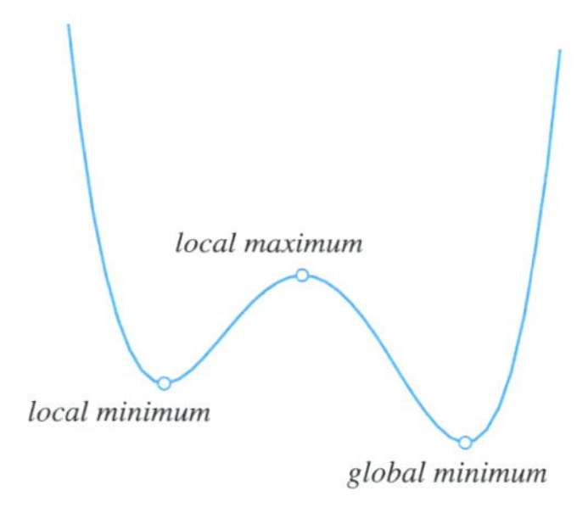


A term that is also frequently used, is *boundary extremum*. This is an extremum that occurs at the boundary of the domain of a function. For instance, take $g(x) = \sqrt{x}$ with domain $[0, \infty\rangle$. Its global minimum $g(0) = 0$ occurs for $x = 0$ at the boundary of the domain.

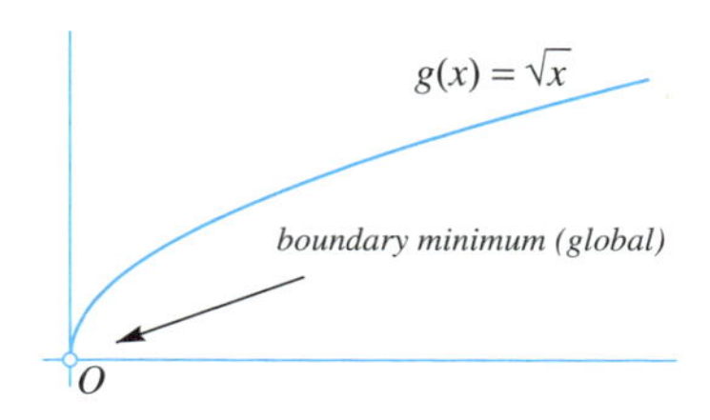


Differentiability doesn't play any role in these definitions. For instance, the global minimum of the function $f(x) = |x|$ is equal to $f(0) = 0$, but the function is not differentiable at this point (see also page 181).

However, if the function $f(x)$ is differentiable at a point $x$ where a local (or global) maximum or minimum occurs, the derivative at that point is equal to zero.

> **Theorem:** If a function $f(x)$ attains a local maximum or minimum value at $x = a$ and it is differentiable at $a$, then $f'(a) = 0$.

N.B.: The *converse* is not true: if $f'(a) = 0$, then $f(x)$ does not necessarily take a minimum or maximum value at $a$. Think of the function $f(x) = x^3$, which doesn't take a maximum or minimum value at $x = 0$, although $f'(0) = 0$.

Find all stationary points and all inflection points of the following functions:

20.47

a. $x^3$

b. $x^3 - x$

c. $x^4 - x^2 - 2x + 1$

d. $x^5 + 10x^2 + 2$

e. $\dfrac{1}{1+x^2}$

20.48

a. $\sin x$

b. $\arctan x$

c. $x^2 \ln x$

d. $x\mathrm{e}^{-x}$

e. $\mathrm{e}^{-x^2}$

20.49 The figures below show graphs of the function

$$f(x) = \begin{cases} x^2 \sin \dfrac{\pi}{x} & \text{als } x \neq 0 \\ 0 & \text{als } x = 0 \end{cases}$$

and its derivative $f'(x)$.

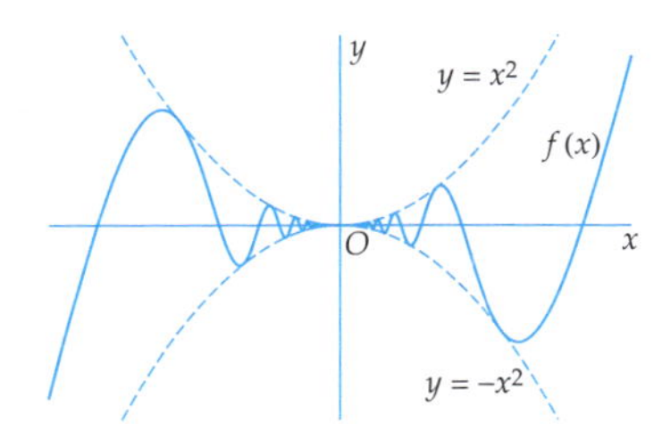


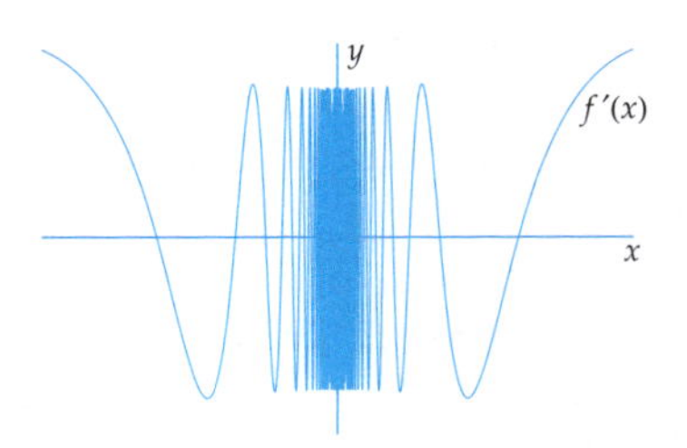


a. Show that $-x^2 \leq f(x) \leq x^2$ for all $x$ and find out for which values of $x$ $f(x) = -x^2$ or $f(x) = x^2$ holds.

b. Find the formula of $f'(x)$ if $x \neq 0$.

c. Show that $\lim\limits_{x \to 0} \dfrac{f(x)}{x} = 0$.
(This means that $f(x)$ is differentiable in $x = 0$ and that $f'(0) = 0$.)

d. Calculate $f'(\frac{1}{2k})$ and $f'(\frac{1}{2k+1})$ for integer $k$.

e. Show that $\lim\limits_{x \to 0} f'(x)$ doesn't exist.
(This implies that $f'(x)$ is not continuous at $x = 0$.)

f. Does $f(x)$ take a local minimum or maximum value at $x = 0$?

g. Is $x = 0$ an inflection point of $f(x)$?

h. Let $g(x) = f(x) + x$. Then $g'(0) = 1$. Is there a $c > 0$ such that $g(x)$ is monotonically increasing on the interval $(-c, c)$?

## Stationary points and inflection points

If $f(x)$ is differentiable in $a$ and $f'(a) = 0$, the tangent line to the graph at $(a, f(a))$ is horizontal, so for $x$ close to $a$ $f(x)$ is nearly constant. Therefore, the point $a$ is called a *stationary point*.

If $f'(a) = 0$ then $a$ is called a *stationary point* of $f(x)$.

Local extremal points of differentiable functions occur at stationary points, but a stationary point does not need to be a local maximum or minimum, as is shown by the function $f(x) = x^3$ (see page 187).

The local extremal points of the derivative $f'(x)$ are also special points of the original function $f(x)$. They are the *inflection points*.

If $f(x)$ is a differentiable function, then each point where the derivative $f'(x)$ attains a maximum or minimum value is called an *inflection point* of the function $f(x)$.

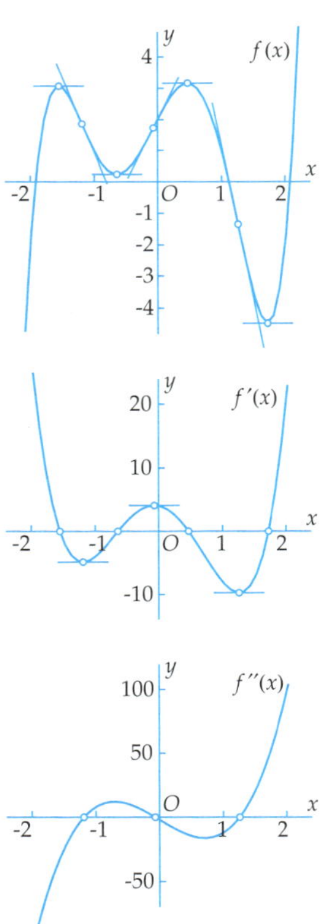


As an example, the graphs of the function $f(x) = x^5 - 5x^3 - x^2 + 4x + 2$, the derivative $f'(x) = 5x^4 - 15x^2 - 2x + 4$ and the second derivative $f''(x) = 20x^3 - 30x - 2$ are drawn (mind the scales on the $y$-axis; they are chosen in such a way as to obtain clear pictures).
In the graphs, the local maximum en minimum values of $f(x)$ and $f'(x)$ are indicated with their corresponding horizontal tangent lines. Also, the tangent lines are drawn in the inflection points of $f(x)$, i.e. in the points where $f'(x)$ attains a local maximum or minimum value. Since $f'(x)$ is also differentiable, in these points $f''(x) = 0$ holds.
As shown, in the inflection points, the graph of $f(x)$ intersects the inflectional tangent line. It is also clear that the 'curvature' of the graph at these points is reversed. This corresponds to the fact that, in these points, $f''(x)$ changes sign. If $f''(x) > 0$, then $f'(x)$ is an increasing function, so the slope of the tangent line of $f(x)$ increases. If $f''(x) < 0$, then $f'(x)$ is a decreasing function, so the slope of the tangent line of $f(x)$ decreases.

20.50 The graph of the polynomial $f(x) = x^5 - 5x^3 - x^2 + 4x + 2$ is drawn on page 189. You can see three zeroes. Are there more? If yes, where are they located? If no, why are there no more zeroes?

20.51 Below, the graphs of two functions $f(x)$ and $g(x)$, their derivatives $f'(x)$ and $g'(x)$ and their second derivatives $f''(x)$ and $g''(x)$ are drawn in a random order. Identify them.

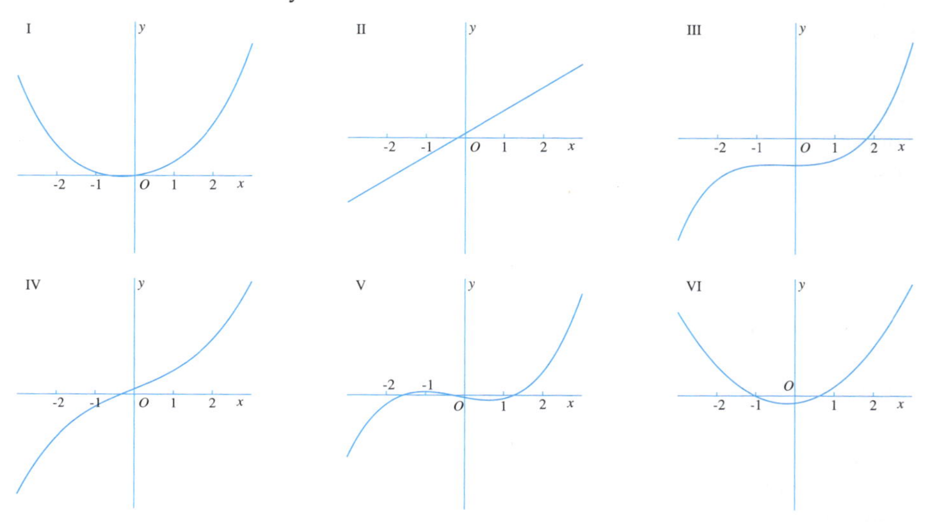


20.52 The same question applies to the following graphs. Again, we have two functions $f(x)$ and $g(x)$, their derivatives $f'(x)$ and $g'(x)$ and their second derivatives $f''(x)$ and $g''(x)$. Identify them.

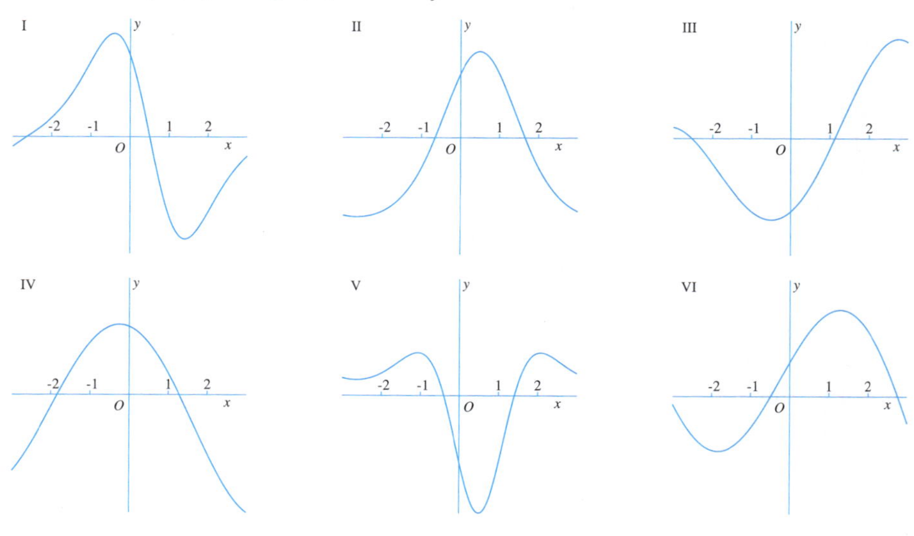

## Playing with functions and their derivatives

When investigating functions, their derivatives can be very useful. Indeed, there is a relation between the increasing and decreasing behaviour of a function and the sign (plus or minus) of its derivative (see the theorem on page 185). The zeroes of the derivative can help to find extremal values of the function, while the zeroes of the second derivative can help to identify inflection points. On the previous pages, you have seen lots of exercises to practice your skills in this field.

We are concluding this chapter with a few challenging problems (see the opposite page) concerning functions and their derivatives. You will need all your ingenuity to solve them!

# 21 Differentials and integrals

Write the following differentials in the form $f'(x)\,dx$:

21.1

a. $d(3x^2+2x+2)$
b. $d(x+\sin 2x+9)$
c. $d(4x^2\sin(x+1))$
d. $d(x^3\sqrt{x^3+1})$
e. $d(\cos(x^2)+5)$
f. $d(3-2x)$

21.2

a. $d(5+x)$
b. $d(\ln(x^2+1))$
c. $d(2-\mathrm{e}^{-x^2})$
d. $d(\mathrm{e}^{\cos x})$
e. $d\left(x-\frac{1}{x}\right)$

21.3

a. $d\left(5x^3+\frac{3}{x^2+1}\right)$
b. $d(x+4)^4$
c. $d((x^4-1)\sin 2x)$
d. $d(\sqrt[4]{x+1})$
e. $d(\tan(x+5))$

21.4

a. $d(x^{2/3}+x^{-2/3})$
b. $d(x-\ln(x^2+1))$
c. $d(\mathrm{e}^{-\sin 2x})$
d. $d\left(\frac{1+x^2}{1-x^2}\right)$

Write the following differentials in the form $d(f(x))$:

21.5

a. $(x^2+2x+2)\,dx$
b. $(x^3-4x)\,dx$
c. $(x^4-4x+5)\,dx$
d. $\sqrt{x}\,dx$ (Hint: $\sqrt{x}=x^{1/2}$)
e. $\frac{4}{x^2}\,dx$

21.6

a. $x\sqrt{x}\,dx$ (Hint: $x\sqrt{x}=x^{3/2}$)
b. $\frac{1}{2\sqrt{x}}\,dx$
c. $(x+1)^4\,dx$
d. $\sin x\,dx$
e. $\sin 5x\,dx$

21.7

a. $(3x^2+2x+2)\,dx$
b. $(x-\sqrt{x})\,dx$
c. $(x^4-4x^3+2x-5)\,dx$
d. $\sqrt{x+1}\,dx$
e. $\left(\frac{1}{x}\right)\,dx$ $(x>0)$

21.8

a. $\sqrt[3]{x}\,dx$
b. $(3+x+\sin 2x)\,dx$
c. $\sin(x+1)\,dx$
d. $\cos(2x+1)\,dx$
e. $\left(\frac{1}{x}\right)\,dx$ $(x<0)$

## Differentials – definition and rules

Suppose that the function $y = f(x)$ is differentiable in the point $x$. For a small increase $dx$ of $x$, the difference $f(x + dx) - f(x)$ is approximately equal to $f'(x)\,dx$, and this approximation gets better as $dx$ gets smaller. The expression $f'(x)\,dx$ is called the *differential* of $f$, with notations $dy$, $df$ or $d(f(x))$. In short:

$$\text{If } y = f(x), \text{ then } dy = f'(x)\,dx$$

On the one hand, the differential thus depends on the choice of the point $x$, and on the other hand on the value of $dx$. If $x$ is given, the differential $dy$ can be seen as the increase of $y$ corresponding to an increase $dx$ of $x$ after *linearization* of the function. Here, 'increase' should be taken in a broad sense: of course, $dx$ and $dy$ can be negative as well.

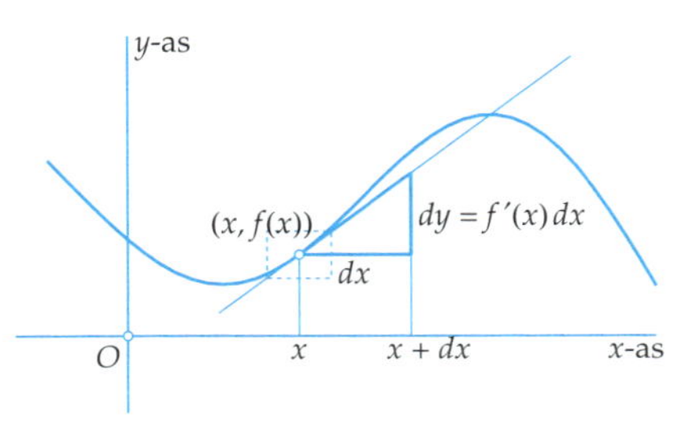


The underlying idea, again, is that the graph of a differentiable function $f(x)$ around a point $(x, f(x))$ almost becomes a straight line when you zoom in far enough. That line is the tangent line to the graph at the point $x$.

The figure to the right shows the result after zooming in. Here, a much smaller value of $dx$ is taken. Again, $dy = f'(x)\,dx$ is the increase of $y$ along the tangent line to the graph at the point $(x, f(x))$ corresponding to the increase $dx$ of $x$. It is also clear that, as $dx$ gets smaller, $dy$ will be a better approximation of the corresponding increment $f(x + dx) - f(x)$ of the function.

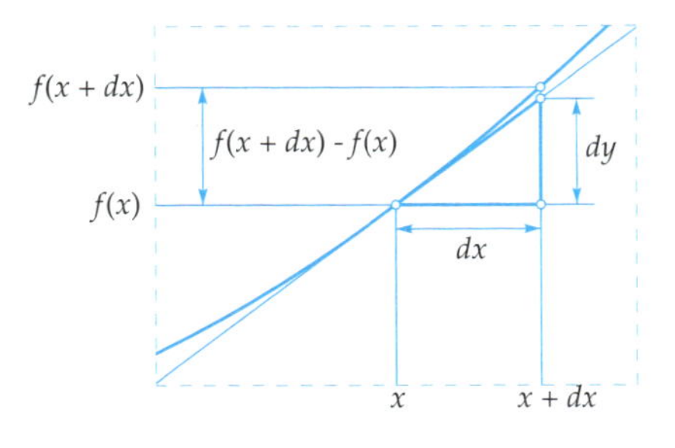


Differentials satisfy the following rules. In fact, these are the well-known differentiation rules, translated in terms of differentials:

$$\begin{aligned}
d(c\,f(x)) &= c\,d(f(x)) \quad \text{for every constant } c \\
d(f(x)+g(x)) &= d(f(x))+d(g(x)) \\
d(f(x)\,g(x)) &= g(x)\,d(f(x))+f(x)\,d(g(x)) \quad \text{(product rule)} \\
d\left(\frac{f(x)}{g(x)}\right) &= \frac{g(x)\,d(f(x))-f(x)\,d(g(x))}{(g(x))^2} \quad \text{(quotient rule)} \\
d(f(g(x))) &= f'(g(x))\,d(g(x)) = f'(g(x))\,g'(x)\,dx \quad \text{(chain rule)}
\end{aligned}$$

21.9 In the following exercises, a measured value $x_m$, an upper bound $h$ for the measurement error and a differentiable function $f(x)$ are given. Calculate $f(x_m)$ and, using the differential, find an upper bound for the error $k$ in this calculation. You can use a calculator. Round up $k$ to an in your eyes reasonable number of decimals (which doesn't need to be the same in each item):

a. $x_m = 2.124, \quad f(x) = 1 + x^2, \quad h = 0.01$
b. $x_m = 0.2124, \quad f(x) = 1 + x^2, \quad h = 0.001$
c. $x_m = 1.284, \quad f(x) = 1 - x^2, \quad h = 0.003$
d. $x_m = 12.84, \quad f(x) = 1 - x^2, \quad h = 0.03$
e. $x_m = 8.372, \quad f(x) = \sin x, \quad h = 0.01$
f. $x_m = 0.672, \quad f(x) = \tan 2x, \quad h = 0.005$
g. $x_m = 0.4394, \quad f(x) = \ln x, \quad h = 0.001$
h. $x_m = 4.394, \quad f(x) = \ln x, \quad h = 0.01$
i. $x_m = 43.94, \quad f(x) = \ln x, \quad h = 0.1$
j. $x_m = 2.984, \quad f(x) = \frac{1}{x}, \quad h = 0.01$

21.10 In the exercises above, you can check your estimate $k = |f'(x_m)|h$ by also calculating $f(x_m + h)$ and $f(x_m - h)$. The difference should approximately be equal to $2k$. Check this for a number of cases, using your calculator.

21.11 If $x_m$ is a measured value and $x_w$ is the real value of a magnitude $x$, the expression $|x_w - x_m|$ is called the *absolute error* and the quotient $\frac{|x_w - x_m|}{|x_w|}$ is called the *relative error*. Since in most cases $x_w$ is unknown, we can only give estimates. If $h$ is an estimate for the maximum absolute error, we usually take the quotient $q = \frac{h}{|x_m|}$ as an estimate for the maximum relative error. Explain the following frequently used rules of thumb, where $h_x, q_x, h_y, q_y$ are estimates for the maximum absolute and relatieve measurement errors in measured values $x_m$ and $y_m$, respectively:

a. $h_x + h_y$ is an estimate for the maximum absolute error in $x_m + y_m$.
b. $h_x + h_y$ is an estimate for the maximum absolute error in $x_m - y_m$.
c. $q_x + q_y$ is an estimate for the maximum relative error in $x_m y_m$.
d. $q_x + q_y$ is an estimate for the maximum relative error in $\frac{x_m}{y_m}$.

## Error estimates

Differentials are often used in error estimates. Suppose that $x_m$ is a measured value of a magnitude $x$, and that the (small) measurement error is estimated to be less than $h$. Thus, we expect that the unknown real value $x_w$ of $x$ satisfies $|x_w - x_m| < h$.

Also, suppose that we don't need $x$ itself, but a function value $f(x)$. We calculate $f(x_m)$, but we would like to known $f(x_w)$. Can we give a reasonable estimate $k$ for the error $f(x_w) - f(x_m)$ if we use $f(x_m)$ instead of $f(x_w)$?

Let $x_w = x_m + dx$. If $dx$ is small, we have

$$f(x_w) - f(x_m) = f(x_m + dx) - f(x_m) \approx f'(x_m)\,dx$$

Since we supposed that $|dx| < h$, the number $k = |f'(x_m)|\,h$ should be a reasonable estimate for the maximum error in taking $f(x_m)$ as an approximation for the unknown function value $f(x_w)$.

For instance, suppose that $x_m = 0.847$ and $f(x) = x^3$, and suppose that we may take $h = 0.02$. Then $f(x_m) = (0.847)^3 = 0.607645423$ and $f'(x_m) = 3(x_m)^2 = 3(0.847)^2 = 2.152227$, so $k = f'(x_m)\,h = 0.04304454$. A reasonable error estimate therefore is $|f(x_w) - f(x_m)| < 0.044$, or, expressed directly in terms of the unknown value $f(x_w)$,

$$0.563 < f(x_w) < 0.651$$

Stated in a rather sloppy way:

*When calculating the function value $f(x_m)$ of a measured result $x_m$, the inaccuracy in $x_m$ is multiplied by the absolute value of the derivative $f'(x_m)$.*

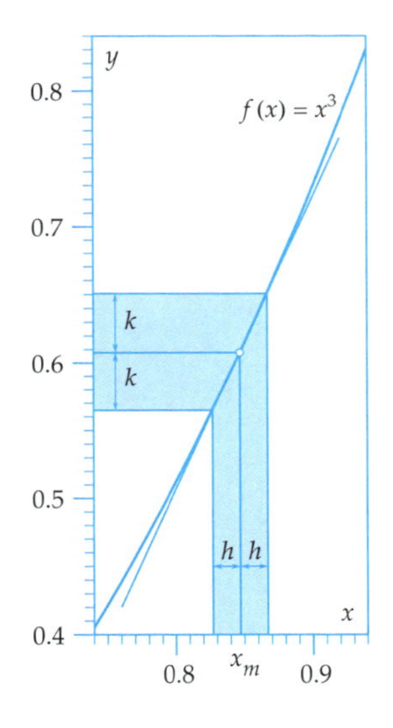


Here again, everything boils down to *linearization* of the function $f(x)$ around $x_m$, i.e. substituting its graph by the tangent line at the measured value $x_m$. A small derivative $f'(x_m)$ implies that the inaccuracy decreases, while a big derivative results in greater inaccuracy. Of course, we always assume that $h$ is small and that $f(x)$ is a differentiable function.

21.12 In each of the exercises below, a point $x$ and a function $f(x)$ are given. Using a calculator or a computer, produce a table like the one on the opposite page, taking for $dx$ the values $dx = 0.1$, $dx = 0.01$, $dx = 0.001$ and $dx = 0.0001$:

a. $x = 2$, $f(x) = 1 + x^2$
b. $x = 1$, $f(x) = \ln x$
c. $x = \frac{1}{4}\pi$, $f(x) = \tan x$
d. $x = 2$, $f(x) = \arctan x$
e. $x = 0$, $f(x) = \cos x$
f. $x = 0$, $f(x) = \sin x$ (explain what you see in this case!)

## Are differentials good approximations?

For small values of $dx$, the differential $dy = f'(x)\,dx$ approximates the difference $\Delta f = f(x+dx) - f(x)$ rather well. But how well is this approximation really? This depends on the function $y = f(x)$, the chosen point $x$ and the size of $dx$.

Here, we concentrate on the dependence of $dx$. The error made by replacing the difference $\Delta f = f(x+dx) - f(x)$ by the differential $df = f'(x)dx$ for 'well-behaved' functions[1] is of order $(dx)^2$. In fact, it can be shown that

$$\Delta f - df = (f(x+dx) - f(x)) - f'(x)\,dx \approx \frac{1}{2} f''(x)\,(dx)^2$$

for $dx$ small. In this respect, you should realize that for a small $dx$, the square $(dx)^2$ is much smaller than $dx$ itself. For instance, when $dx = 0.01$, then $(dx)^2 = 0.0001$. Hence, the difference between $\Delta f$ and the differential $df$ goes to zero 'much faster' than $dx$ itself when $dx$ goes to zero.

A proof of the above is beyond the scope of this book. However, we will give an illustration in the form of a table where we take $f(x) = \sin x$ and $x = \frac{1}{4}\pi$. For $dx$ we substitute the values $dx = 0.1$, $dx = 0.01$, $dx = 0.001$ and $dx = 0.0001$.

| $dx$ | $df = f'(x)\,dx$ | $\Delta f$ | $\Delta f - df$ | $\frac{1}{2} f''(x)\,(dx)^2$ |
|---|---|---|---|---|
| 0.1 | 0.0707106781 | 0.0670603 | −0.0036504 | −0.0035355 |
| 0.01 | 0.00707106781 | 0.007035595 | −0.000035473 | −0.000035355 |
| 0.001 | 0.000707106781 | 0.00070675311 | −0.00000035367 | −0.00000035355 |
| 0.0001 | 0.0000707106781 | 0.0000707071425 | −0.0000000035357 | −0.0000000035355 |

Clearly, if $dx$ is made smaller by a factor *ten*, then $\Delta f - df$ (fourth column) becomes smaller by a factor of approximately *one hundred*! Also remark that, indeed, as $dx$ gets smaller and smaller, the difference $\Delta f - df$ (fourth column) gets closer and closer to $\frac{1}{2} f''(x)\,(dx)^2$ (fifth column).

For later use, we notice that we can estimate the difference $\Delta f - df$ if we know an upper bound $M$ for $|f''(x)|$ on the interval where the $x$-values are taken. Then, the following is true:

$$|\Delta f - df| \leq \frac{1}{2} M |dx|^2$$

In the table above, $f(x) = \sin x$ holds, so $f''(x) = -\sin x$. Then we can take $M = 1$. Verify that the resulting estimate $|\Delta f - df| \leq \frac{1}{2}|dx|^2$ is satisfied by the entries in the fourth column of the table.

[1]The function $f(x)$ is called 'well-behaved' in this respect if $f''(x)$ exists and is continuous in a neighbourhood of $x$.

21.13 Calculate the area of the region bounded by the $x$-axis and the graph of the function $f(x) = x - x^3$ between the points $(0,0)$ and $(1,0)$.

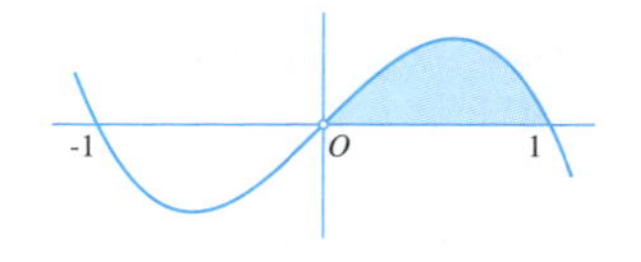


21.14 Calculate the area of the region bounded by the $x$-axis and the graph of the function $f(x) = x^3 - x^4$ between the points $(0,0)$ and $(1,0)$.

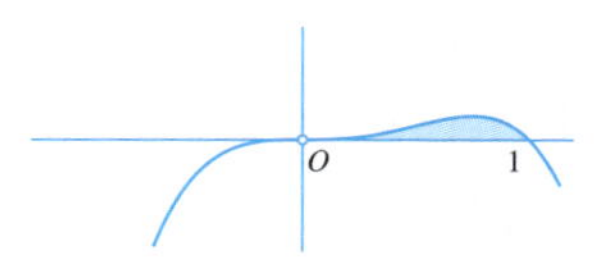


21.15 Calculate the area of the region bounded by the $x$-axis and the graph of the function $f(x) = \sin x$ between the points $(0,0)$ and $(\pi,0)$.

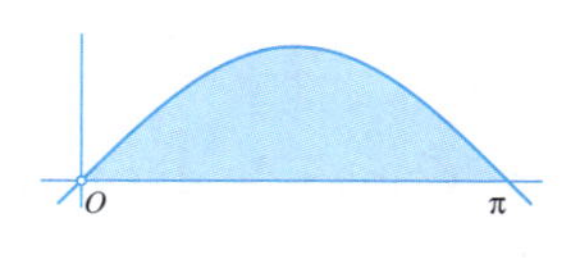


21.16 Calculate the area of the region bounded by the $x$-axis and the graph of the function $f(x) = 2\cos 2x$ between the points $(-\frac{\pi}{4},0)$ and $(\frac{\pi}{4},0)$.

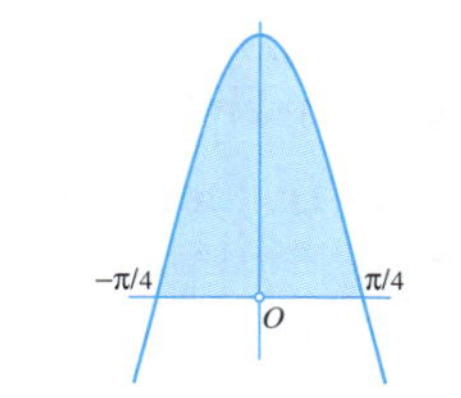


21.17 Calculate the area of the region bounded by the $x$-axis and the graph of the function $f(x) = (\sin x)\,e^{\cos x}$ between the points $(0,0)$ and $(\pi,0)$.

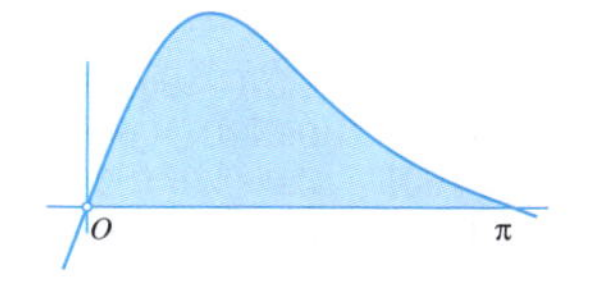

## Calculating an area

Suppose that you are asked to calculate the area of the region $V$ enclosed by the $x$-axis and the graph of $f(x) = 1 - x^2$ between the points $(-1,0)$ and $(1,0)$. This can be done as follows.
Choose a number $x$ between $-1$ and 1, and denote the area of the part of $V$ to the left of the vertical line through $(x,0)$ by $O(x)$.

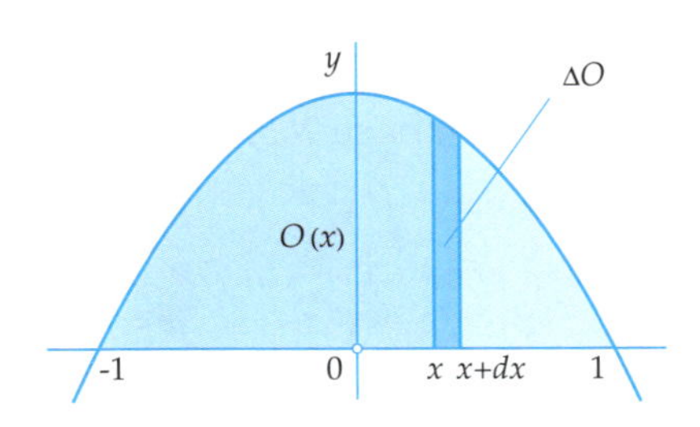


As soon as we know a formula for $O(x)$ as a function of $x$, we also know the area of $V$, since it is equal to $O(1)$.

For small positive values of $dx$, the difference $O(x + dx) - O(x)$, which we will call $\Delta O$ for short, equals the area of the small strip of $V$ between the vertical lines through the points $(x,0)$ and $(x + dx, 0)$.

That strip almost coincides with the small rectangular strip with base $dx$ and height $f(x) = 1 - x^2$, as shown to the right. The area of this strip is equal to $f(x) \times dx = (1 - x^2)\,dx$. As $dx$ gets smaller, the resemblance gets better, which leads to the idea that $(1 - x^2)\,dx$ might be the differential of the function $O(x)$ that we are looking for.

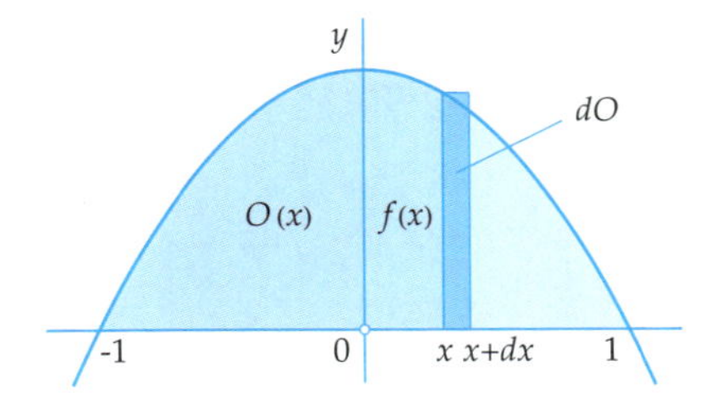


It can be proven that this indeed is true, in other words, that $dO = (1 - x^2)dx$. The derivative of the function $O(x)$ then is $1 - x^2$, and it follows that $O(x) = x - \frac{1}{3}x^3 + c$ for some constant $c$. Since $O(-1) = 0$ (for $x = -1$ the area is zero) $c = \frac{2}{3}$ must hold, so $O(x) = x - \frac{1}{3}x^3 + \frac{2}{3}$. It follows that the requested area of $V$ is equal to $O(1) = 1 - \frac{1}{3} + \frac{2}{3} = \frac{4}{3}$.

21.18 Calculate the area of the region below the graph of the given function $f(x)$ between the lines $x = a$ and $x = b$. To get an idea, first make a rough sketch of the graph of $f(x)$. Then, find a primitive function $F(x)$ of $f(x)$, and finally calculate the requested area. If necessary, you can use the list of primitives on page 205:

a. $f(x) = 1 + x^2$ $\quad a = -1,\ b = 1$
b. $f(x) = x^3 + x^2$ $\quad a = 0,\ b = 2$
c. $f(x) = 1 + \sqrt{x}$ $\quad a = 0,\ b = 1$
d. $f(x) = 2 + \cos x$ $\quad a = 0,\ b = \pi$
e. $f(x) = \mathrm{e}^x$ $\quad a = -1,\ b = 1$
f. $f(x) = x^{\frac{3}{2}}$ $\quad a = 1,\ b = 4$
g. $f(x) = \dfrac{1}{x}$ $\quad a = 1,\ b = \mathrm{e}$
h. $f(x) = \dfrac{1}{1+x^2}$ $\quad a = -1,\ b = 1$

Calculate the following integrals. To get an idea of what you are calculating, first sketch a graph of the integrand:

21.19

a. $\displaystyle\int_0^2 (2x + x^3 + 1)\,dx$

b. $\displaystyle\int_1^4 (x + \sqrt{x})\,dx$

c. $\displaystyle\int_0^1 (1 + x^{-\frac{3}{4}})\,dx$

d. $\displaystyle\int_{-\pi}^{\pi} (1 - \sin x)\,dx$

21.20

a. $\displaystyle\int_0^2 \mathrm{e}^{-2x}\,dx$

b. $\displaystyle\int_0^1 (\mathrm{e}^x + \mathrm{e}^{-x})\,dx$

c. $\displaystyle\int_0^1 \frac{1}{x+1}\,dx$

d. $\displaystyle\int_0^{\frac{1}{2}} \frac{1}{\sqrt{1-x^2}}\,dx$

## Area and primitive function

If $F'(x) = f(x)$ for all $x$ in an interval $I$, $F(x)$ is called a *primitive function* of $f(x)$ on $I$. Such a primitive function is not unique: for each constant $c$, $G(x) = F(x) + c$ also is a primitive of $f(x)$, and *every* primitive of $f(x)$ on $I$ can be written in this form. Therefore, primitive functions are determined up to a constant: if we know one primitive, we know them all.

What we have done on page 199 with the area below the graph of the function $f(x) = 1 - x^2$, can be repeated for an arbitrary function $f$ that is continuous and non-negative on an interval $[a, b]$.

Choose a number $x$ between $a$ and $b$. The area below the graph of $f(x)$ between the vertical lines through $(a, 0)$ and $(x, 0)$ is denoted by $O(x)$. Just as on page 199, you can show that the differential $dO$ is equal to $f(x)\,dx$ (in the proof, the continuity of $f(x)$ is used). This means that $O'(x) = f(x)$, in other words, that $O(x)$ is a primitive of $f(x)$ on $[a, b]$.

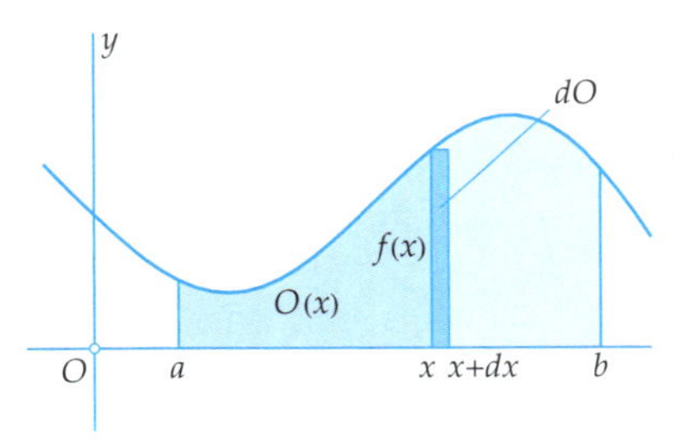


Now, suppose that $F(x)$ is an arbitrary other primitive function of $f(x)$. Then, for some constant $c$, we have $F(x) = O(x) + c$. Since $O(a) = 0$, it follows that $F(a) = c$, so $O(x) = F(x) - F(a)$. In particular $O(b) = F(b) - F(a)$.

The number $F(b) - F(a)$ is called the *integral* of $f(x)$ on the interval $[a, b]$, notation

$$\int_a^b f(x)\,dx = F(b) - F(a)$$

The function $f(x)$ is called the *integrand*. The sign $\int$ before the differential $f(x)\,dx$ is called the *integral sign*. It was invented by G.W. Leibniz (1646-1716), one of the pioneers of the differential and integral calculus. Originally, it was an elongated capital 'S', the first letter of the Latin word *summa*, which means sum. Later, we will expand more on the connection between integration and summation.

As an aside, the *integration variable* $x$ can be replaced by any other letter, unless that letter is already used in a different sense. It is a 'dummy variable', comparable to the summation index in a summation formula.

Calculate the following integrals:

21.21

a. $\displaystyle\int_0^2 (x^4 - 5x^3 - 1)\, dx$

b. $\displaystyle\int_0^2 (x - \sqrt{x})\, dx$

c. $\displaystyle\int_{\frac{1}{2}}^2 \left(x + \frac{1}{x}\right) dx$

d. $\displaystyle\int_{-\pi}^{\pi} \sin x\, dx$

e. $\displaystyle\int_{\frac{\pi}{2}}^{\pi} \cos x\, dx$

21.22

a. $\displaystyle\int_{-2}^2 (3x^2 - 2x^3)\, dx$

b. $\displaystyle\int_1^2 \sqrt[5]{x}\, dx$

c. $\displaystyle\int_{-1}^1 \left(x - \mathrm{e}^{-x}\right) dx$

d. $\displaystyle\int_{-\pi}^0 (x + \sin x)\, dx$

e. $\displaystyle\int_{\frac{\pi}{4}}^{\frac{\pi}{2}} (\sin x - \cos x)\, dx$

21.23

a. $\displaystyle\int_0^2 2^x\, dx$

b. $\displaystyle\int_1^2 \mathrm{e}^{x-1}\, dx$

c. $\displaystyle\int_{-1}^1 \left(\mathrm{e}^x - \mathrm{e}^{-x}\right) dx$

d. $\displaystyle\int_{-\frac{\pi}{2}}^0 \sin 2x\, dx$

e. $\displaystyle\int_0^1 \cos \pi x\, dx$

21.24

a. $\displaystyle\int_1^0 \frac{1}{1+x^2}\, dx$

b. $\displaystyle\int_{-1}^0 \frac{1}{x+2}\, dx$

c. $\displaystyle\int_0^{-2} \frac{1}{1-2x+x^2}\, dx$

d. $\displaystyle\int_{\frac{1}{2}}^{-\frac{1}{2}} \frac{1}{\sqrt{1-x^2}}\, dx$

e. $\displaystyle\int_{\frac{\pi}{3}}^0 \frac{1}{\cos^2 x}\, dx$

Calculate:

21.25

a. $\displaystyle\frac{d}{dx}\int_0^x t^2\, dt$

b. $\displaystyle\frac{d}{dx}\int_{-5}^x t^2\, dt$

c. $\displaystyle\frac{d}{dx}\int_x^3 t^2\, dt$

d. $\displaystyle\frac{d}{dx}\int_{-x}^x t^2\, dt$

21.26

a. $\displaystyle\frac{d}{dx}\int_x^0 \sin t\, dt$

b. $\displaystyle\frac{d}{dx}\int_{-x}^0 \sin t\, dt$

c. $\displaystyle\frac{d}{dx}\int_0^{2x} \cos t\, dt$

d. $\displaystyle\frac{d}{dx}\int_{-x}^x \cos t\, dt$

## Integrals – general definition and properties

Suppose that $F(x)$ is a primitive function of the function $f(x)$ on an interval $I$. Such a primitive is determined by $f(x)$ up to a constant. For arbitrary points $a$ and $b$ in $I$ the difference $F(b) - F(a)$ then is *independent* of the choice of the primitive. This difference is called the *integral* of $f(x)$ with *lower bound* $a$ and *upper bound* $b$, notation $\int_a^b f(x)\,dx$. For $F(b) - F(a)$ the shorter notation $[F(x)]_a^b$ is also used, so

$$\int_a^b f(x)\,dx = [F(x)]_a^b = F(b) - F(a)$$

This integral definition is an extension of the definition on page 201, which was restricted to the case that $a < b$ and that $f(x)$ is continuous and non-negative on $[a, b]$. There, we showed that the area formula $O(x) = \int_a^x f(t)\,dt$ is a primitive function of $f(x)$ and that for any other primitive $F(x)$ of $f(x)$ we have $F(x) = F(a) + \int_a^x f(t)\,dt$.

From the more general integral definition, the following properties immediately follow:

a. $\displaystyle\int_a^b c\,f(x)\,dx = c\int_a^b f(x)\,dx$ for each constant $c$

b. $\displaystyle\int_a^b (f(x) + g(x))\,dx = \int_a^b f(x)\,dx + \int_a^b g(x)\,dx$

c. $\displaystyle\int_b^a f(x)\,dx = -\int_a^b f(x)\,dx$

d. $\displaystyle\int_a^c f(x)\,dx = \int_a^b f(x)\,dx + \int_b^c f(x)\,dx$ for each $a, b, c$ in $I$

e. $\displaystyle\frac{d}{dx}\int_a^x f(t)\,dt = f(x)$ and $\displaystyle\frac{d}{dx}\int_x^b f(t)\,dt = -f(x)$

Property (e.) shows what happens if you differentiate an integral with respect to its upper or lower bound. Since in this case we use the letter $x$ in one of the bounds, we must use a different letter as integration variable.

Obvious questions are: does any function $f(x)$ have primitive functions? And: if a function has a primitive function, how do you find it? The answer to the first question is *no*. There are functions which don't have primitive functions. But in the next section, we will see that there are always primitive functions if $f(x)$ is *continuous*. The next chapter is completely devoted to answering the second question.

Calculate:

21.27

a. $\displaystyle\int_1^2 \frac{3}{x^2}\,dx$

b. $\displaystyle\int_0^2 10^x\,dx$

c. $\displaystyle\int_0^8 \sqrt[3]{x}\,dx$

d. $\displaystyle\int_{-\frac{\pi}{4}}^0 \cos x\,dx$

e. $\displaystyle\int_{-2}^{-1} \frac{1}{x}\,dx$

21.28

a. $\displaystyle\int_{-1}^1 \frac{1}{1+x^2}\,dx$

b. $\displaystyle\int_{-1}^0 \frac{1}{x-3}\,dx$

c. $\displaystyle\int_1^2 \frac{4}{3x^5}\,dx$

d. $\displaystyle\int_{-\frac{1}{2}}^{\frac{1}{2}\sqrt{3}} \frac{1}{\sqrt{1-x^2}}\,dx$

e. $\displaystyle\int_1^2 \frac{2}{x\sqrt{x}}\,dx$

## Primitives of standard functions

If $F(x)$ is a primitive function of $f(x)$ on an interval $[a,b]$ (so $F'(x) = f(x)$), then

$$\int_a^b f(x)\,dx = F(b) - F(a)$$

Hence, it is convenient to have a list of primitives of standard functions. To this end, we could simply refer to the list of standard derivatives on page 179: just read it in reverse. But, for easier reference, we are giving you a fresh list below. It is a bit shorter than the list on page 179.

| $f(x)$ | $F(x)$ | |
|---|---|---|
| $x^p$ | $\frac{1}{p+1}\,x^{p+1}$ | provided $p \neq -1$ |
| $a^x$ | $\frac{1}{\ln a}\,a^x$ | for each $a > 0$, $a \neq 1$ |
| $e^x$ | $e^x$ | |
| $\frac{1}{x}$ | $\ln\lvert x\rvert$ | |
| $\sin x$ | $-\cos x$ | |
| $\cos x$ | $\sin x$ | |
| $\frac{1}{\cos^2 x}$ | $\tan x$ | |
| $\frac{1}{\sqrt{1-x^2}}$ | $\arcsin x$ | |
| $\frac{1}{1+x^2}$ | $\arctan x$ | |

As remarked earlier, a primitive function is not unique: for each constant $c$, $F(x) + c$ is also a primitive of $f(x)$.

Special attention should be payed to the primitive $F(x) = \ln|x|$ of the function $f(x) = \frac{1}{x}$. Note that $f(x)$ is defined for all $x \neq 0$, while the function $\ln x$ is only defined for $x > 0$. But, if $x < 0$, then, according to the chain rule, the derivative of the function $\ln(-x)$ is $-\frac{1}{-x} = \frac{1}{x}$. So, if $x < 0$, then the function $\ln(-x)$ is a primitive function of $\frac{1}{x}$. We can combine the two cases by writing $\ln|x|$. This, indeed, is a primitive function of $\frac{1}{x}$ for $x > 0$ as well as for $x < 0$. On page 211, however, we will slightly modify and extend this observation.

21.29

a. Using the graphs of $\sin x$ and $\cos x$, show that

$$\int_0^{2\pi} \sin x\,dx = \int_0^{2\pi} \cos x\,dx = 0$$

b. Plot the graphs of $\sin^2 x$ and $\cos^2 x$ and show that they are the same up to a horizontal translation.

c. From the above result conclude that

$$\int_0^{\pi} \sin^2 x\,dx = \int_0^{\pi} \cos^2 x\,dx$$

d. Now, using the relation $\cos^2 x + \sin^2 x = 1$, calculate both integrals.

e. Also, calculate the integrals

$$\int_0^{\frac{\pi}{2}} \sin^2 x\,dx \quad \text{en} \quad \int_0^{\frac{\pi}{2}} \cos^2 x\,dx$$

21.30 Calculate the area of the region enclosed between the following curves. First make a (rough) sketch of the situation and then use $\int_a^b f(x)\,dx - \int_a^b g(x)\,dx = \int_a^b (f(x) - g(x))\,dx$:

a. The parabola $y = x^2$ and the parabola $y = 1 - x^2$.
b. The parabola $y = x^2 - 2$ and the line $y = x$.
c. The graphs of the functions $f(x) = \sqrt{x}$ and $g(x) = x^3$.
d. The graphs of the functions $f(x) = \cos \frac{\pi}{2}x$ and $g(x) = x^2 - 1$.
e. The graph of the function $f(x) = \mathrm{e}^{x}$, the $y$-axis and the line $y = \mathrm{e}$.

21.31 On the graph of the function $f(x) = x^3$ the point $P = (1, 1)$ is chosen. The tangent line at $P$ intersects the graph of $f(x)$ in a point $Q$ different from $P$. Calculate the coordinates of $Q$ and calculate the area of the region bounded by the graph of $f(x)$ and the segment $PQ$.

*Hint:* first, find an equation of the tangent line. Intersecting it with the graph of $f(x)$ yields a cubic (degree 3) equation in $x$. Remember that you already know a double root of this equation, namely $x = 1$! The third root is the $x$-coordinate of $Q$.

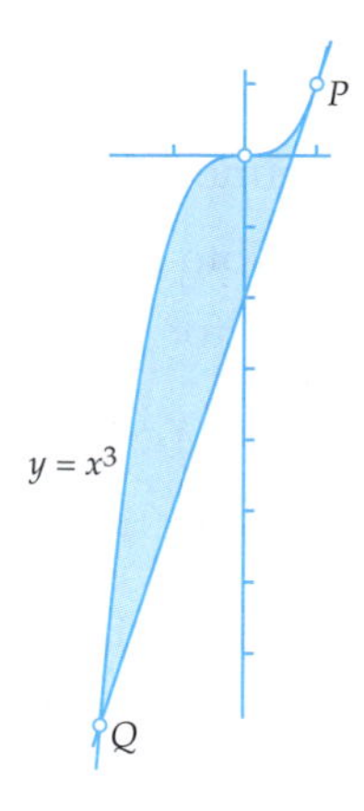

## The relation between area and integral once again

Suppose that $f(x)$ is continuous on the interval $[a, b]$. As shown on page 201, if $f(x) \geq 0$ on $[a, b]$, a primitive function of $f(x)$ is given by $O(x)$, the function that represents the area below the graph of $f$ on the interval $[a, x]$.

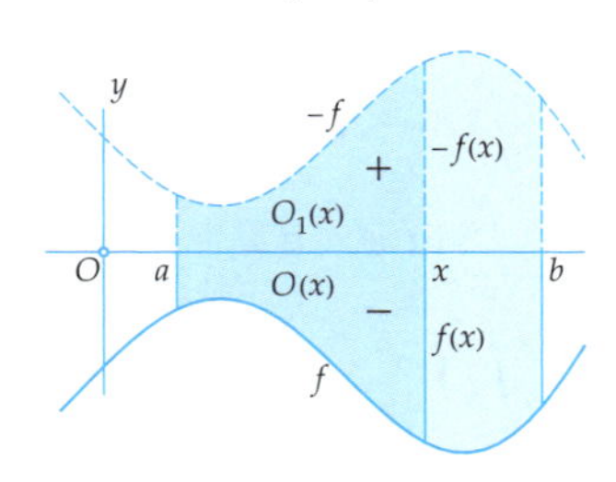


If $f(x) \leq 0$ on $[a, b]$, then $-f(x) \geq 0$, so the (non-negative) function $O_1(x)$ representing the area below the graph of $-f(x)$ on $[a, x]$, is a primitive function of $-f(x)$. A primitive function of $f(x)$ therefore is $O(x) = -O_1(x)$. At the same time, this is the area on $[a, x]$ between the graph of $f$ and the horizontal axis, but taken with a minus sign.

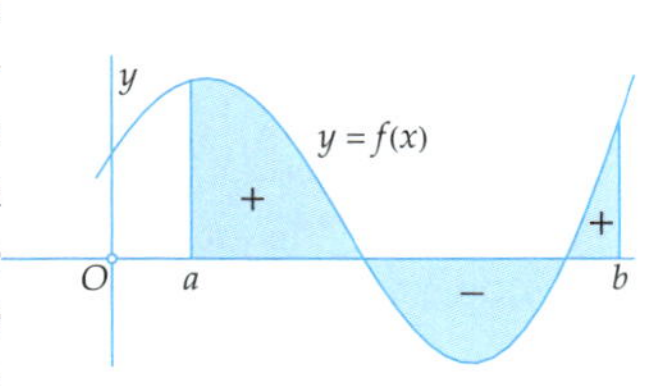


If $f(x)$ changes sign on $[a, b]$, we must combine positive and negative contributions to the integral. In the area function, the area's of the regions where $f(x) \leq 0$, must be taken with a minus sign. In the situation sketched to the right, on the interval $[a, b]$ two regions must be taken with a plus sign, while one region gets a minus sign.

With this sign convention, in all cases the area function $O(x)$ yields a primitive function of $f(x)$. Every other primitive function $F(x)$ can then be written as $F(x) = O(x) + c$. Since $O(a) = 0$, we have $\int_a^b f(x)\,dx = F(b) - F(a) = O(b) - O(a) = O(b)$. But $O(b)$ also is the 'area' between the graph of $f(x)$ and the horizontal axis, so we have:

*The integral $\int_a^b f(x)\,dx$ is equal to the area between the graph, the horizontal axis and the vertical lines through a and b, where the area of the regions below the horizontal axis should be taken with a minus sign.*

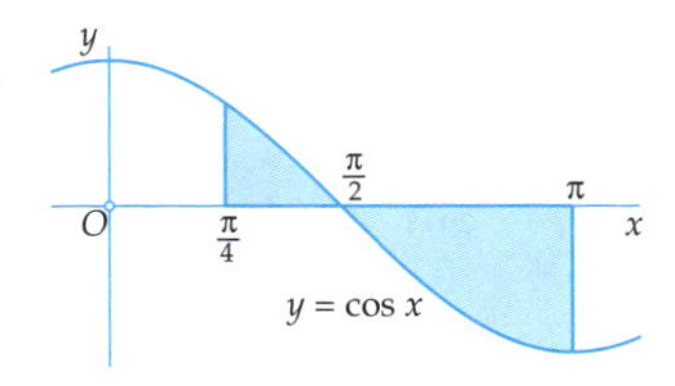


Example: take $f(x) = \cos x$ on $[\frac{\pi}{4}, \pi]$. Then $F(x) = \sin x$ is a primitive function, so

$$\int_{\frac{\pi}{4}}^{\pi} \cos x\,dx = \sin \pi - \sin \frac{\pi}{4} = -\frac{1}{2}\sqrt{2}$$

Write the following indefinite integrals in the form $F(x) + c$. For the second set, use suitably chosen trigonometric formulas (see page 313 and the example on the opposite page):

21.32

a. $\int (4x^3 - 2x^2 + x + 1)\,dx$

b. $\int (3 - 2x^3)\,dx$

c. $\int \sin(3x)\,dx$

d. $\int (\sqrt{x} - \frac{2}{x^2})\,dx$

21.33

a. $\int \sin^2 x\,dx$

b. $\int \cos^2 3x\,dx$

c. $\int \sin^2 5x\,dx$

d. $\int \cos^2 \frac{x}{2}\,dx$

21.34 Find the following indefinite integrals using suitably chosen trigonometric formulas (see page 313):

a. $\int \sin x \cos 5x\,dx$

b. $\int \cos 3x \cos 2x\,dx$

c. $\int \sin 2x \sin 4x\,dx$

21.35 In *Fourier analysis* (a branch of mathematics with lots of applications, for example in signal theory), the following results play an important role. Here, $m$ and $n$ are positive integers with $m \neq n$. Prove these results, using suitably chosen trigonometric formulas:

a. $\int_0^{2\pi} \sin mx \cos nx\,dx = 0$

b. $\int_0^{2\pi} \sin mx \sin nx\,dx = 0$

c. $\int_0^{2\pi} \cos mx \cos nx\,dx = 0$

d. $\int_0^{2\pi} \sin^2 mx\,dx = \pi$

e. $\int_0^{2\pi} \cos^2 mx\,dx = \pi$

## Indefinite integrals

If $F(x)$ is a primitive function of $f(x)$ on the interval $I$, then for each $a$ and each $x$ in $I$

$$F(x) = F(a) + \int_a^x f(t)\,dt$$

Every primitive function of $f(x)$ on $I$ can be written in this form. Such a primitive function of $f(x)$ is ofen denoted by

$$\int f(x)dx$$

(so without integration bounds). Then, we speak of an *indefinite integral* (also: *anti-derivative*) of $f(x)$. Every other primitive function on $I$ then can be obtained by adding a constant.

Example, with $I = \langle -\infty, \infty \rangle$:

$$\int x^5 dx = \frac{1}{6}x^6 + c$$

Each choice of the so-called *integration constant* $c$ yields a primitive function of $f(x) = x^5$. Many computer algebra packages, when being asked for an indefinite integral, only return one primitive function, leaving out the integration constant $c$.

Below is another example. Here, we use the trigonometric formula $\cos 2x = 2\cos^2 x - 1$.

$$\int \cos^2 x\,dx = \int \frac{\cos 2x + 1}{2}\,dx = \frac{1}{4}\sin 2x + \frac{1}{2}x + c$$

Describe all the primitive functions of the following functions:

21.36

a. $f(x) = \dfrac{1}{x-1}$

b. $f(x) = \dfrac{1}{2-x}$

c. $f(x) = \dfrac{3}{2x-1}$

d. $f(x) = \dfrac{4}{2-3x}$

e. $f(x) = \dfrac{1}{x^2}$

f. $f(x) = \dfrac{1}{(x-1)^3}$

21.37

a. $f(x) = \dfrac{1}{\sqrt{|x|}}$

b. $f(x) = \dfrac{1}{\sqrt[3]{x}}$

c. $f(x) = \dfrac{1}{\sqrt[5]{x-1}}$

d. $f(x) = \dfrac{1}{\sqrt{|2-x|}}$

e. $f(x) = \dfrac{1}{\cos^2 x}$

f. $f(x) = \dfrac{1}{\cos^2 \pi x}$

## The primitive functions of $f(x) = \frac{1}{x}$

The domain of the function $f(x) = \frac{1}{x}$ consists of two intervals: $\langle -\infty, 0\rangle$ and $\langle 0, \infty\rangle$. On the interval $\langle 0, \infty\rangle$, the function $F(x) = \ln x$ is a primitive function, while on $\langle -\infty, 0\rangle$, the function $F(x) = \ln(-x)$ is a primitive function. Indeed, according to the chain rule $\frac{d}{dx}\ln(-x) = \frac{1}{-x}(-1) = \frac{1}{x}$. Both formulas for $F(x)$ can be combined to $F(x) = \ln|x|$ since $|x| = x$ if $x > 0$ and $|x| = -x$ if $x < 0$. Other primitive functions are $\ln|x| + c$ for an arbitrary constant $c$. But this doesn't give us *all* the primitive functions, because the integration constant on $\langle -\infty, 0\rangle$ isn't necessarily the same as the one on $\langle 0, \infty\rangle$. So, strictly speaking, it not correct to state that

$$\int \frac{1}{x}\,dx = \ln|x| + c$$

as is frequently seen. A primitive function that is not included, is, e.g.,

$$F(x) = \begin{cases} \ln x + 1 & \text{als} \quad x > 0 \\ \ln(-x) - 1 & \text{als} \quad x < 0 \end{cases}$$

Below, the graphs of $f(x) = \dfrac{1}{x}$ and this primitive function $F(x)$ are drawn.

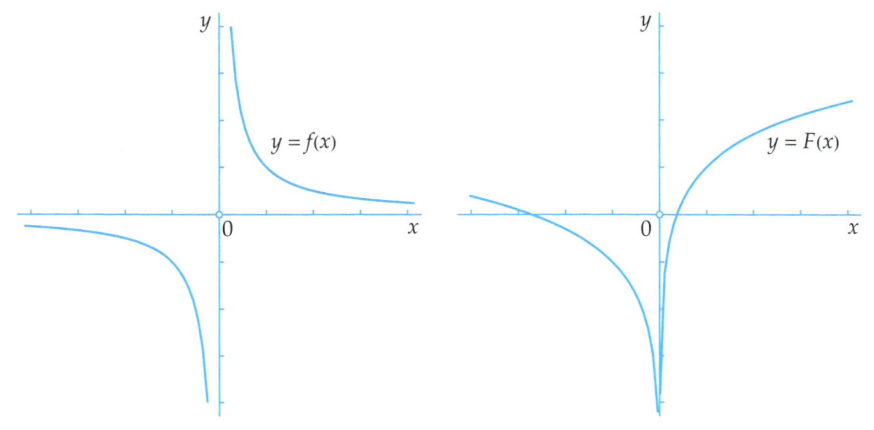


Hence, *all* the primitive functions of $f(x) = \dfrac{1}{x}$ are given by

$$F(x) = \begin{cases} \ln x + c_1 & \text{if} \quad x > 0 \\ \ln(-x) + c_2 & \text{if} \quad x < 0 \end{cases}$$

The same phenomenon occurs for all functions $f(x)$ for which the domain separates into at least two distinct intervals, for example because there are vertical asymptotes.

# 22 Integration techniques

Calculate the following integrals by using the substitution rule. Use the technique of 'bringing a function behind the $d$'. Write down all the intermediate steps, just like in the examples on the opposite page. If you get stuck, first study the examples in full detail!

22.1

a. $\displaystyle\int_0^1 (1+x)^9\,dx$

b. $\displaystyle\int_{-1}^1 (2+3x)^5\,dx$

c. $\displaystyle\int_0^2 (3-x)^6\,dx$

d. $\displaystyle\int_{-1}^0 (5-2x)^5\,dx$

e. $\displaystyle\int_1^{\mathrm{e}} \frac{\ln x}{x}\,dx$

f. $\displaystyle\int_{\mathrm{e}}^{\mathrm{e}^2} \frac{1}{x(1+\ln x)}\,dx$

22.2

a. $\displaystyle\int_{-1}^1 \mathrm{e}^{2x+1}\,dx$

b. $\displaystyle\int_0^1 x\mathrm{e}^{x^2}\,dx$

c. $\displaystyle\int_{-1}^1 x\mathrm{e}^{-x^2}\,dx$

d. $\displaystyle\int_0^1 x^2\mathrm{e}^{x^3}\,dx$

e. $\displaystyle\int_0^1 \frac{\mathrm{e}^x}{1+\mathrm{e}^{2x}}\,dx$

f. $\displaystyle\int_{-1}^0 \frac{\mathrm{e}^x}{\sqrt{1+\mathrm{e}^x}}\,dx$

22.3

a. $\displaystyle\int_0^{\pi} \cos\frac{1}{3}x\,dx$

b. $\displaystyle\int_{-1}^0 \sin\pi x\,dx$

c. $\displaystyle\int_0^{\sqrt[3]{\pi}} x^2\cos x^3\,dx$

d. $\displaystyle\int_0^{\frac{3\pi}{4}} \sin x\cos x\,dx$

e. $\displaystyle\int_0^{\frac{2\pi}{3}} \sin x\cos^2 x\,dx$

f. $\displaystyle\int_0^{\frac{\pi}{6}} \cos^3 x\,dx$

22.4

a. $\displaystyle\int_0^{\pi} \sin^5 x\,dx$

b. $\displaystyle\int_0^{\frac{5\pi}{6}} \cos x\sqrt{1+\sin x}\,dx$

c. $\displaystyle\int_{-\frac{\pi}{2}}^0 \tan\frac{x}{2}\,dx$

d. $\displaystyle\int_0^{\pi} \frac{\cos\sqrt{x}}{\sqrt{x}}\,dx$

e. $\displaystyle\int_0^{\frac{3\pi}{4}} \cos x\,\sin(\sin x)\,dx$

f. $\displaystyle\int_0^{\frac{\pi}{2}} \frac{\sin x}{2+\cos x}\,dx$

## The substitution rule

Suppose that $\frac{d}{dy}F(y) = f(y)$. In the language of differentials, we then have $d(F(y)) = f(y)dy$. Substituting a function $g(x)$ for $y$, we get

$$d(F(g(x))) = f(g(x))d(g(x)) = f(g(x))g'(x)\,dx$$

The last step boils down to 'removing from behind the $d$' the function $g(x)$ by means of the well-known rule $d(g(x)) = g'(x)dx$ for differentials.

We can translate this rule in terms of integrals. When using it to calculate integrals, however, the point is not 'removing a function from behind the $d$' but 'bringing a function behind the $d$' by using the same rule $g'(x)dx = d(g(x))$ in reverse order:

$$\begin{aligned}\int_a^b f(g(x))g'(x)\,dx &= \int_a^b f(g(x))d(g(x)) = \int_a^b d(F(g(x))) \\ &= \big[F(g(x))\big]_a^b = F(g(b)) - F(g(a))\end{aligned}$$

This rule is known as the *substitution rule* for integrals. Example:

$$\begin{aligned}\int_0^{\pi/2} \cos x\, \mathrm{e}^{\sin x} dx &= \int_0^{\pi/2} \mathrm{e}^{\sin x} d(\sin x) = \int_0^{\pi/2} d(\mathrm{e}^{\sin x}) \\ &= \big[\mathrm{e}^{\sin x}\big]_0^{\pi/2} = \mathrm{e}^1 - \mathrm{e}^0 = \mathrm{e} - 1\end{aligned}$$

Here, we have taken $f(y) = \mathrm{e}^y$ and $y = g(x) = \sin x$, so $d(g(x)) = \cos x\,dx$. Note that in this case $F(y) = \mathrm{e}^y$ as well.

Below are a few more examples. In each case, identify the function $g(x)$ and verify the calculation:

$$\int_2^3 \frac{1}{x \ln x}\,dx = \int_2^3 \frac{1}{\ln x}\,d(\ln x) = [\ln(\ln x)]_2^3 = \ln(\ln 3) - \ln(\ln 2)$$

$$\begin{aligned}\int_{-\pi/4}^{\pi/3} \tan x\,dx = \int_{-\pi/4}^{\pi/3} \frac{\sin x}{\cos x}\,dx &= -\int_{-\pi/4}^{\pi/3} \frac{1}{\cos x}\,d(\cos x) \\ = -\big[\ln(\cos x)\big]_{-\pi/4}^{\pi/3} &= -\ln\frac{1}{2} + \ln\frac{1}{\sqrt{2}} = \frac{1}{2}\ln 2\end{aligned}$$

$$\int_0^3 \sqrt{x+1}\,dx = \int_0^3 \sqrt{x+1}\,d(x+1) = \left[\frac{2}{3}(x+1)^{3/2}\right]_0^3 = \frac{14}{3}$$

$$\int_0^{\sqrt{\pi}} x\sin(x^2)\,dx = \frac{1}{2}\int_0^{\sqrt{\pi}} \sin(x^2)\,d(x^2) = \frac{1}{2}\left[-\cos(x^2)\right]_0^{\sqrt{\pi}} = 1$$

$$\int_2^4 (2x-5)^5 dx = \frac{1}{2}\int_2^4 (2x-5)^5 d(2x-5) = \frac{1}{12}\left[(2x-5)^6\right]_2^4 = \frac{182}{3}$$

Calculate the following integrals, either by an explicit substitution or by the method of 'bringing a function behind the $d$'. In some cases, the first method will be easier, in other cases the other. If you use an explicit substitution, always clearly state the used substitution in terms of differentials, and also in the integration bounds. See the examples on the opposite page.

You will see that calculating integrals is not a matter of simply 'follow the recipe', because you have to think a few steps ahead, realizing which substitution will make the integral more manageable. Integration, unlike differentiation, is not just 'following the rules'. You need experience and creativity!

22.5

a. $\int_0^1 x(1+x)^9\,dx$

b. $\int_0^1 x(1-x)^4\,dx$

c. $\int_{-1}^0 x(2x+3)^5\,dx$

d. $\int_{-1}^1 x(1-x^2)^7\,dx$

e. $\int_0^1 x(1+x^2)^8\,dx$

f. $\int_0^1 (x+1)(2x+x^2)^3\,dx$

22.6

a. $\int_0^1 \frac{x}{1+x}\,dx$

b. $\int_{-1}^1 \frac{x-1}{x+2}\,dx$

c. $\int_0^{\frac{1}{4}} \frac{x}{1-2x}\,dx$

d. $\int_{-1}^0 \frac{x}{1+x^2}\,dx$

e. $\int_0^1 \frac{x}{4-x^2}\,dx$

f. $\int_0^{\frac{1}{2}} \frac{x}{1+4x^2}\,dx$

22.7

a. $\int_0^5 x\sqrt{x+4}\,dx$

b. $\int_{-1}^1 x\sqrt{x^2+1}\,dx$

c. $\int_2^6 x\sqrt{2x-3}\,dx$

d. $\int_2^3 \frac{x}{\sqrt{x-1}}\,dx$

e. $\int_{-1}^0 \frac{x}{\sqrt{1-x}}\,dx$

22.8

a. $\int_1^2 x\sqrt[3]{x-1}\,dx$

b. $\int_0^1 \frac{\sqrt{x}}{1+x}\,dx$

c. $\int_0^1 \frac{x}{1+\sqrt{x}}\,dx$

d. $\int_0^1 \frac{\sqrt{x}}{1+\sqrt{x}}\,dx$

e. $\int_0^{\sqrt{3}} \frac{1}{\sqrt{4-x^2}}\,dx$

## Explicit substitutions

Sometimes, it is convenient to perform a substitution explicitly, by choosing a new variable. The integration bounds then also have to be stated in terms of the new variable. In the following example, we choose the substitution $y = x - 1$ so $x = y + 1$ and $dx = d(y+1) = dy$ by the rules for differentials.

$$\begin{aligned}\int_0^2 x(x-1)^5\,dx &= \int_{y=-1}^{y=1} (y+1)y^5\,dy = \int_{y=-1}^{y=1} (y^6+y^5)\,dy \\ &= \left[\frac{1}{7}y^7 + \frac{1}{6}y^6\right]_{-1}^{1} = \frac{2}{7}\end{aligned}$$

In the next example, we choose $y = \sqrt{x+1}$, which yields $x = y^2 - 1$ and $dx = 2y\,dy$. Note that also in this case, we transform the integration bounds, writing down them explicitly in terms of the new variable $y$.

$$\begin{aligned}\int_0^3 x\sqrt{x+1}\,dx &= \int_{y=1}^{y=2} (y^2-1)y\cdot 2y\,dy = \int_{y=1}^{y=2} (2y^4-2y^2)\,dy \\ &= \left[\frac{2}{5}y^5 - \frac{2}{3}y^3\right]_{1}^{2} = \frac{116}{15}\end{aligned}$$

To calculate the following integral, we use the substitution $y = \mathrm{e}^x$. Then $x = \ln y$, so $dx = \frac{1}{y}\,dy$.

$$\begin{aligned}\int_{-1}^{1} \frac{1}{\mathrm{e}^x + \mathrm{e}^{-x}}\,dx &= \int_{y=1/\mathrm{e}}^{y=\mathrm{e}} \frac{1}{y+\frac{1}{y}}\,\frac{1}{y}\,dy = \int_{y=1/\mathrm{e}}^{y=\mathrm{e}} \frac{1}{y^2+1}\,dy \\ &= \big[\arctan y\big]_{1/\mathrm{e}}^{\mathrm{e}} = \arctan \mathrm{e} - \arctan\frac{1}{\mathrm{e}}\end{aligned}$$

In the next example, we put $x = 3\sin t$. Then $dx = 3\cos t\,dt$ and $\sqrt{9-x^2} = \sqrt{9 - 9\sin^2 t} = 3\cos t$.

$$\begin{aligned}\int_0^{\frac{3}{2}} \frac{1}{\sqrt{9-x^2}}\,dx &= \int_{t=0}^{t=\pi/6} \frac{1}{3\cos t}\,3\cos t\,dt \\ &= \int_{t=0}^{t=\pi/6} dt = \big[\,t\,\big]_0^{\pi/6} = \frac{\pi}{6}\end{aligned}$$

To calculate the next integral, we choose $y = \sqrt{x}$. Then $x = y^2$ and $dx = 2y\,dy$.

$$\begin{aligned}\int_1^4 \frac{1}{x+2\sqrt{x}}\,dx &= \int_{y=1}^{y=2} \frac{2y}{y^2+2y}\,dy = \int_{y=1}^{y=2} \frac{2}{y+2}\,dy \\ &= \big[\,2\ln(y+2)\,\big]_1^2 = 2(\ln 4 - \ln 3)\end{aligned}$$

Calculate the following integrals by partial integration. If necessary, also consult the examples on page 219:

22.9

a. $\int_{-\pi}^{0} x \cos x \, dx$

b. $\int_{1}^{e} x \ln x \, dx$

c. $\int_{1}^{e} \ln x \, dx$

d. $\int_{0}^{1} x \arctan x \, dx$

e. $\int_{0}^{\pi} x^2 \sin x \, dx$

f. $\int_{0}^{\frac{1}{2}} \arcsin x \, dx$

22.10

a. $\int_{0}^{1} x \, e^{3x} dx$

b. $\int_{0}^{1} x^2 \, e^{-x} \, dx$

c. $\int_{-2}^{0} (2x+1) e^{x} \, dx$

d. $\int_{0}^{1} x^3 e^{-x^2} \, dx$

e. $\int_{-\pi}^{\pi} e^{2x} \cos x \, dx$

f. $\int_{1}^{e} \sqrt{x} \ln x \, dx$

## Partial integration

The product rule for differentials (see page 193) can be written as $f(x)\, d(g(x)) = d(f(x)\, g(x)) - g(x)\, d(f(x))$. Hence

$$\int_a^b f(x)\, d(g(x)) = \int_a^b d(f(x)\, g(x)) - \int_a^b g(x)\, d(f(x))$$

which means that

$$\begin{aligned}\int_a^b f(x)\, d(g(x)) &= [f(x)\, g(x)]_a^b - \int_a^b g(x)\, d(f(x)) \\ &= f(b)\, g(b) - f(a)\, g(a) - \int_a^b g(x)\, d(f(x))\end{aligned}$$

This is known as *partial integration*. In combination with the substitution rule, this rule sometimes makes it possible to calculate integrals that, at first sight, seem forbidding. Example:

$$\begin{aligned}\int_1^2 x^3 \ln x\, dx &= \int_1^2 \ln x\, d\left(\frac{1}{4}x^4\right) \\ &= \left[(\ln x)\left(\frac{1}{4}x^4\right)\right]_1^2 - \int_1^2 \frac{1}{4}x^4\, d(\ln x) \\ &= 4\ln 2 - \int_1^2 \frac{1}{4}x^3\, dx = 4\ln 2 - \left[\frac{1}{16}x^4\right]_1^2 = 4\ln 2 - \frac{15}{16}\end{aligned}$$

As you see, we first brought the factor $x^3$ behind the $d$. After using partial integration, we then removed $\ln x$ from behind the $d$ using the differential rule $d(\ln x) = \frac{1}{x}\, dx$, leading to an integral that could be calculated easily.

The question which factor should be brought behind the $d$ in a given situation, cannot be answered in general. Sometimes, an inadequate choice will only lead you further astray. Example:

$$\int_0^1 x\mathrm{e}^x dx = \int_0^1 x d(\mathrm{e}^x) = [x\mathrm{e}^x]_0^1 - \int_0^1 \mathrm{e}^x dx = \mathrm{e} - [\mathrm{e}^x]_0^1 = \mathrm{e} - (\mathrm{e} - 1) = 1$$

but

$$\int_0^1 x\mathrm{e}^x dx = \int_0^1 \mathrm{e}^x d\left(\frac{1}{2}x^2\right) = \left[\frac{1}{2}x^2\mathrm{e}^x\right]_0^1 - \int_0^1 \frac{1}{2}x^2\mathrm{e}^x dx = ??$$

## Mixed exercises

Calculate the following integrals:

22.11

a. $\int_0^{\pi^2} \sin\sqrt{x}\,dx$

b. $\int_1^4 \mathrm{e}^{\sqrt{x}}\,dx$

c. $\int_1^{\mathrm{e}} x^{\frac{2}{3}} \ln x\,dx$

d. $\int_{-\frac{1}{2}}^{\frac{1}{2}} \arccos x\,dx$

e. $\int_0^{\frac{\pi}{3}} \sin 2x\,dx$

f. $\int_0^{\frac{1}{2}} \arctan 2x\,dx$

22.12

a. $\int_{-2}^{1} |x|\,dx$

b. $\int_0^1 (1 - x + x^3)\,dx$

c. $\int_0^1 x\sqrt{4 - x^2}\,dx$

d. $\int_1^{\mathrm{e}} \ln 2x\,dx$

e. $\int_{-\pi}^{\pi} \sin^2 x\,dx$

f. $\int_0^1 \frac{1}{1+\sqrt{x}}\,dx$

22.13

a. $\int_0^1 \sin^2(\pi x)\,dx$

b. $\int_0^2 x^3\,\mathrm{e}^{-x^2}\,dx$

c. $\int_0^1 \sqrt{x+1}\,dx$

d. $\int_0^{2\pi} |\cos x|\,dx$

e. $\int_0^{\frac{\pi}{3}} \sin 2x\,\mathrm{e}^{\cos x}\,dx$

f. $\int_0^1 x \arctan 2x\,dx$

22.14

a. $\int_0^2 \frac{x^2}{1+x^3}\,dx$

b. $\int_0^1 x(1-x)^{20}\,dx$

c. $\int_{-1}^{1} \sin^3(\pi x)\,dx$

d. $\int_0^{\frac{\pi}{3}} \cos^3 x\,dx$

e. $\int_0^{\mathrm{e}} \ln(1+3x)\,dx$

f. $\int_0^{\frac{\pi}{4}} \cos 2x\,\mathrm{e}^{-\sin 2x}\,dx$

## Three more examples of partial integration

1. $$\int_0^{\pi} x \sin x \, dx = -\int_0^{\pi} x \, d(\cos x) = -\big[x \cos x\big]_0^{\pi} + \int_0^{\pi} \cos x \, dx$$
$$= \pi + \big[\sin x\big]_0^{\pi} = \pi$$

2. $$\int_0^1 \arctan x \, dx = \big[x \arctan x\big]_0^1 - \int_0^1 x \, d(\arctan x) = \frac{\pi}{4} - \int_0^1 \frac{x}{1+x^2} \, dx$$
$$= \frac{\pi}{4} - \frac{1}{2}\int_0^1 \frac{1}{1+x^2} \, d(1+x^2) = \frac{\pi}{4} - \frac{1}{2}\Big[\ln(1+x^2)\Big]_0^1$$
$$= \frac{\pi}{4} - \frac{1}{2}\ln 2$$

In the next example, we apply partial integration twice, each time by bringing the factor $\mathrm{e}^{x}$ behind the $d$.

3. $$\int_0^{\pi} \mathrm{e}^{x} \sin x \, dx = \int_0^{\pi} \sin x \, d(\mathrm{e}^{x}) = [\mathrm{e}^{x} \sin x]_0^{\pi} - \int_0^{\pi} \mathrm{e}^{x} d(\sin x)$$
$$= -\int_0^{\pi} \mathrm{e}^{x} \cos x \, dx = -\int_0^{\pi} \cos x \, d(\mathrm{e}^{x})$$
$$= -[\mathrm{e}^{x} \cos x]_0^{\pi} + \int_0^{\pi} \mathrm{e}^{x} d(\cos x)$$
$$= \mathrm{e}^{\pi} + 1 - \int_0^{\pi} \mathrm{e}^{x} \sin x \, dx$$

Now it seems that we are back where we started, since we are again faced with the original integral. However, on second view there is a difference: denoting the original integral for a moment by $I$, we have derived $I = \mathrm{e}^{\pi} + 1 - I$, and this is an equation from which $I$ can be solved! The result is

$$I = \int_0^{\pi} \mathrm{e}^{x} \sin x \, dx = \frac{\mathrm{e}^{\pi} + 1}{2}$$

Calculate the following improper integrals, if they exist:

22.15

a. $\int_1^\infty \frac{1}{x^2}\,dx$

b. $\int_2^\infty \frac{1}{x^3}\,dx$

c. $\int_1^\infty \frac{1}{x^{10}}\,dx$

d. $\int_2^\infty \frac{1}{x^p}\,dx \quad (p > 1)$

e. $\int_{-\infty}^{-1} \frac{1}{x}\,dx$

22.16

a. $\int_1^\infty \frac{1}{\sqrt{x}}\,dx$

b. $\int_2^\infty \frac{1}{\sqrt{x+1}}\,dx$

c. $\int_{-\infty}^\infty \frac{1}{4+x^2}\,dx$

d. $\int_4^\infty \frac{x}{1+x^2}\,dx$

e. $\int_{-\infty}^\infty \frac{1}{1+|x|}\,dx$

22.17

a. $\int_0^\infty \mathrm{e}^{-x}\,dx$

b. $\int_{-\infty}^0 \mathrm{e}^{3x}\,dx$

c. $\int_1^\infty x\,\mathrm{e}^{-x}\,dx$

d. $\int_{-\infty}^0 x^2\,\mathrm{e}^{x}\,dx$

e. $\int_{-\infty}^\infty \mathrm{e}^{-|x|}\,dx$

22.18

a. $\int_1^\infty \frac{\ln x}{x^2}\,dx$

b. $\int_1^\infty \frac{\ln x}{x}\,dx$

c. $\int_{-\infty}^0 \frac{\arctan x}{1+x^2}\,dx$

d. $\int_{-\infty}^\infty \sin x\,dx$

e. $\int_0^\infty \sin x\,\mathrm{e}^{-x}\,dx$

## Improper integrals – type 1

Some definite integrals cannot be calculated directly, but only via a limit process. These integrals are called *improper integrals*. We distinguish two types: a type where the interval of integration has an infinite length and a type where the integrand is not continuous at a boundary point of the integration interval, for instance because of a vertical asymptote. For both, we assume that the integrand $f(x)$ is continuous on the integration interval, with the possible exception of one or two of its boundary points. Then, primitive functions of the integrand $f(x)$ do exist; let's call one of them $F(x)$.

Type 1:

If $f(x)$ is continuous on the interval $[a, \infty\rangle$ and if $\lim_{M\to\infty} F(M)$ exists, then

$$\int_a^\infty f(x)\,dx = \lim_{M\to\infty}\int_a^M f(x)\,dx = \lim_{M\to\infty} F(M) - F(a)$$

If $f(x)$ is continuous on the interval $\langle -\infty, b]$ and if $\lim_{N\to -\infty} F(N)$ exists, then

$$\int_{-\infty}^b f(x)\,dx = \lim_{N\to -\infty}\int_N^b f(x)\,dx = F(b) - \lim_{N\to -\infty} F(N)$$

If $f(x)$ is continuous on the interval $\langle -\infty, \infty\rangle$ and if the limits $\lim_{M\to\infty} F(M)$ and $\lim_{N\to -\infty} F(N)$ both exist and at least one of them is finite, then

$$\int_{-\infty}^\infty f(x)\,dx = \lim_{M\to\infty}\lim_{N\to -\infty}\int_N^M f(x)\,dx = \lim_{M\to\infty} F(M) - \lim_{N\to -\infty} F(N)$$

Example:

$$\begin{aligned}\int_{-\infty}^\infty \frac{1}{1+x^2}\,dx &= \lim_{M\to\infty}\lim_{N\to -\infty}\int_N^M \frac{1}{1+x^2}\,dx = \lim_{M\to\infty}\lim_{N\to -\infty}\left[\arctan x\right]_N^M \\ &= \lim_{M\to\infty}\arctan M - \lim_{N\to -\infty}\arctan N = \frac{\pi}{2} - (-\frac{\pi}{2}) = \pi\end{aligned}$$

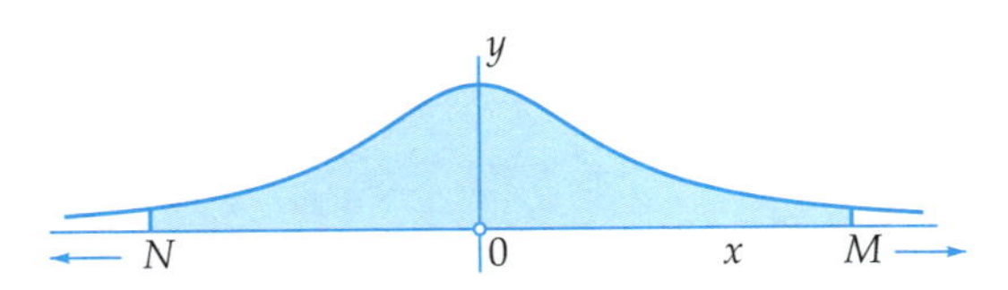

Calculate the following improper integrals, if they exist:

22.19

a. $\int_0^1 \frac{1}{\sqrt{x}}\,dx$

b. $\int_0^2 \frac{1}{\sqrt[3]{x}}\,dx$

c. $\int_0^1 \frac{1}{\sqrt[10]{x}}\,dx$

d. $\int_0^2 \frac{1}{x^p}\,dx \quad (0 < p < 1)$

e. $\int_0^1 \frac{1}{x}\,dx$

22.20

a. $\int_0^1 \frac{1}{x^2}\,dx$

b. $\int_{-1}^0 \frac{1}{\sqrt{x+1}}\,dx$

c. $\int_{-1}^0 \frac{1}{\sqrt[3]{x}}\,dx$

d. $\int_{-2}^2 \frac{1}{\sqrt{2-x}}\,dx$

e. $\int_{-1}^2 \frac{1}{\sqrt{|x|}}\,dx$

22.21

a. $\int_0^1 \ln x\,dx$

b. $\int_0^{\frac{\pi}{2}} \tan x\,dx$

c. $\int_{-1}^1 \ln(1-x)\,dx$

d. $\int_0^1 \ln 3x\,dx$

e. $\int_0^4 \frac{\ln x}{\sqrt{x}}\,dx$

22.22

a. $\int_0^1 \frac{x}{1-x}\,dx$

b. $\int_0^{\frac{\pi}{2}} \frac{\cos x}{\sin x}\,dx$

c. $\int_{-2}^2 \frac{1}{\sqrt{4-x^2}}\,dx$

d. $\int_0^2 \frac{x}{\sqrt{4-x^2}}\,dx$

e. $\int_0^1 x \ln x\,dx$

## Improper integrals – type 2

Type 2:

If $f(x)$ is continuous on the interval $[a, b\rangle$ and if $\lim_{t\uparrow b} F(t)$ exists, then

$$\int_a^b f(x)\,dx = \lim_{t\uparrow b} \int_a^t f(x)\,dx = \lim_{t\uparrow b} F(t) - F(a)$$

If $f(x)$ is continuous on the interval $\langle a, b]$ and if $\lim_{u\downarrow a} F(u)$ exists, then

$$\int_a^b f(x)\,dx = \lim_{u\downarrow a} \int_u^b f(x)\,dx = F(b) - \lim_{u\downarrow a} F(u)$$

If $f(x)$ is continuous on the interval $\langle a, b\rangle$ and if the two limits $\lim_{t\uparrow b} F(t)$ and $\lim_{u\downarrow a} F(u)$ exist and at least one of them is finite, then

$$\int_a^b f(x)\,dx = \lim_{u\downarrow a}\lim_{t\uparrow b} \int_u^t f(x)\,dx = \lim_{t\uparrow b} F(t) - \lim_{u\downarrow a} F(u)$$

Example:

$$\begin{aligned}\int_{-1}^1 \frac{1}{\sqrt{1-x^2}}\,dx &= \lim_{u\downarrow -1}\lim_{t\uparrow 1} \int_u^t \frac{1}{\sqrt{1-x^2}}\,dx = \lim_{u\downarrow -1}\lim_{t\uparrow 1} \big[\arcsin x\big]_u^t \\ &= \lim_{t\uparrow 1} \arcsin t - \lim_{u\downarrow -1} \arcsin u = \frac{\pi}{2} - (-\frac{\pi}{2}) = \pi\end{aligned}$$

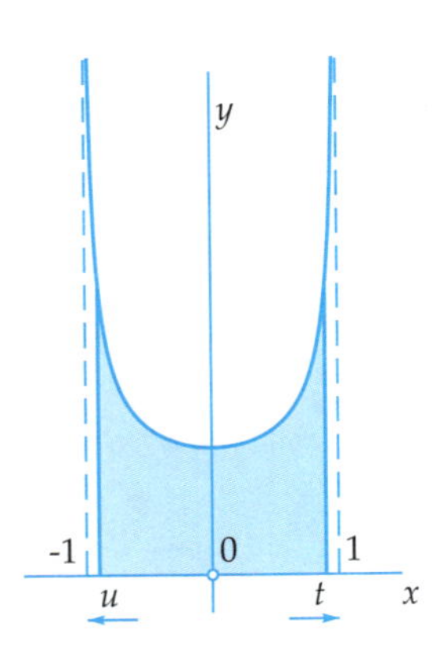

22.23 For these exercises, you need a programmable calculator or a computer with a computer algebra package.

For $N = 10$, $N = 100$ and $N = 1000$ calculate the sum $\sum_{i=0}^{N-1} f(x_i)dx$, which is an approximation of the given integral. To do this, partition the integration interval in $N$ subintervals of equal length. Enter the results in a table and compare them to the exact value of the integral:

a. $\int_{-1}^{1} (1 - x^2)\, dx$

b. $\int_{0}^{1} 2^x\, dx$

c. $\int_{1}^{10} \log_{10} x\, dx$

22.24 For each of the integrals above, find an upper bound $M$ for $|f'(x)|$ on the integration interval and use this bound to verify the estimate $\frac{1}{2}M(b-a)dx$ (see the opposite page) for the error in approximating the integral by each of the calculated sums.

## Sums and integrals

Suppose that $F(x)$ is a primitive function of $f(x)$ on the interval $[a, b]$. Partition $[a, b]$ into $N$ subintervals by means of partition points $x_0, x_1, x_2, \ldots, x_N$ with $a = x_0 < x_1 < \cdots < x_N = b$. Then

$$\begin{aligned}\int_a^b f(x)dx &= \int_{x_0}^{x_1} f(x)dx + \int_{x_1}^{x_2} f(x)dx + \cdots + \int_{x_{N-1}}^{x_N} f(x)dx \\ &= \sum_{i=0}^{N-1} \int_{x_i}^{x_{i+1}} f(x)dx = \sum_{i=0}^{N-1} F(x_{i+1}) - F(x_i)\end{aligned}$$

Put $dx_i = x_{i+1} - x_i$, $\Delta F_i = F(x_{i+1}) - F(x_i)$ and $dF_i = F'(x_i)dx_i = f(x_i)dx_i$. Then $\Delta F_i \approx dF_i$, provided $dx_i$ is small, so

$$\int_a^b f(x)dx \approx \sum_{i=0}^{N-1} f(x_i)dx_i$$

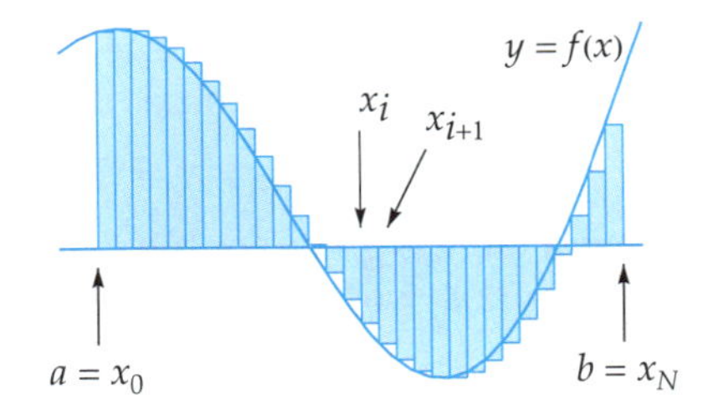


We see that such an integral can be approximated by a *sum* (hence the integral sign $\int$, which is an elongated capital S) of differentials, one for each subinterval.

Note that $f(x_i)\,dx_i$ is the area of the rectangle with base $dx_i$ and height $f(x_i)$, with a minus sign if $f(x_i) < 0$. If the number of subintervals increases while their length goes to zero, the sum will gradually approximate the integral.

This last statement, illustrated by the figure, can be made more plausible in the following way. Suppose that $f(x)$ has a continuous derivative on $[a, b]$ and that an upper bound $M$ exists such that $|f'(x)| < M$ for all $x$ in $[a, b]$. Most 'decent' functions satisfy this condition. Furthermore, for simplicity, we assume that all subintervals $[x_i, x_{i+1}]$ are of equal length. Their length $dx = x_{i+1} - x_i$ then satisfies $dx = (b - a)/N$. Since $F'(x) = f(x)$, we have $F''(x) = f'(x)$ and according to the estimate on page 197, $|\Delta F_i - dF_i| \leq \frac{1}{2}M(dx)^2$ holds, so

$$\begin{aligned}\left|\int_a^b f(x)dx - \sum_{i=0}^{N-1} f(x_i)dx\right| &= \left|F(b) - F(a) - \sum_{i=0}^{N-1} f(x_i)dx\right| = \left|\sum_{i=0}^{N-1} \Delta F_i - \sum_{i=0}^{N-1} dF_i\right| \\ &= \left|\sum_{i=0}^{N-1} (\Delta F_i - dF_i)\right| \leq N \times \frac{1}{2}M(dx)^2 = \frac{1}{2}M(b-a)dx\end{aligned}$$

since $Ndx = (b - a)$. From this, it follows that the difference between the integral and the sum indeed goes to zero if $dx$ goes to zero.

22.25 For these exercises, you need a programmable calculator or a computer with a computer algebra package. If using a programmable calculator, use the highest precision mode; if using a calculater algebra package, work with at least 15 significant digits.

We will always partition the integration interval into $n$ *equal* subintervals. By $M(n)$ we understand the result of the midpoint rule and by $T(n)$ the result of the trapezium rule (see the definitions on the opposite page). Furthermore, we define

$$S(n) = \frac{2M(n) + T(n)}{3}$$

This rule is known as *Simpson's rule*. It can be proven that $S(n)$ yields a much better approximation to the integral than each of $M(n)$ and $T(n)$.

*Example:* since $\int_0^1 \frac{4}{1+x^2}\,dx = \left[4\arctan x\right]_0^1 = \pi$ , we can use numerical methods for approximating the integral to find approxximations to $\pi$. For comparison: the first fifteen decimals of $\pi$ are given by

$$\pi = 3.141592653589793\ldots$$

Below, a table is given for the values of $M(n)$, $T(n)$ and $S(n)$ for $n = 8$, $n = 16$, $n = 32$ and $n = 64$.

| $n$ | $M(n)$ | $T(n)$ | $S(n)$ |
|---|---|---|---|
| 8 | 3.142894729591688 | 3.138988494491089 | 3.141592651224821 |
| 16 | 3.141918174308560 | 3.140941612041388 | 3.141592653552836 |
| 32 | 3.141674033796337 | 3.141429893174975 | 3.141592653589217 |
| 64 | 3.141612998641850 | 3.141551963485654 | 3.141592653589784 |

Make a similar table for the numerical approximation of the following integrals. For comparison, we also give the 'exact' values, rounded off to fifteen decimals.

a. $\int_0^4 \mathrm{e}^{-x^2}\,dx \approx 0.886226911789569$

b. $\int_0^1 \sin\left(\mathrm{e}^x\right)\,dx \approx 0.874957198780384$

c. $\int_0^{\sqrt{\pi}} \sin\left(x^2\right)\,dx \approx 0.894831469484145$

## Numerical integration methods

Suppose that $f(x)$ is continuous on the interval $[a,b]$. If $F(x)$ is a primitive function of $f(x)$, then $\int_a^b f(x)\,dx = F(b) - F(a)$. However, on the following pages, we will see that it is not always possible to find a formula expressed in 'elementary' functions for a primitive function of $f(x)$, even if $f(x)$ itself is given by a simple formula. In such cases, numerical approximations of the integral can be obtained by dividing $[a,b]$ into small subintervals. The formula on page 225 then yields the following numerical approximation of the integral:

$$\int_a^b f(x)dx \approx \sum_{i=0}^{N-1} f(x_i)dx_i$$

Let's call this approximation $L$. In calculating $L$, we have multiplied the length $dx_i = x_{i+1} - x_i$ of each subinterval by the function value $f(x_i)$ in the *left-hand* end point $x_i$.

But, instead of the left-hand end point, we can also take another point in each subinterval $[x_i, x_{i+1}]$, for instance the midpoint $\frac{1}{2}(x_i + x_{i+1})$, or the right-hand end point $x_{i+1}$. This results in the *midpoint rule* $M$ and the *right-hand rule* $R$ with formulas

$$M = \sum_{i=0}^{N-1} f\left(\frac{x_i + x_{i+1}}{2}\right) dx_i \quad \text{and} \quad R = \sum_{i=0}^{N-1} f(x_{i+1})\,dx_i$$

Both are also approximations of the integral. In the figure below, they are illustrated with subintervals of equal length.

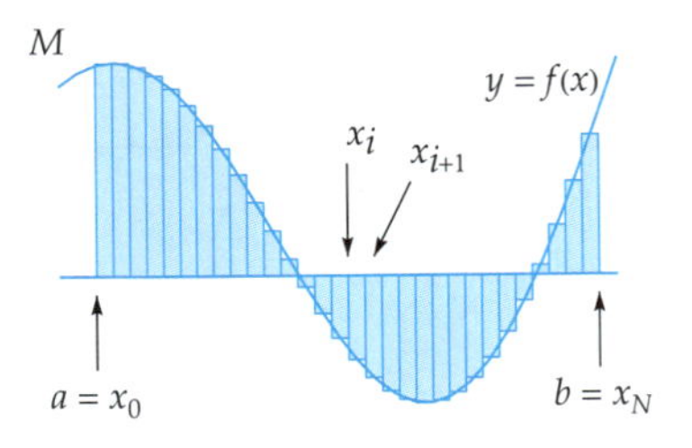


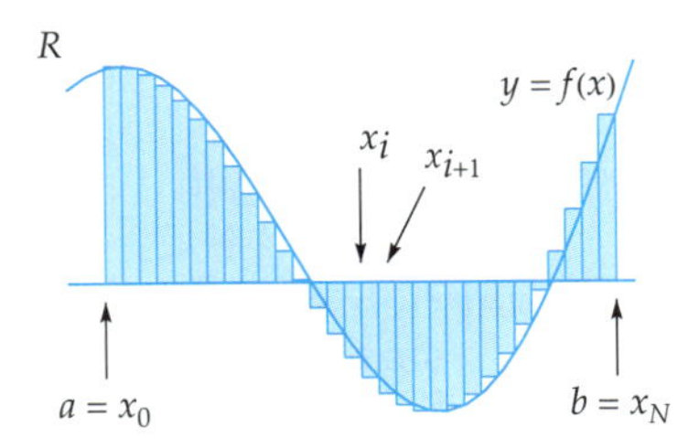


A better result than $L$ or $R$ is given by the *mean value* $\frac{1}{2}(L+R)$ of $L$ and $R$. This rule is called the *trapezium rule*. If all the subintervals have the same length $dx$, it can be written as

$$T = \left(\frac{1}{2}f(x_0) + f(x_1) + f(x_2) + \cdots + f(x_{N-2}) + f(x_{N-1}) + \frac{1}{2}f(x_N)\right) dx$$

Even better results can be obtained by taking a suitable combination of $M$ and $T$, the so-called *Simpson's rule*, given on the opposite page.

22.26 The figure to the right shows the graphs of the function $\varphi(t) = \frac{1}{\sqrt{2\pi}}\mathrm{e}^{-\frac{1}{2}t^2}$, the *probability density function* of the standard normal distribution from statistics, and the corresponding *distribution function*

$$\Phi(x) = \frac{1}{\sqrt{2\pi}}\int_{-\infty}^{x} \mathrm{e}^{-\frac{1}{2}t^2}\,dt$$

y

y = φ(t)

Φ(x)

0 x t

It can be proven (but this is not easy!) that $\lim_{x\to\infty}\Phi(x) = 1$.

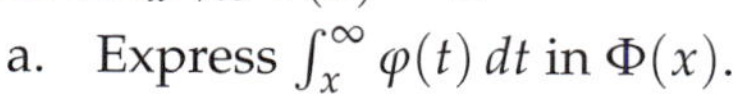

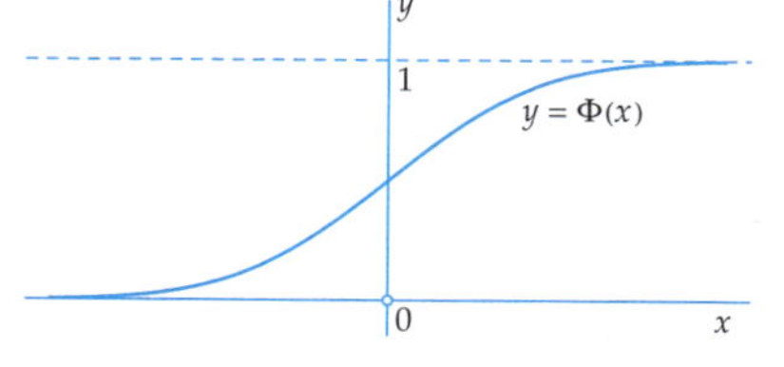


a. Express $\int_x^{\infty}\varphi(t)\,dt$ in $\Phi(x)$.
b. Show that $\Phi(-x) = 1 - \Phi(x)$.
c. Find $\Phi(0)$.
d. Calculate $\int_{-\infty}^{\infty}\mathrm{e}^{-x^2}\,dx$.

22.27 The *error function* $\mathrm{Erf}(x)$ is defined by the integral

$$\mathrm{Erf}(x) = \frac{2}{\sqrt{\pi}}\int_0^x \mathrm{e}^{-t^2}\,dt$$

a. Find $\mathrm{Erf}(0)$, $\lim_{x\to\infty}\mathrm{Erf}(x)$ and $\lim_{x\to-\infty}\mathrm{Erf}(x)$.
b. Plot the graph of $\mathrm{Erf}(x)$.
c. Write $\mathrm{Erf}(x)$ in terms of $\Phi(x)$.

22.28 The *sine integral function* $\mathrm{Si}(x)$ is defined by the integral

$$\mathrm{Si}(x) = \int_0^x \frac{\sin t}{t}\,dt$$

$y = \frac{\sin t}{t}$ y Si(x)

0 x t

The figure to the right show the graphs of the functions $\frac{\sin t}{t}$ and $\mathrm{Si}(x)$. It can be proven (but this is not easy!) that $\lim_{x\to\infty}\mathrm{Si}(x) = \frac{\pi}{2}$.

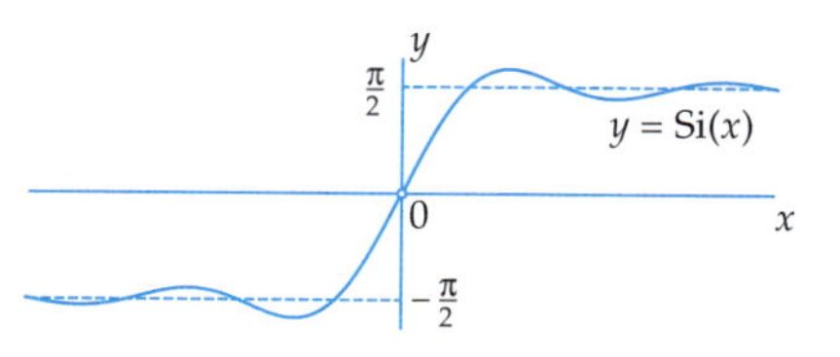


a. Show that $\mathrm{Si}(-x) = -\mathrm{Si}(x)$.
b. Find the $x$-values of the local maximum and minimum values of $\mathrm{Si}(x)$.
c. Calculate $\displaystyle\int_{-\infty}^{\infty}\frac{\sin t}{t}\,dt$.
d. Write $\displaystyle\int_a^b \frac{\sin mt}{t}\,dt$ (where $m$ is a constant) in terms of $\mathrm{Si}(x)$.

## Is finding a formula for primitive functions always possible?

Suppose that we want to calculate the integral $\int_a^b f(x)\,dx$. The integrand $f(x)$ is often given by a formula in terms of standard functions (powers, sines, cosines, exponential functions, logarithmic functions et cetera). We would like to have the primitive function $F(x)$ in such a formula as well, since we could then write down the integral as $F(b) - F(a)$. We have seen many examples in which it was indeed done this way.

However, while the reverse process, *differentiation* of functions in formula form, never leads to problems – the differentiation rules are always applicable – finding a primitive function often leads to serious problems, even if the function $f(x)$ looks innocent.

For instance, take $\varphi(x) = \frac{1}{\sqrt{2\pi}}\mathrm{e}^{-\frac{1}{2}x^2}$. This function occurs frequently in statistics as the probability density function of the standard normal distribution. Its graph is the well-known Gaussian bell curve. The probability that a standard normally distributed stochastic variable takes a value in the interval $[a, b]$ is equal to the integral $\int_a^b \frac{1}{\sqrt{2\pi}}\mathrm{e}^{-\frac{1}{2}x^2}\,dx$. Thus, it would be nice if we had a formula for a primitive function of $\varphi(x)$. However, nobody has succeeded in finding such a formula in terms of well-known standard functions, and this can even be *proven*! Such a formula simply doesn't exist!

However, primitive functions *do* exist, for instance the area function

$$O(x) = \int_0^x \frac{1}{\sqrt{2\pi}}\mathrm{e}^{-\frac{1}{2}t^2}\,dt$$

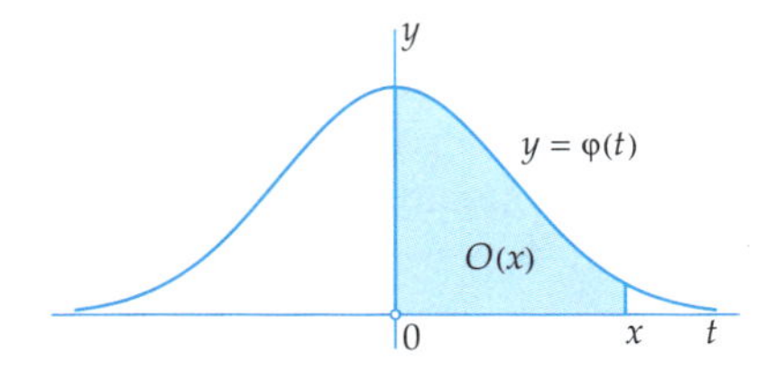


Indeed, since the integrand is obviously continuous, the area function is a primitive function (also see page 201).

Of course, the lower bound 0 of the integral can be replaced by any other number, since the difference is only a constant. In statistics, one usually takes $-\infty$ as the lower bound. The corresponding primitive function is denoted by $\Phi(x)$. In the first exercise on the opposite page, this function is investigated further.

What can be done in such cases, is *numerically* calculate the values of $O(x)$ (or $\Phi(x)$) by means of approximation methods. This is how tables of the distribution function $\Phi(x)$ of the standard normal distribution were compiled in former days. They can still be found in many books on statistics and probability theory. But nowadays, these values are calculated on the fly using suitable software in calculators and computers.

# 23 Applications

23.1 Find the tangent vector to the following parametric curves. If necessary, consult the illustrations of these parameter curves in the exercises in chapter 19 on page 166. Also find out whether there are points on the curve where the tangent vector is the zero vector:

a. $(\cos 3t, \sin 2t)$
b. $(\cos 2t, \sin 3t)$
c. $(\cos^3 t, \sin^3 t)$
d. $(\cos^3 t, \sin t)$
e. $(\cos^3 t, \sin 2t)$
f. $(\cos \frac{1}{2}t, \sin^3 t)$
g. $(\sqrt[3]{\cos t}, \sqrt[3]{\sin t})$
h. $(\sqrt[3]{\cos t}, \sin^3 t)$

23.2 Find the tangent vector to the following curves given in polar coordinates. Take $\varphi$ as parameter; the parametrization is then $(r(\varphi)\cos\varphi, r(\varphi)\sin\varphi)$. If necessary, consult the illustrations of the curves on page 168.

a. $r = \cos\varphi$
b. $r = \cos 2\varphi$
c. $r = \cos 3\varphi$
d. $r = \sin \frac{1}{2}\varphi$
e. $r = \cos \frac{3}{2}\varphi$
f. $r^2 = \cos 2\varphi$
g. $r = 1 + \cos\varphi$
h. $r = 1 + 3\cos 7\varphi$

23.3 The equation $r = e^{c\varphi}$ in polar coordinates is the equation of a logarithmic spiral (see page 169). Show that the radius vector and the tangent vector include a constant angle (i.e. the angle only depends on $c$). Calculate this angle for $c = 1$.

23.4 Find the tangent vector to the following parametric space curves. If necessary, consult the exercises on page 170 for illustrations of these curves.

a. $(t, 2t^2 - 1, t^3)$
b. $(\sin t, \sin 2t, \cos t)$
c. $(\sin t, \sin 2t, \cos 3t)$
d. $(\sin 2\pi t, t, \cos 2\pi t)$
e. $(\sin 2\pi t, t^2 - 1, t^3)$
f. $(\cos t, \sin t, \cos 12t)$

## The tangent vector to a parametric curve

The figure to the right shows the curve with parametrization

$$(x(t), y(t)) = (t^3 - 2t, 2t^2 - 2t - 2.4)$$

The part drawn corresponds to $-1.6 \leq t \leq 2.2$. The points $P = (x(t), y(t))$ and $Q = (x(t+dt), y(t+dt))$ are close to each other on the curve: we took $t = 1.9$ and $dt = 0.1$. Since $dt$ is small, the part of the curve between $P$ and $Q$ is nearly straight; it almost coincides with a part of the tangent line to the curve at the point $P$.

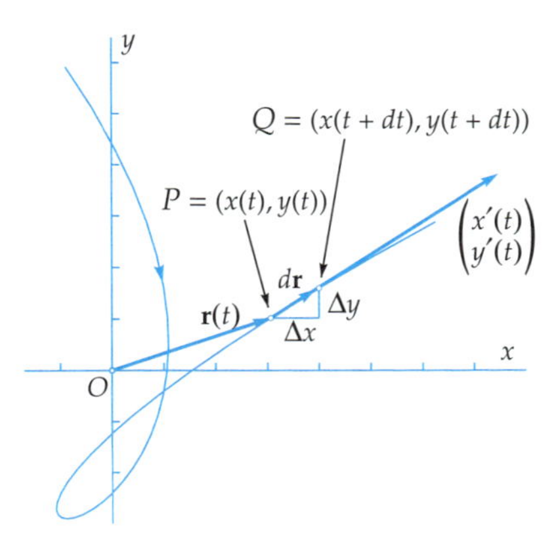


The vector from the origin $O$ to the point $P$ on the curve is usually called the *radius vector*, and denoted by $\mathbf{r}(t)$. The difference vector $\Delta\mathbf{r} = \mathbf{r}(t+dt) - \mathbf{r}(t)$ is the vector from $P$ to $Q$. Its components are $\Delta x = x(t+dt) - x(t)$ and $\Delta y = y(t+dt) - y(t)$. Since $dt$ is small, $\Delta x \approx dx = x'(t)\,dt$ and $\Delta y \approx dy = y'(t)\,dt$ hold, so

$$\Delta\mathbf{r} = \begin{pmatrix} \Delta x \\ \Delta y \end{pmatrix} \approx d\mathbf{r} = \begin{pmatrix} dx \\ dy \end{pmatrix} = \begin{pmatrix} x'(t)\,dt \\ y'(t)\,dt \end{pmatrix} = \begin{pmatrix} x'(t) \\ y'(t) \end{pmatrix} dt$$

The vector $\mathbf{v}(t) = \begin{pmatrix} x'(t) \\ y'(t) \end{pmatrix}$ is called the *tangent vector* at $P$ to the curve. It is usually drawn as an arrow starting at $P$. The tangent vector is then part of the tangent line to the curve at $P$.

The vector $d\mathbf{r}$, with the differentials $dx$ and $dy$ as components, is equal to the tangent vector multiplied by the factor $dt$. If $dt$ is small, $d\mathbf{r}$ almost coincides with $\Delta\mathbf{r}$. In the drawing above, $t = 1.9$, $dt = 0.1$, $P = (3.059, 1.02)$ and $\Delta\mathbf{r} = \begin{pmatrix} 0.941 \\ 0.58 \end{pmatrix}$, $\mathbf{v}(t) = \begin{pmatrix} 8.83 \\ 5.6 \end{pmatrix}$ and $d\mathbf{r} = \begin{pmatrix} 0.883 \\ 0.56 \end{pmatrix} = \mathbf{v}(t)\,dt$ (we have drawn the tangent vector $\mathbf{v}(t)$ at half the size to keep the drawing in reasonable proportions).

In many applications, the parameter $t$ represents time. The radius vector $\mathbf{r}(t)$ then denotes the position of $P$ at time $t$ and the tangent vector $\mathbf{v}(t)$ is the *velocity vector*, the vector that represents the velocity of $P$ at time $t$.

Space curves can be treated in a similar way, but all vectors then have three components instead of two.

23.5 Find the length of the following curves:

a. The circle $(\cos t, \sin t), 0 \leq t \leq 2\pi$.
b. The circle $(R \sin t, R \cos t), 0 \leq t \leq 2\pi$.
c. The helix $(R \cos 2\pi t, R \sin 2\pi t, at), 0 \leq t \leq 1$.
d. The logarithmic spiral $(\mathrm{e}^{c\varphi} \cos \varphi, \mathrm{e}^{c\varphi} \sin \varphi), 0 \leq \varphi \leq 2\pi$.

23.6 The parametrization

$$(x, y) = (t - \sin t, 1 - \cos t)$$

represents a curve that is traced by a point $P$ on the boundary of a circle that rolls over the $x$-axis. Such a curve is called a *cycloid*. The radius of the circle is 1 and its centre $M$ moves according to the parametrization $(t, 1)$ along the line $y = 1$. With respect to $M$, the point $P$ describes the circular movement $(-\sin t, -\cos t)$. The superposition of both movements yields the parametrization of the cycloid.

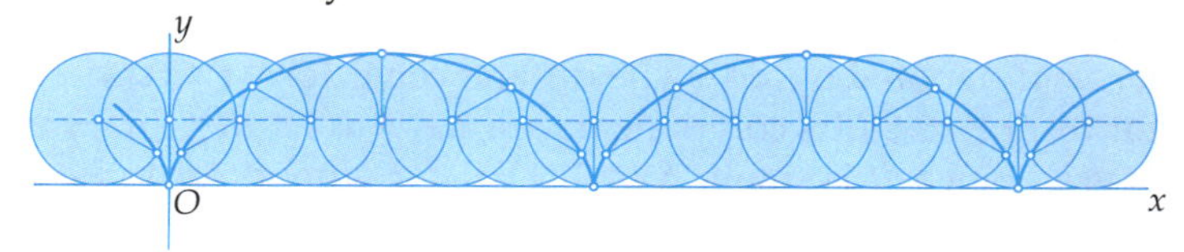


a. Find the coordinates of $P$ for $t = 0, t = \frac{\pi}{2}, t = \pi, t = \frac{3\pi}{2}$ and $t = 2\pi$.

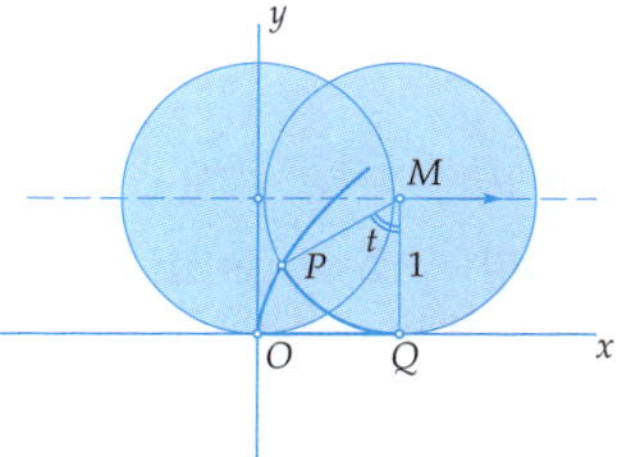


b. Suppose that $Q$ is the point of contact of the circle with the $x$-axis at time $t$. Show that the segment $OQ$ and the circular arc $PQ$ have the same length. (This means that, indeed, the circle rolls over the $x$-axis without slipping.)
c. Find the velocity vector (tangent vector) and the scalar velocity. Simplify the expression for the velocity using suitable trigonometric formulas. Find out whether there are points on the cycloid where the velocity is zero. For which points on the curve is the maximum velocity attained?

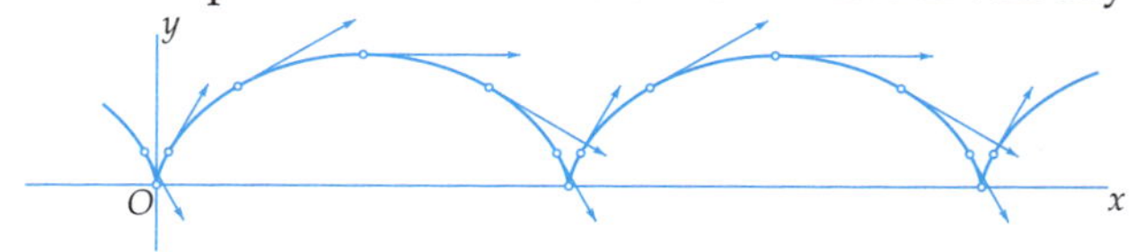


d. Find the length of one arc of the cycloid.

## The length of a curve

Suppose that a planar curve is given with parametrization $(x(t), y(t))$, where $x(t)$ and $y(t)$ are differentiable functions of the variable $t$ on an interval $[a, b]$, and suppose that $x'(t)$ and $y'(t)$ are both continuous functions on $[a, b]$. For each $t$ with $a \leq t \leq b$ the expression $(x(t), y(t))$ represents a point in the plane, and as $t$ runs from $a$ to $b$, the point $(x(t), y(t))$ traces the curve from $A = (x(a), y(a))$ to $B = (x(b), y(b))$. Suppose that along the way, the tangent vector doesn't reverse its direction. The point then always moves in the same direction along the curve. As will be shown, the length $L$ of the curve between $A$ and $B$ is then given by

$$L = \int_a^b \sqrt{(x'(t))^2 + (y'(t))^2}\, dt$$

Let's call $L(t)$ the length of the part of the curve between the points $A = (x(a), y(a))$ and $P = (x(t), y(t))$. For a positive $dt$, the increase $\Delta L = L(t + dt) - L(t)$ is equal to the length of the segment of the curve between the points $P = (x(t), y(t))$ and $Q = (x(t + dt), y(t + dt))$.
For small values of $dt$, the length $\Delta L$ is nearly equal to the distance $d(P, Q)$. According to Pythagoras' theorem, this is equal to $d(P, Q) = \sqrt{(\Delta x)^2 + (\Delta y)^2}$.

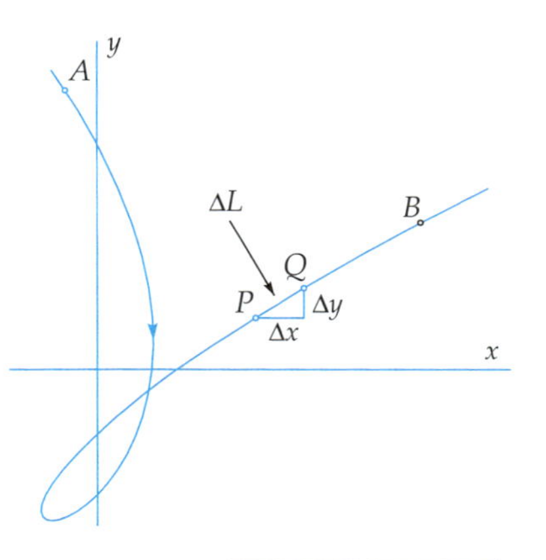


Since $\Delta x \approx x'(t)dt$ and $\Delta y \approx y'(t)dt$, we have $d(P, Q) \approx \sqrt{(x'(t))^2 + (y'(t))^2}\, dt$. This expression is a differential: the differential $dL$ of the function $L(t)$. Hence, the length of the curve between $A$ and $B$ is indeed given by the integral above.

If $t$ represents time, the tangent vector $\mathbf{v}(t) = \begin{pmatrix} x'(t) \\ y'(t) \end{pmatrix}$ is the velocity vector of $P$ at time $t$. The length $|\mathbf{v}(t)| = \sqrt{(x'(t))^2 + (y'(t))^2}$ is then called the *scalar velocity* of $P$ and denoted by $v(t)$. It is clear that the distance traced by $P$ between $t = a$ and $t = b$ is given by the integral $\int_a^b v(t)dt$.

The treatment of space curves is completely analogous. All vectors then have three coordinates, $v(t) = \sqrt{(x'(t))^2 + (y'(t))^2 + (z'(t))^2}$ and

$$L = \int_a^b v(t)\, dt = \int_a^b \sqrt{(x'(t))^2 + (y'(t))^2 + (z'(t))^2}\, dt$$

23.7 The volume of a right circular cone with height $h$ and radius of its base circle $r$ is equal to $\frac{1}{3}\pi r^2 h$ ($\frac{1}{3}\times$ base area $\times$ height). Verify this formula by calculating the volume of the solid of revolution that results when rotating the graph of the function $z = \frac{r}{h}y$, $(0 \leq y \leq h)$ around the $y$-axis.

23.8 The graph of the function $f(x) = \sin x, 0 \leq x \leq \pi$, is rotated around the $x$-axis. Calculate the volume of the resulting solid of revolution.

23.9 The graph of the function $f(x) = x^2, -1 \leq x \leq 1$, is rotated around the $y$-axis. Calculate the volume of the resulting solid of revolution.

23.10 The graph of the function $f(x) = x^4, -1 \leq x \leq 1$, is rotated around the $y$-axis. Calculate the volume of the resulting solid of revolution.

23.11 The graph of the function $f(x) = \frac{1}{x}, 1 \leq x < \infty$, is rotated around the $x$-axis. Calculate the volume of the resulting solid of revolution.

23.12 The graph of the function $f(x) = \frac{1}{\sqrt{x}}, 1 \leq x < \infty$, is rotated around the $x$-axis. Calculate the volume of the resulting solid of revolution.

23.13 The graph of the function $f(x) = \mathrm{e}^{-x}, 0 \leq x < \infty$, is rotated around the $x$-axis. Calculate the volume of the resulting solid of revolution.

23.14 The graph of the function $f(x) = \ln 2x, 0 \leq x \leq 1$, is rotated around the $y$-axis. Calculate the volume of the resulting solid of revolution.
*Hint:* interchange $x$ and $y$ and use the inverse function.

23.15 The planar region $G$ that is bounded by the parabolas $y = x^2$ and $x = y^2$ is rotated around the $x$-axis. Calculate the volume of the resulting solid of revolution.

23.16 Calculate the volume of the part of the sphere $x^2 + y^2 + z^2 \leq R^2$ between the planes $z = h$ and $z = R$ (with $-R \leq h \leq R$).

## The volume of a solid of revolution

Suppose that the function $z = f(y)$ is continuous and non-negative on the interval $[a, b]$. Let the solid that is bounded by the planes $y = a$, $y = b$ and the surface resulting from rotating the graph of this function around the $y$-axis, be called $K$. In the figure below, only the part of $K$ that is situated in the first octant is drawn. (The first octant in space is given by $x \geq 0, y \geq 0$ and $z \geq 0$.) What is the volume of $K$?

Choose a number $y$ between $a$ and $b$. If the volume of the part of $K$ to the left of the vertical plane through $(0, y, 0)$ is called $I(y)$, the requested volume of $K$ is equal to $I(b)$.

For a small positive $dy$, the increase $\Delta I = I(y + dy) - I(y)$ is equal to the thin slice of $K$ between the vertical planes through the points $(0, y, 0)$ and $(0, y + dy, 0)$.
For a small positive $dy$, that slice almost coincides with the drawn thin cylindrical slice with thickness $dy$ and circles with radius $f(y)$ as its vertical bounding planes.

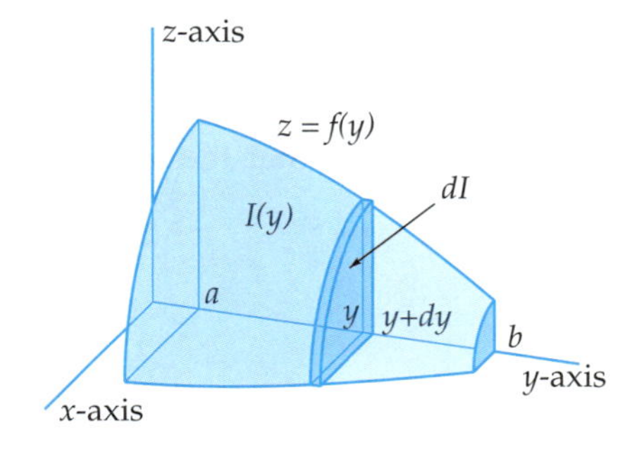


These circles have an area $\pi \times f(y)^2$, so the volume of the cylindrical slice is equal to $\pi f(y)^2 dy$. The expression $\pi f(y)^2 \, dy$ is a differential, namely the differential $dI$ of the volume function $I(y)$, so $dI = \pi f(y)^2 \, dy$. Therefore, the volume of $K$ is given by the integral

$$\text{Volume}(K) = I(b) = \int_a^b \pi f(y)^2 \, dy$$

If $G(y)$ is an arbitrary primitive function of $\pi f(y)^2$, this integral is equal to $G(b) - G(a)$.

For instance, take the function $z = \sqrt{R^2 - y^2}$ on the interval $[-R, R]$. Its graph is a semicircle with radius $R$, and the corresponding solid of revolution is the sphere with radius $R$ and the origin as its centre. Hence, its volume is given by

$$\int_{-R}^{R} \pi (R^2 - y^2) \, dy$$

A primitive function of $\pi(R^2 - y^2)$ is $G(y) = \pi(R^2 y - \frac{1}{3} y^3)$, so the volume of the sphere is $G(R) - G(-R) = \frac{4}{3} \pi R^3$.

23.17 Show that the surface area of the curved part of a right circular cone with height $h$ and base circle of radius $r$ is equal to $\pi r \sqrt{r^2 + h^2}$.

23.18 Show that the surface area of a sphere with radius $R$ is equal to $4\pi R^2$.

23.19 Find the surface area of the part of the sphere $x^2 + y^2 + z^2 = R^2$ between the planes $z = h$ and $z = R$ (with $-R \leq h \leq R$).

23.20 The parabola segment $y = x^2$ with $-1 \leq x \leq 1$ is rotated around the $y$-axis. Find the surface area of the resulting solid of revolution.

23.21 The graph of the function $f(x) = \dfrac{1}{x}$ with $1 \leq x < \infty$ is rotated around the $x$-axis. Show that the surface area of the resulting solid of revolution is infinite.

## The surface area of a solid of revolution

Suppose that the function $z = f(y)$ is non-negative and differentiable on the interval $[a,b]$. We now look at the *surface area* of the solid of revolution resulting from rotating the graph of this function around the $y$-axis. We will show that this area is equal to the integral

$$\int_a^b 2\pi f(y)\sqrt{1+(f'(y))^2}\,dy$$

Choose $y$ between $a$ and $b$. Let the surface area of the part located to the left of the vertical plane through $(0,y,0)$ be called $O(y)$. The requested surface area is then equal to $O(b)$.
For small positive values of $dy$, the increase $\Delta O = O(y+dy) - O(y)$ is equal to the area of the thin dark coloured ribbon between the vertical planes through the points $(0,y,0)$ and $(0,y+dy,0)$.

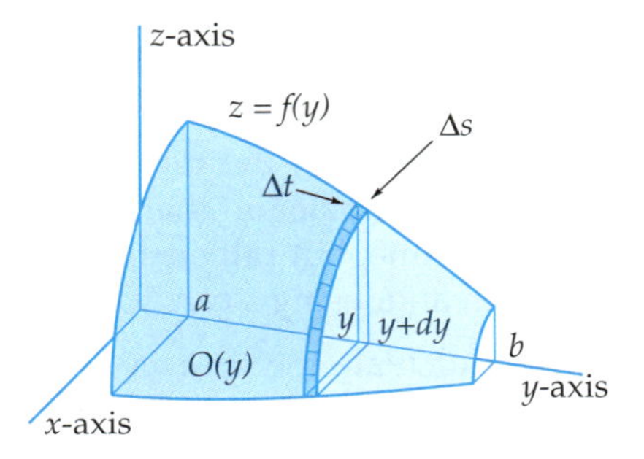


Divide this ribbon into $N$ equal parts, as shown in the figure. There, $N = 40$ is taken, giving ten parts in the first octant. Each part is nearly a rectangle. Let its sides be called $\Delta s$ and $\Delta t$ as shown. Since the left-hand boundary of the dark ribbon is a circle with radius $f(y)$, its perimeter is $2\pi f(y)$. Hence, $\Delta t = \frac{2\pi}{N} f(y)$ holds. For small values $dy$, the arc $\Delta s$ is nearly a straight line segment with length

$$\sqrt{dy^2 + (f(y+dy) - f(y))^2} \approx \sqrt{dy^2 + (f'(y)dy)^2} = \sqrt{1+f'(y)^2}\,dy$$

Each part therefore has an area of approximately $\Delta t \times \Delta s$. In total, there are $N$ parts, so the dark ribbon has an area that is approximately equal to

$$2\pi f(y)\sqrt{1+(f'(y))^2}dy$$

It can be proven that this is the differential $dO$ of $O(y)$. It follows that the surface area of the solid of revolution is indeed equal to the integral above.

If the curve that is rotated around the $y$-axis, is given by the parametrization $(y(t), z(t))$, where the parameter $t$ runs through an interval $[c,d]$, then the surface area of the solid of revolution is equal to

$$\int_c^d 2\pi z(t)\sqrt{(y'(t))^2+(z'(t))^2}\,dt$$

23.22 In the example on the opposite page, $\lambda = 0.3$ and $P_0 = 1300$. Using a calculator, find the time $t$ for which $P(t) = 2000$ holds.

23.23 Show that in each exponential growth model, i.e. each model for which the solution functions $P = P(t)$ are given by a differential equation that can be written as

$$dP = \lambda P \, dt$$

a *doubling time* $t_d$ exist, i.e. a number $t_d$ satisfying $P(t + t_d) = 2P(t)$ for each $t$. Express the doubling time $t_d$ in terms of $\lambda$. (We assume $\lambda > 0$.)

23.24 Let $P(t)$ be a solution function of an exponential growth model with doubling time $t_d = 5$. Suppose that $P_0 = 100$. Calculate the time $t$ for which $P(t) = 1\,000\,000$ using a calculator. Take a wild guess first.

23.25 For $\lambda < 0$, the differential equation $dP = \lambda P \, dt$ models *exponential decline*, a phenomenon that occurs for, e.g., radioactive decay. Here, not the doubling time, but rather the *halving time* is used. Explain what is meant by this term and, using a calculator, calculate the halving time for $\lambda = -0.2$.

23.26 Calculate the time needed for exponential decline with halving time $t_h = 3$ until a magnitude $P_0 = 100$ is reduced to $P = 0.001$. Use a calculator. Again, take a wild guess first.

23.27 The differential equation

$$\frac{1}{P^{1+a}} \, dP = \lambda \, dt$$

(with $\lambda > 0$ and $a > 0$) is called a *doomsday equation*. For small positive values $a$, this differential equation may be seen as a variation of the differential equation for exponential growth. The solution functions, however, behave quite differently, as will be shown.

a. Show that the differential equation can be written as $d(P^{-a}) = d(-a\lambda t)$.
b. Show that all the solutions can be written as $P(t) = (a\lambda(T - t))^{-1/a}$ for a certain constant $T$. In particular, for $t = 0$ we have $P_0 = P(0) = (a\lambda T)^{-1/a}$.
c. Express $T$ in terms of $a$, $\lambda$ and $P_0$ and show that $\lim_{t \uparrow T} P(t) = \infty$. This is the reason why time $T$ is called *doomsday*.
d. Calculate *doomsday* using a calculator for $a = 0.2$, $\lambda = 0.05$ and $P_0 = 100$ and also find a formula for $P(t)$ in this case.

## Exponential growth

Many applications use mathematical models for growth processes, for instance when you study populations varying in size with time. Often, a differentiable function $P(t)$ is used whichs models the size of the population at time $t$. In one of the most simple models, we assume that for small increases $dt$ of time the *relative* population growth $\dfrac{\Delta P}{P} = \dfrac{P(t+dt)-P(t)}{P(t)}$ is approximately proportional to $dt$. This means that there is a constant $\lambda$ such that $\dfrac{\Delta P}{P} \approx \lambda\, dt$. This leads to the *differential equation*

$$\frac{1}{P}\, dP = \lambda\, dt$$

as a mathematical model. This must be satisfied by the differentiable function $P = P(t)$, representing the population size at time $t$.

The left-hand side of the differential equation is equal to the differential $d(\ln P)$ and the right-hand side is equal to the differential $d(\lambda t)$. These differentials are equal, so for a certain constant $c$ we must have

$$\ln P(t) = \lambda t + c$$

or, with $P_0 = \mathrm{e}^{\,c}$,

$$P(t) = \mathrm{e}^{\,\lambda t + c} = P_0\, \mathrm{e}^{\,\lambda t}$$

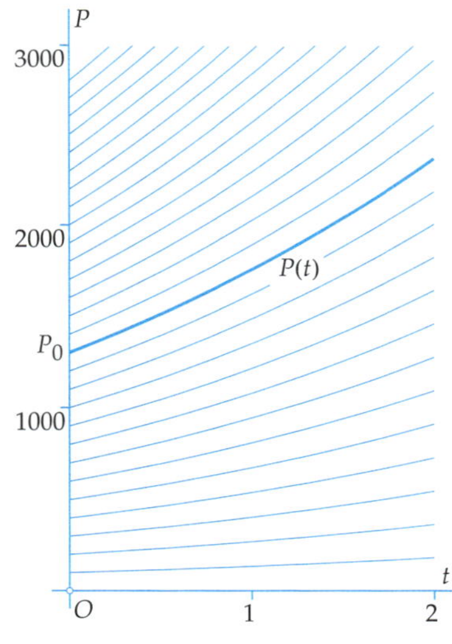


This model is called an *exponential growth model*. Apparently, $P_0$ represents the size of the population at time $t = 0$ (substitute $t = 0$ into the formula).
In the situation drawn in the figure, the growth rate $\lambda$ was taken as $\lambda = 0.3$. The graph of the function $P(t)$ with $P_0 = 1300$ is drawn as a thick line. For a number of other values of $P_0$ the graphs are drawn as thin lines.

The value of $\lambda$ is determined by external circumstances, e.g. fertility, food supply and so on. After $\lambda$ has been chosen, the starting value $P_0$ completely determines the solution function $P(t)$.

23.28 In the example on the opposite page, the line element field of the logistic differential equation

$$dP = \mu(M - P)P\,dt$$

is given for $M = 4000$ and $\mu = 0.0004$. The scales on the axes are not equal: one unit on the $t$-axis corresponds with one thousand units on the $P$-axis.

a. Show that all line elements on a horizontal line $P = c$ have the same direction.

b. Calculate the tangent of the slope angle of a line element on the line $P = 2000$. (N.B.: since the scales are not equal, the tangent is not equal to the derivative $\frac{dP}{dt}$.)

c. Calculate the tangent of the slope angle of the line elements on the lines $P = 1000$ and $P = 3000$. Compare these calculated values with the tangents of the angles in the drawing (use a protractor).

d. Plot a rough sketch of the line element field in the region of the $(t, P)$-plane for which $P > 4000$ holds. Can you explain this in terms of the growth model?

e. Plot a rough sketch of the line element field in the region of the $(t, P)$-plane for which $P < 0$ holds.

23.29 For each line element field drawn below, find the corresponding differential equation. Explain your answers:

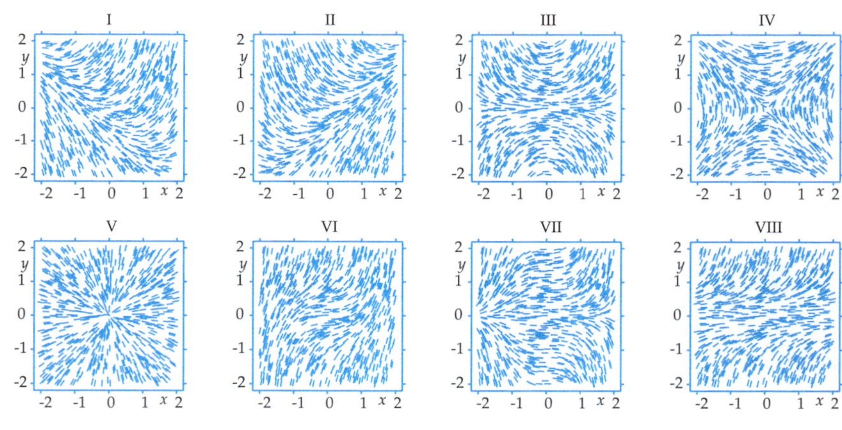


a. $y\,dy = x\,dx$

b. $dy = xy\,dx$

c. $dy = y^2\,dx$

d. $dy = (x^2 + y^2)\,dx$

e. $dy = (x - y)\,dx$

f. $dy = (x + y)\,dx$

g. $dy = x^2 y\,dx$

h. $x\,dy = y\,dx$

## Logistic growth – the line element field

For any positive $\lambda$, the differential equation $dP = \lambda P dt$ models exponential growth. The population size $P(t)$ at time $t$ is then given by the formula $P(t) = P_0\,e^{\lambda t}$, where $P_0$ represents the population size at time $t = 0$.

For big values of $t$, however, such growth models cannot be realistic, since the exponential function goes to infinity very fast. Food shortage and other restrictions will slow down the growth. We are now looking for models where this also is expressed in a reasonable way. The simplest solution is to let the growth factor $\lambda$ no longer be a constant, but make it dependent on the population size $P$. As $P$ approximates a 'saturation value' $M$, the growth has to decrease to zero.

The function $\lambda(P) = \mu(M - P)$ satisfies this condition if $\mu$ is positive. The modified model is then given by the differential equation

$$dP = \mu(M - P)P\,dt$$

where $P = P(t)$ is a (still unknown) function of time $t$ and $\mu$ and $M$ are positive constants that depend on the specific circumstances. This growth model is known as *logistic growth*.

As an example, we choose $M = 4000$ and $\mu = 0.0004$. To get an idea of the solution functions $P(t)$, we sketch the *line element field* with, in this case, $0 \leq t \leq 10$ and $0 \leq P \leq 4000$.

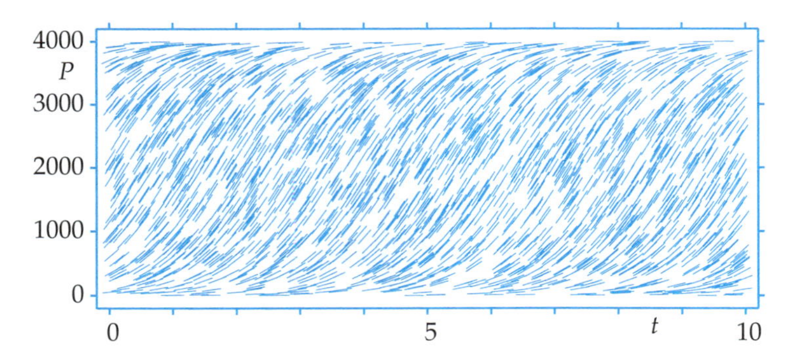


In this region we have chosen (using a computer) 2500 points $(t, P)$ *at random*. In each point, a small line is drawn with slope $\frac{dP}{dt} = \mu(M - P)P$. Just as iron filings make the field lines of a magnetic field visible, the line element field makes the solution curves visible, i.e. the graphs of the solution functions $P = P(t)$ of the differential equation.

23.30 Suppose that $P(t)$ is the solution function of the logistic differential equation

$$dP = \mu(M - P)P\,dt$$

with $M = 4000$ and $\mu = 0.0004$ (as in the figure on the opposite page) satisfying $P(0) = 2000$.

a. Calculate for which $t$ the function value $P(t) = 3000$ is reached.
b. Calculate for which $t$ the function value $P(t) = 1000$ was reached.
c. Show that $P(-t) + P(t) = M$. Wat does this mean geometrically?

23.31 Show that for each solution function $P(t)$ of the logistic differential equation $P(t_0 + t) + P(t_0 - t) = M$ holds. Here (as on the opposite page), $t_0$ is the time for which $P(t_0) = \frac{1}{2}M$. Verify that this means that the graph of $P(t)$ is point symmetric with respect to the point $(t_0, \frac{1}{2}M)$.

23.32 To derive the formula for the solution functions of the logistic differential equation, we used $0 < P < M$, in other words, we assumed that the population size $P$ is positive and smaller than the saturation value $M$.

Now assume that $P > M$. Verify that, in this case,

$$\frac{d}{dP}\left(\ln\frac{P}{P-M}\right) = \frac{M}{(M-P)P}$$

can be used to find a formula for the solution functions, and find such a formula. In particular, show that the solution function $P(t)$ satisfying the condition $P(0) = 2M$ is given by the formula $P(t) = \dfrac{2M}{2 - \mathrm{e}^{-\mu M t}}$. Also calculate $\lim_{t\to\infty} P(t)$.

23.33 Find equations for all the solution curves of the following differential equations. As on the opposite page, use the method of *separation of the variables*, i.e. bring all the expressions involving $x$ to one side, and all the expressions involving $y$ to the other side of the equality sign:

a. $y\,dy = x\,dx$
b. $x\,dy = y\,dx$
c. $dy = xy\,dx$
d. $dy = y^2\,dx$
e. $dy = x^2 y\,dx$

## Logistic growth – the solution functions

The line element field of the logistic growth model described by the differential equation

$$dP = \mu(M - P)P\,dt$$

provides a good qualitative insight into the behaviour of the solution curves, but in this case it is also possible to find an exact formula for the solution functions $P = P(t)$.

As a first step, we write the differential equation in the form

$$\frac{M}{(M-P)P}\,dP = \mu M\,dt$$

It is easy to verify that $\dfrac{d}{dP}\left(\ln\dfrac{P}{M-P}\right) = \dfrac{M}{(M-P)P}$. So we can write the differential equation as

$$d\left(\ln\frac{P}{M-P}\right) = d(\mu M t)$$

from which it follows that $\ln\dfrac{P}{M-P} = \mu M t + c$ for some constant $c$. Putting $c = -\mu M t_0$ and solving $P$ from this equation yields:

$$P = P(t) = \frac{M\mathrm{e}^{\mu M(t-t_0)}}{\mathrm{e}^{\mu M(t-t_0)} + 1} = \frac{M}{1 + \mathrm{e}^{-\mu M(t-t_0)}}$$

For $t = t_0$, the population size $P(t)$ equals $M/2$, half the saturation value $M$. The graph of $P(t)$ looks like an elongated letter S with $\lim_{t\to-\infty} P(t) = 0$ and $\lim_{t\to\infty} P(t) = M$.

Below, some of the solution curves are sketched, again with $M = 4000$ and $\mu = 0.0004$, with the line element field as a background.

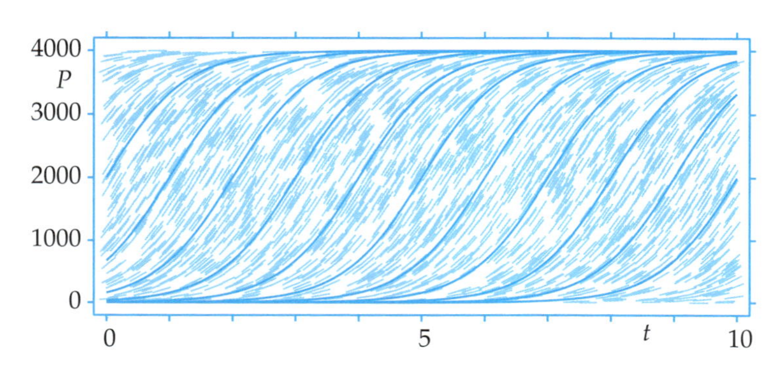

# VIII Mathematical Background

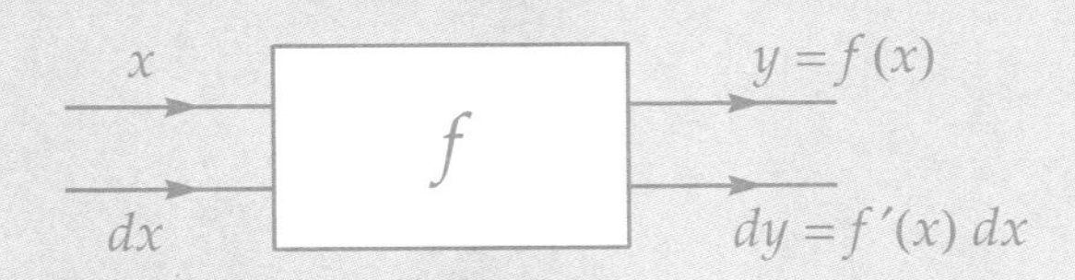


In this part, we present – without exercises – additional background material to a number of subjects covered in the previous chapters. You may consult them as required. Among others, we discuss the real number line, various kinds of intervals, the symbols $\infty$ and $-\infty$, coordinate systems in the plane and in space, the function concept, graphs, limits and continuity. In the last chapter, we present proofs for the various rules for differentiation and the formulas for the derivatives of the standard functions.

# 24 Real numbers and coordinates

## The real number line

The set of all real numbers is denoted by $\mathbb{R}$. The positive real numbers are the numbers greater than 0, the negative real numbers are the numbers less than 0. The *absolute value* $|r|$ of a real number $r$ is equal to $r$ if $r \geq 0$, and equal to $-r$ if $r < 0$.

A geometric image of the set of all real numbers is obtained by choosing two different points on a straight line, calling them 0 and 1, and then positioning all the other numbers accordingly (see below). You can imagine that the line continues both ways, without bounds. This yields the *real number line*, a line where each point corresponds to a real number. Apart from the integer numbers

$$\ldots, -5, -4, -3, -2, -1, 0, 1, 2, 3, 4, 5, \ldots$$

and the non-integer rational numbers (the numbers that can be written as a fraction), the real number line also contains irrational numbers, e.g. the numbers $\sqrt{2}$, e and $\pi$.



*The real number line*

Since we can identify real numbers with points on the real number line, we often speak (somewhat sloppy) of 'the point $r$' instead of 'the real number $r$'.

Each real number can be written as a finite or infinite decimal progression. For example:

$$\begin{aligned}
\sqrt{2} &= 1.41421356237309504 88\ldots \\
\frac{3}{16} &= 0.1875 \\
\pi &= 3.14159265358979323 85\ldots \\
\frac{22}{7} &= 3.14285714285714285 71\ldots \\
\mathrm{e} &= 2.71828182845904523 54\ldots \\
\frac{271801}{99990} &= 2.71828182818281828 18\ldots
\end{aligned}$$

## The braces notation for sets

It is often convenient to describe sets of real numbers using the *braces notation*. The elements of such a set (the numbers that are included in the set) then are enumerated between a pair of braces, or they are described between the braces in words or by a formula. For instance,

$$A = \{1, 2, 3, 4, 5, 6\}$$

is the set of all possible outcomes of throwing a dice. And

$$B = \{x \in \mathbb{R} \mid x > 0\}$$

is the set of all positive real numbers. In the latter notation, the symbol $\in$ means 'is element of'. To the left of the vertical bar it says that the set $B$ consists of real numbers, and to the right of the vertical bar the conditions are given for the numbers to belong to $B$. The notation thus can be read as: '$B$ is the set of all $x$ that are element of $\mathbb{R}$ and that satisfy $x > 0$'.

## Intervals

An *interval* is a set of real numbers corresponding to an uninterrupted part of the real number line. We distinguish the following kinds of intervals (we always assume that $a$ and $b$ are real numbers with $a < b$):

| *Notation:* | *Braces form:* | *name:* |
|---|---|---|
| $\langle a, b \rangle$ | $\{x \in \mathbb{R} \mid a < x < b\}$ | bounded open interval |
| $[a, b]$ | $\{x \in \mathbb{R} \mid a \leq x \leq b\}$ | bounded closed interval |
| $\langle a, b]$ | $\{x \in \mathbb{R} \mid a < x \leq b\}$ | bounded half-open interval |
| $[a, b\rangle$ | $\{x \in \mathbb{R} \mid a \leq x < b\}$ | bounded half-open interval |
| $[a, \infty\rangle$ | $\{x \in \mathbb{R} \mid a \leq x\}$ | unbounded interval |
| $\langle a, \infty\rangle$ | $\{x \in \mathbb{R} \mid a < x\}$ | unbounded open interval |
| $\langle -\infty, b]$ | $\{x \in \mathbb{R} \mid x \leq b\}$ | unbounded interval |
| $\langle -\infty, b\rangle$ | $\{x \in \mathbb{R} \mid x < b\}$ | unbounded open interval |
| $\langle -\infty, \infty\rangle$ | { all real numbers } | unbounded open interval |

The first four intervals are *bounded* intervals. Their *length* in each case is equal to $b - a$. The other intervals all are unbounded; they have 'infinite length'. The symbols $\infty$ and $-\infty$ (which *do not* belong to $\mathbb{R}$!) are pronounced 'infinity' and 'minus infinity'. Sometimes, $+\infty$ ('plus infinity') is used instead of $\infty$.

An *open neighbourhood* of a point $r \in \mathbb{R}$ is an open interval $\langle a, b\rangle$ containing $r$, so with $a < r < b$.

## Mathematics and reality

The real number line is a good illustration of the tension that exists between reality and the mathematics that is used to get a grip on it. Mathematics is always an idealization, and the same holds true for the real number line. Notwithstanding its name (*real* number line), it is a mathematical model, an ideal image, invented by humans.

In reality, a straight line is never completely straight, and never of unbounded length. In reality, we see a stroke drawn by a pencil on a piece of paper, or a material straightedge, or a tape measure divided into centimeters, inches or whatever other units of measurement used. On such a straightedge or tape measure there is almost no difference between $\pi$ and $22/7$, or between e and $271801/99990$ (see page 247).

Nevertheless, if we apply mathematics in real-life situations, we often prefer to work in an idealized mathematical model, since it gives us complete control over all derivations, formulas and calculations: they are 'exact'. Applying mathematics can often be divided into four distinct steps:

1. Choose a mathematical model.
2. In the model, perform derivations and calculations.
3. Interpret the results in terms of the real-life situation.
4. Investigate whether these interpreted results are good approximations of the observations.

Steps 1, 3 and 4 belong to the domain of the person who applies the mathematical tools. He or she should have a thorough knowledge of the field of application, and, moreover, should know what kind of mathematical models can be used in the situation at hand. This book almost exclusively deals with step 2. It describes the tool kit that everyone applying mathematics at university or college level should have mastered during his or her education.

## Coordinates in the plane

In the plane, a *coordinate system* can be established in the following way. Choose a point $O$, called the *origin*, and two straight lines through $O$, the *coordinate axes*. Choose scales on the axes to make them real number lines where the origin corresponds to zero. Through any point $P$, draw lines parallel to the axes. The coordinates of $P$ are now the real numbers corresponding to the intersection points of the lines through $P$ with the coordinate axes.

Usually, the coordinate axes are chosen perpendicular. The coordinate system is then called *rectangular* (or *orthogonal*). In most cases, the first axis is drawn horizontally, and since the corresponding first coordinate usually is denoted by the letter $x$, the axis is then called the $x$-axis. The vertical axis is usually called the $y$-axis, and the coordinate system is called an $Oxy$-system.

Furthermore, it is usual to choose the scale on the $x$-axis ascending from left to right, and from the bottom up on the $y$-axis. All these conventions, however, are not rigid. Sometimes, other letters are used, and sometimes the axes are drawn differently.

If $x$ and $y$ are the two coordinates of a point $P$, they are denoted by $(x, y)$. One frequently writes $P = (x, y)$, identifying the point $P$ with its coordinate pair. Because the coordinates of each point consist of two real numbers, the system is called $\mathbb{R}^2$ (pronounced R-two). Again, this is a mathematical idealization: one imagines the plane unbounded into all directions, and imagines that each pair of real numbers $(x, y)$ corresponds to exactly one point in this idealized plane.

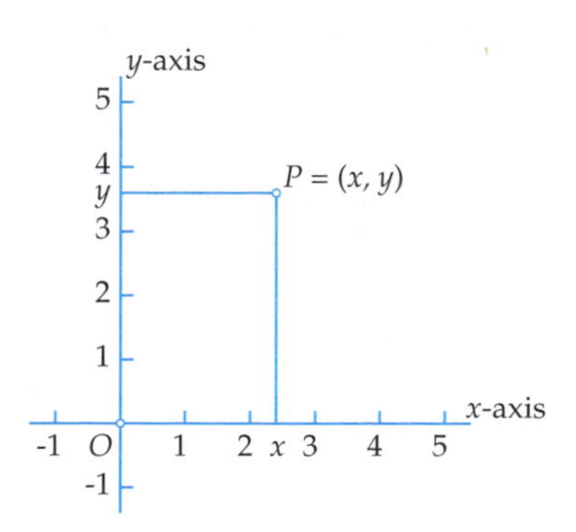


When drawing a graph of a function, it might be convenient to choose unequal scales on the $x$-axis and the $y$-axis (see, e.g., the graph on page 135). In geometric situations, however, the scales are usually equal. In that case, the coordinate system is called *orthonormal* or *cartesian*, after René Descartes (1596-1650), also called Cartesius, one of the pioneers of working with coordinates in geometry.

In an orthonormal coordinate system, the distance $d(P_1, P_2)$ of the points $P_1$ and $P_2$ with coordinates $(x_1, y_1)$ and $(x_2, y_2)$, respectively, is given by $d(P_1, P_2) = \sqrt{(x_1 - x_2)^2 + (y_1 - y_2)^2}$ (by Pythagoras' theorem).

## Pythagoras' theorem

The Greek Pythagoras lived around 500 BC, originally on Samos and later in southern Italy. The theorem that bears his name, however, is much older. We know from clay tablets that the Babylonians in Mesopotamia (present-day Iraq) already knew it around 1800 BC.

**Theorem of Pythagoras:** In any right-angled triangle with sides $a$ and $b$, the hypotenuse $c$ satisfies $a^2 + b^2 = c^2$.

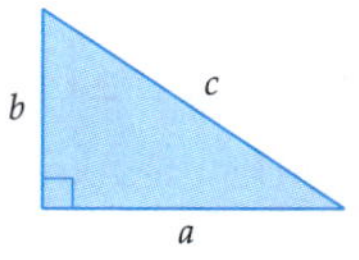


**Proof:** Draw eight copies of the given right-angled triangle, four white and four blue ones, inside a square with sides $a + b$ as shown in the leftmost figure below. The small square in the middle is coloured blue as well. This yields a tilted blue square in that figure with sides of length $c$, so its area equals $c^2$. Interchange the two upper blue triangles with the two lower white triangles (middle figure). The resulting blue region consists of two neighbouring squares: one with sides $b$ and one with sides $a$ (see the rightmost figure). It follows that $c^2$, the area of the blue square in the left figure, equals $a^2 + b^2$, the sum of the areas of the two blue squares in the right figure. End of proof!

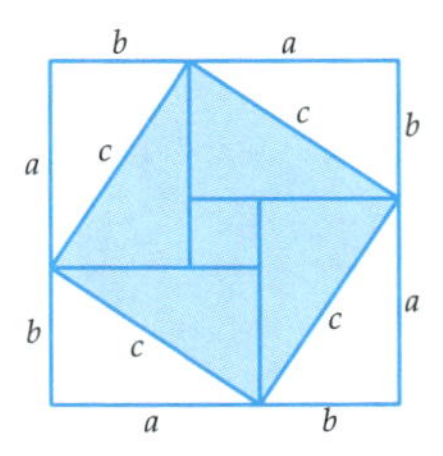


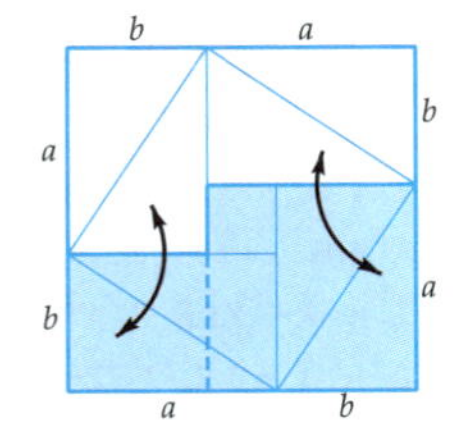


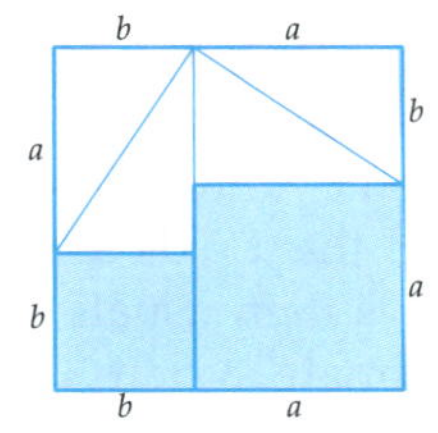


In many old texts, Pythagoras' theorem is formulated as follows:

> **Theorem of Pythagoras (rectangle version):** In any rectangle with sides $a$ and $b$, the diagonal $c$ satisfies $a^2 + b^2 = c^2$.

In space, a similar theorem holds:

> **Theorem of Pythagoras (three-dimensional version):** In any rectangular parallelepiped with edges $a$, $b$ and $c$, the solid diagonal $d$ satisfies $a^2 + b^2 + c^2 = d^2$.

To prove the three-dimensional version, just apply the planar version twice: first in the diagonal plane through a vertical edge $c$ and a solid diagonal $d$. This is a rectangle with diagonal $d$. If the horizontal side of this rectangle is $e$, then $d^2 = e^2 + c^2$ holds, while $e$ itself satisfies $e^2 = a^2 + b^2$. Combining both results yields $d^2 = a^2 + b^2 + c^2$, as desired.

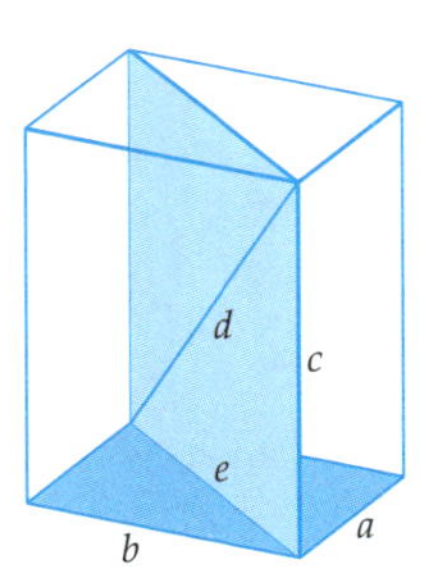


## Coordinates in space

In space, a coordinate system can be established as follows. Choose a point $O$, the *origin*, and three straight lines through $O$, not in the same plane: the *coordinate axes*. Choose scales on the axes making them real number lines with the origin at zero. Each pair of axes determines a plane through $O$. This yields three planes: the *coordinate planes*.

Through any point $P$, choose planes parallel to the coordinate planes. The coordinates of $P$ then are the real numbers corresponding to the intersection points of these planes with the coordinate axes. The three-dimensional space in which such a coordinate system is chosen, is called $\mathbb{R}^3$ (pronounced R-three).

The coordinate axes are usually chosen perpendicular. The coordinate system is then called *rectangular* (or *orthogonal*). Usually, the first two axes are drawn horizontally, and since the corresponding coordinates are frequently designated by the letters $x$ and $y$, the axes then are called the $x$-axis and the $y$-axis. The vertical axis is then usually called the $z$-axis, and the coordinate system is then called an $Oxyz$-system. If the scales on the three axes are equal, the system is called *orthonormal* or *cartesian*.

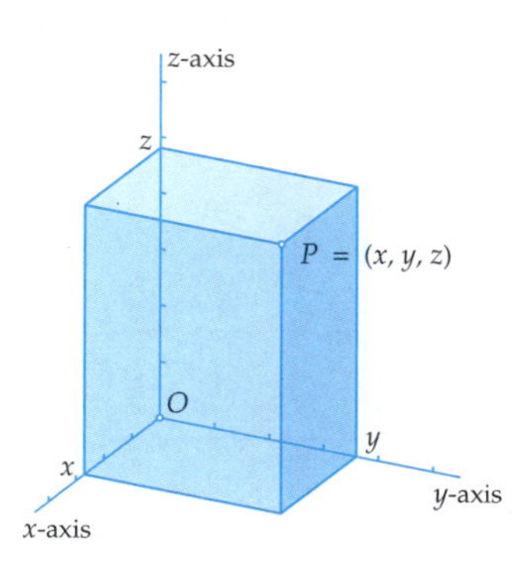


In an orthonormal coordinate system in space, the distance $d(P_1, P_2)$ of the points $P_1$ and $P_2$ with coordinates $(x_1, y_1, z_1)$ and $(x_2, y_2, z_2)$, respectively, is given by

$$d(P_1, P_2) = \sqrt{(x_1 - x_2)^2 + (y_1 - y_2)^2 + (z_1 - z_2)^2}$$

on account of the three-dimensional version of Pythagoras' theorem.

# 25 Functions, limits and continuity

## Function, domain and range

A *function from* $\mathbb{R}$ *to* $\mathbb{R}$ is a rule that transforms real numbers into real numbers in a well-defined manner. This can be done by a formula, a description in words, or any other way. The function concept can be illustrated by the *input output model* below. It is a kind of *black box* that accepts real numbers as input, and for each accepted number produces a real number as output.

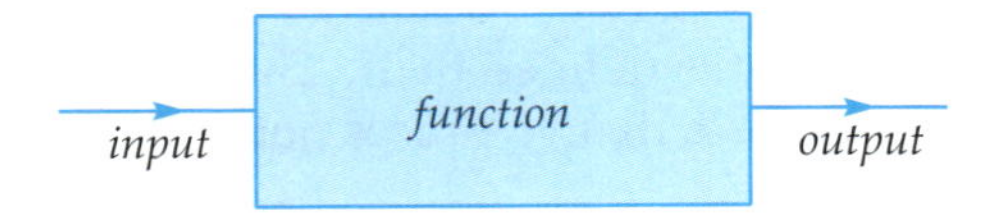


Often, the input is denoted by a letter, for instance $x$, and the resulting output by a different letter, for instance $y$. To show that $y$ is the output corresponding to $x$, one writes $y = f(x)$, where $f$, the *function symbol*, stands for what happens in the *black box*. The number $f(x)$ is called the *function value* belonging to $x$.

Instead of the letter $f$ any other letter or symbol can be chosen. An example is the square root function, given by $y = \sqrt{x}$. As input $x$, any real number $x \geq 0$ can be taken; the corresponding output $y$ then is the square root of $x$. And, of course, instead of the letters $x$ and $y$ any other letters may be taken, provided they are not already used in a different meaning.

In many cases, not all real numbers can be used as input of a given function. For instance, the square root function only accepts numbers greater than or equal to zero as input. In general, one calls the set of all real numbers that are accepted as input, the *domain* of the function. The set of all real numbers that can be generated as output, is called the *range* of the function. Examples:

| *Function:* | *Domain:* | *Range:* |
|---|---|---|
| $f(x) = \sqrt{x}$ | $[0, \infty\rangle$ | $[0, \infty\rangle$ |
| $f(x) = x^2$ | $\mathbb{R}$ | $[0, \infty\rangle$ |
| $f(x) = \sin x$ | $\mathbb{R}$ | $[-1, 1]$ |
| $f(x) = \ln x$ | $\langle 0, \infty\rangle$ | $\mathbb{R}$ |

There are many ways to denote functions. The clearest notation perhaps is the *arrow notation*, which, in case of the square root function, looks as follows:

$$f : x \longrightarrow y = \sqrt{x}$$

To the left of the colon, you find the function symbol $f$. Immediately preceding the arrow is the input variable $x$, and following the arrow a formula shows how the output variable $y$ (the function value) depends on $x$. This notation is generally used in higher mathematics, while the shorter forms $y = \sqrt{x}$ or $f(x) = \sqrt{x}$ are more common in applications. In this book, we also use the latter notation, so we write $f(x) = \sqrt{x}$. The drawback is that it looks like an equation, while it actually is a function rule. The arrow notation avoids this type of confusion. In computer algebra packages, therefore, an arrow notation of some kind is always used for functions.

Note that there is a difference between a *function rule*, such as $f(x) = x^3$, and the *graph* of the function that is defined by it. The graph is a certain curve in the plane $\mathbb{R}^2$, namely the curve that, in braces notation, is given by

$$\{(x,y) \in \mathbb{R}^2 \mid y = x^3\}$$

This, however, is often written in shorter form as *the curve* $y = x^3$ (also in this book). In practice, the context clarifies what is meant: a function rule or the graph of a function.

## Inverse functions

If $f$ is a function with domain $D_f$ and range $R_f$, then for each $x \in D_f$ there is exactly one value $y = f(x) \in R_f$. However, the same value $y$ can occur for more than one $x$. Take, e.g., the function $f(x) = x^2$ that yields the value $y = 4$ for inputs $x = 2$ and $x = -2$.

In case that, for any $y \in R_f$, there is *only one* $x \in D_f$, the function $f$ is called *invertible*. In that case, alongside $f$, the *inverse function* $f^{-1}$ is defined with domain $R_f$ and range $D_f$. The inverse function interchanges the input and output variables, so

$$y = f(x) \qquad \Longleftrightarrow \qquad x = f^{-1}(y)$$

Of course, $f^{-1}(f(x)) = x$ for all $x \in D_f$ and $f(f^{-1}(y)) = y$ for all $y \in R_f$.

If the input variable of the inverse function $f^{-1}$ is again called $x$, and the output variable $y$, the graph of the function $y = f^{-1}(x)$ is obtained from the graph of $y = f(x)$ by reflection in the line $y = x$. Examples of invertible functions and their graphs can be found in chapters 17 and 18.

## Symmetry

A planar figure is called *line symmetric* with respect to a line $\ell$ if the perpendicular reflection in $\ell$ transforms the figure as a whole into itself. A planar figure is called *point symmetric* with respect to a point $P$ if the point reflection in $P$ transforms the figure as a whole into itself.

The graph of a function $y = f(x)$ is *line symmetric* with respect to the vertical line $x = c$ if $f(c - x) = f(c + x)$ for all $x$. The graph of the function $y = f(x)$ is *point symmetric* in a point $P = (p, q)$ if $f(p + x) - q = q - f(p - x)$ for all $x$. For instance, the graph of the function $f(x) = \sin x$ is line symmetric with respect to every vertical line $x = \frac{\pi}{2} + k\pi$ and point symmetric with respect to every point $(k\pi, 0)$ (for integer $k$). The graph of $f(x) = \cos x$ is line symmetric in every vertical line $x = k\pi$ and point symmetric in every point $(\frac{\pi}{2} + k\pi, 0)$.

A function with a graph that is line symmetric with respect to the $y$-axis, is called *even*, a function with a graph that is point symmetric with respect to the origin is called *odd*. For instance, $f(x) = x^n$ is an even function if $n$ is even, and an odd function if $n$ is odd. Furthermore, $f(x) = \cos x$ is an even function and $f(x) = \sin x$ is an odd function.

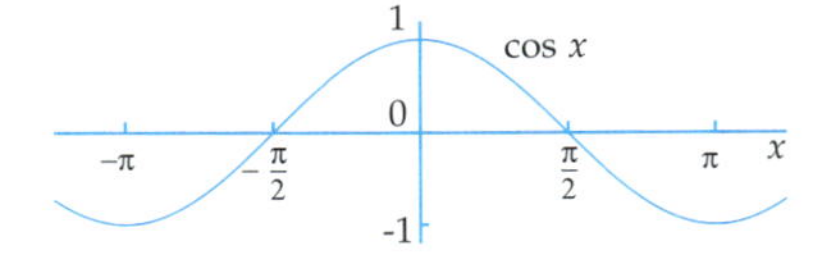


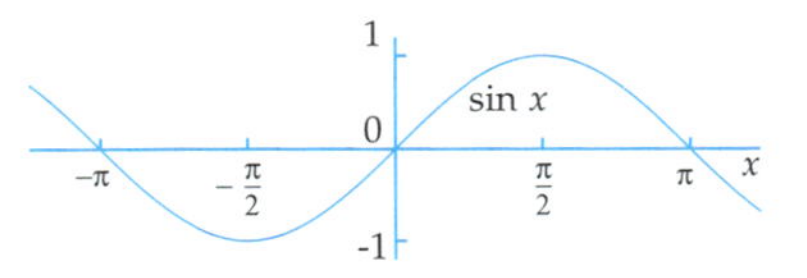


## Periodicity

A function $f$ is called *periodic* if there is a number $p > 0$ such that $f(x + p) = f(x)$ for each $x$. The number $p$ is then called a *period* of the function. For each positive integer $k$, $kp$ then is a period as well. If there is a smallest positive period, this is often called 'the' period of the function. But such a smallest period does not have to exist: every constant function is periodic with period $p$ for each $p > 0$. But this is a trivial example.

More interesting examples are the functions $f(x) = \sin x$ and $f(x) = \cos x$, both with $2\pi$ as their smallest period. But, note that the function $\tan x = \frac{\sin x}{\cos x}$ has the number $\pi$ as its smallest period and not $2\pi$, since for every $x \neq \frac{1}{2}\pi + k\pi$

$$\tan(x + \pi) = \frac{\sin(x + \pi)}{\cos(x + \pi)} = \frac{-\sin x}{-\cos x} = \frac{\sin x}{\cos x} = \tan x$$

## Limits

On page 69, we specified what is understood by the limit of a sequence of real numbers $a_1, a_2, a_3, \ldots$ We distinguished the following cases: $\lim_{n\to\infty} a_n = L$ with $L \in \mathbb{R}$, $\lim_{n\to\infty} a_n = \infty$ and $\lim_{n\to\infty} a_n = -\infty$. Here, we repeat this overview:

| *'lim'-notation:* | *arrow notation:* | *description:* |
|---|---|---|
| $\lim\limits_{n\to\infty} a_n = L$ | $a_n \to L$ if $n \to \infty$ | For each positive number $p$ (no matter how small), there is a term $a_N$ in the sequence for which all terms $a_n$ with $n > N$ satisfy $\lvert a_n - L \rvert < p$. |
| $\lim\limits_{n\to\infty} a_n = \infty$ | $a_n \to \infty$ if $n \to \infty$ | For each positive number $P$ (no matter how big), there is a term $a_N$ in the sequence for which all terms $a_n$ with $n > N$ satisfy $a_n > P$. |
| $\lim\limits_{n\to\infty} a_n = -\infty$ | $a_n \to -\infty$ if $n \to \infty$ | This simply means that $\lim\limits_{n\to\infty} (-a_n) = \infty$. |

Limits of functions were treated in a more intuitive way. For example, on page 153 we argued that $\lim\limits_{x\to 0} \dfrac{\sin x}{x} = 1$. And on page 161, in connection with the definition of the number e, we argued that $\lim\limits_{x\to 0} \dfrac{\mathrm{e}^x - 1}{x} = 1$. Also, we used a limit in defining the derivative of a function in a point $a$, and we did so in a rather intuitive way.

The somewhat vague, intuitive meaning of $\lim_{x\to a} f(x) = L$ is: 'if $x$ gets closer and closer to $a$, then $f(x)$ gets closer and closer to $L$'. With the help of limits of sequences, this can be further specified. The idea is as follows. If such a limit exists and $a_1, a_2, a_3, \ldots$ is a sequence of points in the domain of $f$ with $a_n \neq a$ for all $n$ and $\lim_{n\to\infty} a_n = a$, then $\lim_{n\to\infty} f(a_n) = L$ must hold.

For a precise definition of the concept of a limit for functions, this approach is reversed. Now, we will say that $\lim_{x\to a} f(x) = L$ means that $\lim_{n\to\infty} f(a_n) = L$ for *each* sequence $a_1, a_2, a_3, \ldots$ in the domain of $f$ with $a_n \neq a$ for all $n$ and $\lim_{n\to\infty} a_n = a$. Thus, our definition of the limit for functions is reduced to the definition of limits for sequences that we already have discussed in full detail. But, note the word *each*, marked in italics. The limit $\lim_{n\to\infty} f(a_n) = L$ must hold for *each* sequence with $\lim_{n\to\infty} a_n = a$ and $a_n \neq a$ for all $n$.

Moreover, the condition that all $a_n$ should be unequal to $a$, even if $a$ is in the domain of $f$, arises from the fact that $f(a)$ may be unequal to the limit value $L$.

The value of $f$ in $a$, if it exists, has no influence whatsoever on the limit value of $f(x)$ for $x \to a$, if it exists. Later, we will define *continuity* in $a$ by the condition that $\lim_{x \to a} f(x) = f(a)$.

But first we want to give formal definitions for different kinds of limits of functions. Suppose that $f$ is a function defined on an open neighbourhood of the point $a$, with the possible exception of $a$ itself (think, e.g., of the function $f(x) = \dfrac{\sin x}{x}$ that is not defined for $x = 0$), and suppose that $L$ is a real number.

**Definition:** $\lim_{x \to a} f(x) = L$ means that $f(a_n) \to L$ for each sequence $a_1, a_2, a_3, \ldots$ in the domain of $f$ with $a_n \neq a$ for all $n$ and $a_n \to a$.

In a similar way:

**Definition:** $\lim_{x \to a} f(x) = \infty$ means that $f(a_n) \to \infty$ for each sequence $a_1, a_2, a_3, \ldots$ in the domain of $f$ with $a_n \neq a$ for all $n$ and $a_n \to a$.

**Definition:** $\lim_{x \to a} f(x) = -\infty$ means that $f(a_n) \to -\infty$ for each sequence $a_1, a_2, a_3, \ldots$ in the domain of $f$ with $a_n \neq a$ for all $n$ and $a_n \to a$.

In the last two cases, the graph of $f$ has a vertical asymptote for $x = a$. One-sided limits are defined in a similar way:

**Definition:** $\lim_{x \uparrow a} f(x) = L$ means that $f(a_n) \to L$ for each sequence $a_1, a_2, a_3, \ldots$ in the domain of $f$ with $a_n < a$ for all $n$ and $a_n \uparrow a$.

**Definition:** $\lim_{x \downarrow a} f(x) = L$ means that $f(a_n) \to L$ for each sequence $a_1, a_2, a_3, \ldots$ in the domain of $f$ with $a_n > a$ for all $n$ and $a_n \downarrow a$.

In a similar way, the limits $\lim_{x \uparrow a} f(x) = \infty$, $\lim_{x \downarrow a} f(x) = \infty$, $\lim_{x \uparrow a} f(x) = -\infty$ and $\lim_{x \downarrow a} f(x) = -\infty$ are defined. Then we have a vertical asymptote at $x = a$ as well.

However, this still does not cover all the possibilities. The point $a$ may be situated 'at infinity' as well. We will only give one example; the other possibilities are left to the reader.

**Definition:** $\lim_{x \to \infty} f(x) = L$ means that $f(a_n) \to L$ for each sequence $a_1, a_2, a_3, \ldots$ in the domain of $f$ with $a_n \to \infty$.

In this case, the graph of $f$ has a horizontal asymptote with equation $y = L$. Various examples of limits and exercises with limits are given in chapters 17 and 18.

## Continuity

What does it mean that a function $f(x)$ is continuous in a point $a$ in its domain? Vaguely said, it means that the function values $f(x)$, as $x$ gets closer and closer to $a$, get closer and closer to $f(a)$. We can give precise definitions using limits:

**Definition:** The function $f(x)$ is called *continuous at $a$* if $a$ is in the domain of $f$ and $\lim_{x\to a} f(x) = f(a)$.

**Definition:** The function $f(x)$ is called *continuous from the left at $a$* if $a$ is in the domain of $f$ and $\lim_{x\uparrow a} f(x) = f(a)$.

**Definition:** The function $f(x)$ is called *continuous from the right at $a$* if $a$ is in the domain of $f$ and $\lim_{x\downarrow a} f(x) = f(a)$.

**Definition:** The function $f(x)$ is called *continuous on an interval $I$* if $I$ is in the domain of $f$ and $f(x)$ is continuous at each point of $I$. If such a point is an end point of $I$ (for instance the point $a$ if $I = [a,b]$), then we only require that $f$ is continuous from the right or from the left at that point, depending whether it is a left or right end point of $I$.

Almost all functions given by 'ordinary' formulas and symbols are continuous on all intervals in their domain. It would take us too far to further specify this statement, let alone prove it. As a rule of thumb, you might say that a function is continuous on an interval if you can draw its graph without taking your pencil from the paper. Sometimes, however, you should be careful in this respect, as the special examples of functions in the exercises 20.46 and 20.49 show (pages 186 and 188).

Examples:

- The function $f(x) = \dfrac{1}{x}$ is continuous on the two intervals $\langle -\infty, 0\rangle$ and $\langle 0, \infty\rangle$ that together constitute the domain of $f$.
- The function $f(x) = \tan x$ is continuous on each interval $\langle \frac{k-1}{2}\pi, \frac{k+1}{2}\pi\rangle$ with integer $k$.
- The function $f(x) = \sqrt{x}$ is continuous on $[0, \infty\rangle$.
- The function $f$, defined by $f(x) = 0$ if $x$ is rational and $f(x) = 1$ if $x$ is irrational, is in no point continuous, since each interval contains both rational and irrational points.
- The function

$$f(x) = \begin{cases} \dfrac{\sin x}{x} & \text{if } x \neq 0 \\ 1 & \text{if } x = 0 \end{cases}$$

is continuous at the point 0, since $\lim\limits_{x\to 0} \dfrac{\sin x}{x} = 1$. In all other points $f(x)$ is also continuous, so $f(x)$ is continuous at all points of its domain $\mathbb{R}$.

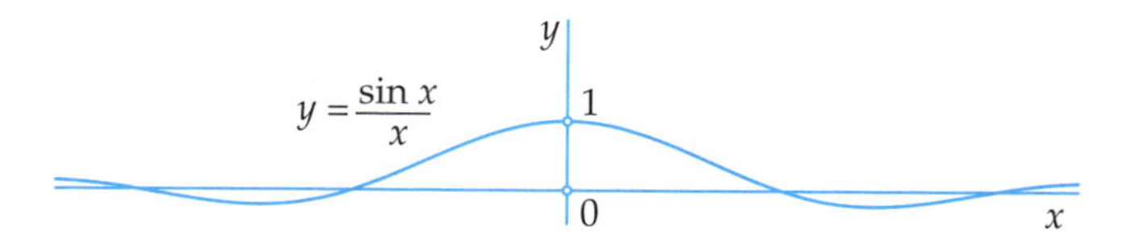

- The *floor function* $f(x) = \lfloor x \rfloor$ is defined by: $\lfloor x \rfloor$ is the greatest integer number less than or equal to $x$. Thus we have, e.g., $\lfloor \pi \rfloor = 3$, $\lfloor 3 \rfloor = 3$, $\lfloor -\sqrt{2} \rfloor = -2$. This function is discontinuous at each integer point, but continuous at all other points. In the integer points, however, the function is continuous from the right, but not continuous from the left. For instance, we have $\lim_{x \downarrow 3} \lfloor x \rfloor = 3 = \lfloor 3 \rfloor$ but $\lim_{x \uparrow 3} \lfloor x \rfloor = 2 \neq \lfloor 3 \rfloor$.

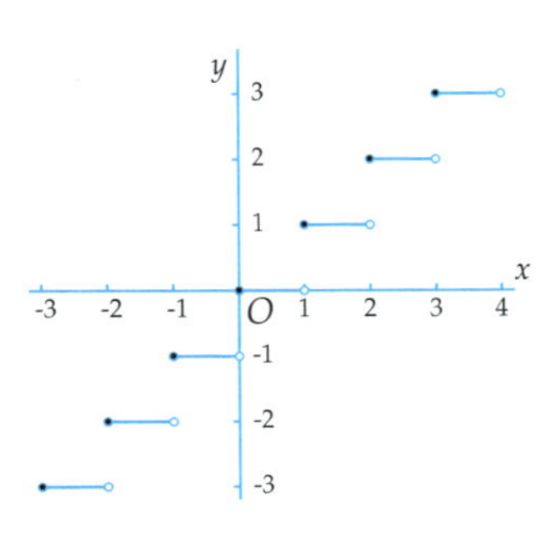


Remark: rounding off a real number to a given number of decimal places can be expressed by means of the floor function. For instance, the number $\pi = 3.1415926\ldots$, is, rounded off to 3 decimal places, equal to

$$\frac{\lfloor 1000\pi + 0.5 \rfloor}{1000} = \frac{\lfloor 3141.5926\ldots + 0.5 \rfloor}{1000} = \frac{3142}{1000} = 3.142$$

The general formula for rounding off the positive number $r$ to $n$ decimal places is $\dfrac{\lfloor 10^n r + 0.5 \rfloor}{10^n}$.

# 26 Additional derivations

## Inner product and the cosine rule

In chapter 13, on page 103, we defined the inner product $\langle \mathbf{a}, \mathbf{b} \rangle$ of two vectors $\mathbf{a} = \begin{pmatrix} a_1 \\ a_2 \end{pmatrix}$ and $\mathbf{b} = \begin{pmatrix} b_1 \\ b_2 \end{pmatrix}$ in the plane by

$$\langle \mathbf{a}, \mathbf{b} \rangle = a_1 b_1 + a_2 b_2$$

We stated that it can be proven that

$$\langle \mathbf{a}, \mathbf{b} \rangle = |\mathbf{a}||\mathbf{b}| \cos \varphi$$

where $\varphi$ is the angle at the origin between the two vectors. Here we present a proof.

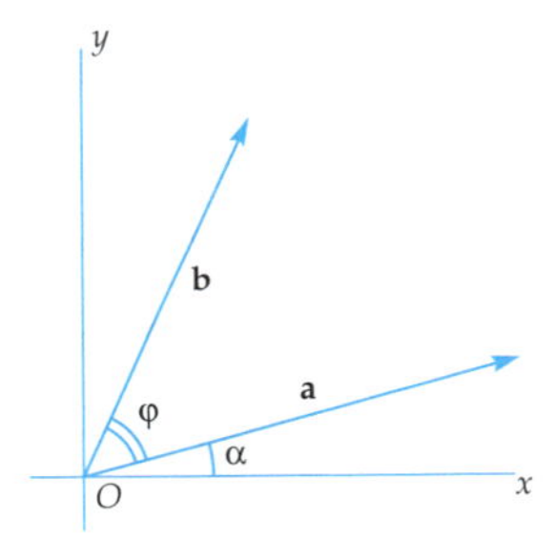


Suppose that $\alpha$ is the angle between the vector $\mathbf{a}$ and the positive $x$-axis, measured anticlockwise. Then (think of polar coordinates) $a_1 = |\mathbf{a}| \cos \alpha$ and $a_2 = |\mathbf{a}| \sin \alpha$. The angle between the vector $\mathbf{b}$ and the positive $x$-axis then is $\alpha + \varphi$, so $b_1 = |\mathbf{b}| \cos(\alpha + \varphi)$ and $b_2 = |\mathbf{b}| \sin(\alpha + \varphi)$. This yields

$$\langle \mathbf{a}, \mathbf{b} \rangle = a_1 b_1 + a_2 b_2 = |\mathbf{a}||\mathbf{b}| \left( \cos \alpha \cos(\alpha + \varphi) + \sin \alpha \sin(\alpha + \varphi) \right)$$

On account of one of the trigonometric formulas, the expression between brackets on the right-hand side is equal to $\cos(\alpha - (\alpha + \varphi)) = \cos(-\varphi) = \cos \varphi$, as desired.

## Exponential and logarithmic functions

In chapter 18, on page 159, we stated that the given properties of logarithmic functions can be derived from the properties of exponential functions. Here we present these derivations. The fundamental relation is always

$$q = a^p \quad \Longleftrightarrow \quad p = \log_a q$$

(here we write $p$ and $q$, since we will use $x$ and $y$ in a different meaning below).

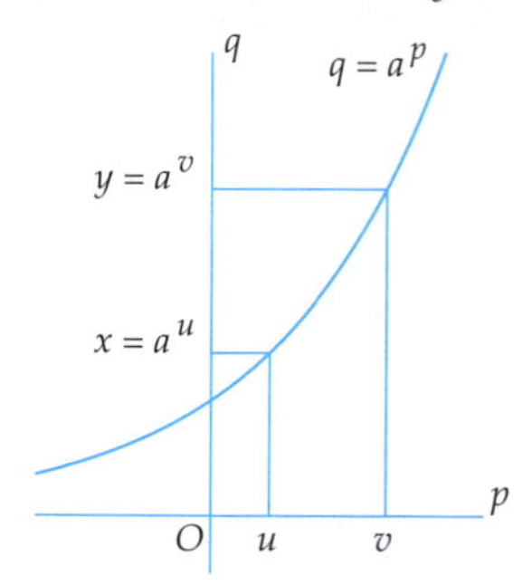


Suppose that $a > 0$ and $a \neq 1$. Let $x = a^u$, $y = a^v$, then $u = \log_a x$ and $v = \log_a y$. On account of the rules for exponential functions on page 157, we have $xy = a^u a^v = a^{u+v}$ so $\log_a(xy) = u + v = \log_a x + \log_a y$. Similarly: $x/y = a^u/a^v = a^{u-v}$ so $\log_a(x/y) = u - v = \log_a x - \log_a y$.

On account of another rule for exponential functions $(a^u)^p = a^{up}$, so, again with $x = a^u$:

$$\log_a(x^p) = \log_a((a^u)^p) = \log_a(a^{up}) = up = pu = p\log_a x$$

It follows from the formulas $a^{\log_a x} = x$ and $p\log_b a = \log_b a^p$ that, using $p = \log_a x$

$$\log_a x \; \log_b a = \log_b\left(a^{\log_a x}\right) = \log_b x$$

Writing this in the form

$$\log_a x = \frac{\log_b x}{\log_b a}$$

we are now able to transform logarithms with base $a$ into logarithms with base $b$. This completes our derivation of the formulas on page 159.

## Rules for differentiation and derivatives

The derivative $f'(x)$ of a function $y = f(x)$ at a point $x$ is defined as

$$f'(x) = \lim_{dx \to 0} \frac{f(x + dx) - f(x)}{dx}$$

provided this limit exists as a finite number. On page 179, a number of differentiation rules and standard derivatives are given. These rules are:

$$\begin{aligned}
(c\,f(x))' &= c\,f'(x) \quad \text{for any constant } c \\
(f(x)+g(x))' &= f'(x)+g'(x) \\
(f(g(x)))' &= f'(g(x))g'(x) \quad \text{(chain rule)} \\
(f(x)g(x))' &= f'(x)g(x)+f(x)g'(x) \quad \text{(product rule)} \\
\left(\frac{f(x)}{g(x)}\right)' &= \frac{f'(x)g(x)-f(x)g'(x)}{(g(x))^2} \quad \text{(quotient rule)}
\end{aligned}$$

The first two rules directly follow from the limit definition. In this section, we prove the product rule and the quotient rule.

A proof of the *product rule* is given using the limit definition of the derivative:

$$\begin{aligned}
(f(x)g(x))' &= \lim_{dx\to 0} \frac{f(x+dx)g(x+dx)-f(x)g(x)}{dx} \\
&= \lim_{dx\to 0} \frac{f(x+dx)g(x+dx)-f(x+dx)g(x)+f(x+dx)g(x)-f(x)g(x)}{dx} \\
&= \lim_{dx\to 0} \left( f(x+dx)\frac{g(x+dx)-g(x)}{dx} + g(x)\frac{f(x+dx)-f(x)}{dx} \right) \\
&= f(x)g'(x)+g(x)f'(x)
\end{aligned}$$

For a proof of the *quotient rule*, we first calculate the derivative of $\frac{1}{g(x)}$.

$$\begin{aligned}
\left(\frac{1}{g(x)}\right)' &= \lim_{dx\to 0} \frac{\frac{1}{g(x+dx)}-\frac{1}{g(x)}}{dx} \\
&= \lim_{dx\to 0} \frac{g(x)-g(x+dx)}{g(x+dx)g(x)dx} \\
&= \lim_{dx\to 0} \left( \frac{1}{g(x+dx)g(x)} \frac{g(x)-g(x+dx)}{dx} \right) \\
&= \frac{-g'(x)}{(g(x))^2}
\end{aligned}$$

The proof of the quotient rule then follows by writing $\frac{f(x)}{g(x)}$ as $f(x)\frac{1}{g(x)}$ and combining the above with the product rule:

$$\left(\frac{f(x)}{g(x)}\right)' = \left(f(x)\frac{1}{g(x)}\right)' = f'(x)\frac{1}{g(x)} + f(x)\frac{-g'(x)}{(g(x))^2} = \frac{f'(x)g(x)-f(x)g'(x)}{(g(x))^2}$$

## Differentials and the chain rule

In chapter 21, on page 193, we defined the differential $dy$ of a differentiable function $y = f(x)$ at the point $x$ corresponding to an increase $dx$ by $dy = f'(x)\,dx$. For such a function and its differential, we can picture the situation as a *black box* model:

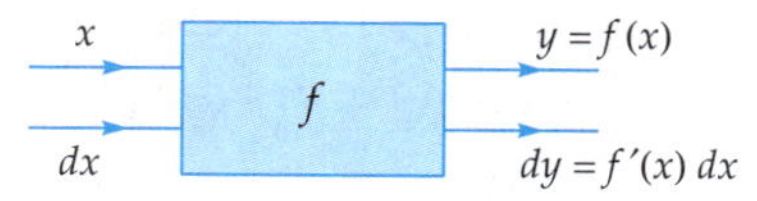


As *input*, we take $x$ and $dx$. The *output* consists of $y$ and $dy$.

A composed function $h(x) = f(g(x))$ can be seen as a concatenation of the functions $y = g(x)$ and $z = f(y)$. Also, the corresponding black boxes can be concatenated, where the outpunt of the $g$ box is taken as the input for the $f$ box. The input of the composed box is the pair $x$, $dx$, while the output is the pair $z = h(x)$, $dz = h'(x)\,dx$.

x
dx
g
y = g(x)
dy = g′(x) dx
f
h
z = f (y) = f (g(x)) = h(x)
dz = f ′(y) dy
= f ′(g(x)) g′(x) dx
= h ′(x) dx

Looking into the $h$ box, we see that $dz = f'(y)\,dy = f'(g(x))\,g'(x)\,dx$, since $y = g(x)$ and $dy = g'(x)\,dx$. Because $dz = h'(x)\,dx$ also holds, it follows that $h'(x) = f'(g(x))\,g'(x)$, which explains the chain rule.

## Standard derivatives

We now present proofs for the formulas for the derivatives of the standard functions on page 179. We will use the limit definition of the derivative, the rules for derivatives and differentials, and also the following two standard limits (see pages 153 and 161):

$$\lim_{dx \to 0} \frac{\mathrm{e}^{dx} - 1}{dx} = 1 \qquad \text{and} \qquad \lim_{dx \to 0} \frac{\sin dx}{dx} = 1$$

- $\dfrac{d}{dx}\mathrm{e}^{x} = \mathrm{e}^{x}$. The proof uses the limit definition:

$$\lim_{dx \to 0} \frac{\mathrm{e}^{x+dx} - \mathrm{e}^{x}}{dx} = \mathrm{e}^{x} \lim_{dx \to 0} \frac{\mathrm{e}^{dx} - 1}{dx} = \mathrm{e}^{x}$$

- $\frac{d}{dx} a^x = a^x \ln a.$ The proof uses the chain rule:

$$\frac{d}{dx}a^x = \frac{d}{dx}\mathrm{e}^{\ln a^x} = \frac{d}{dx}\mathrm{e}^{x\ln a} = \mathrm{e}^{x\ln a}\ln a = a^x \ln a$$

- $\frac{d}{dx}\ln x = \frac{1}{x}.$ The proof uses differentials and the relation

$$y = \ln x \iff x = \mathrm{e}^y$$

From $dx = \mathrm{e}^y\,dy$, it follows that $d(\ln x) = dy = \frac{1}{\mathrm{e}^y}dx = \frac{1}{x}dx.$

- $\frac{d}{dx}\log_a x = \frac{1}{x\ln a}.$ The proof uses the relation $\log_a x = \frac{\ln x}{\ln a}$ :

$$\frac{d}{dx}\log_a x = \frac{d}{dx}\left(\frac{\ln x}{\ln a}\right) = \frac{1}{x\ln a}$$

- $\frac{d}{dx}x^p = p\,x^{p-1}.$ The proof uses the chain rule:

$$\frac{d}{dx}x^p = \frac{d}{dx}\mathrm{e}^{\ln x^p} = \frac{d}{dx}\mathrm{e}^{p\ln x} = \mathrm{e}^{p\ln x}\frac{p}{x} = x^p\frac{p}{x} = p\,x^{p-1}$$

- $\frac{d}{dx}\sin x = \cos x.$ The proof uses the limit definition and the following auxiliary limit, that will be proven first:

$$\lim_{dx\to 0}\frac{1-\cos dx}{dx} = 0$$

Proof of the auxiliary limit:

$$\lim_{dx\to 0}\frac{1-\cos dx}{dx} = \lim_{dx\to 0}\frac{2\sin^2(\frac{1}{2}dx)}{dx} = \lim_{dx\to 0}\sin(\tfrac{1}{2}dx)\frac{\sin(\frac{1}{2}dx)}{\frac{1}{2}dx} = 0$$

The main part of the proof is:

$$\begin{aligned}
\lim_{dx\to 0}\frac{\sin(x+dx)-\sin x}{dx} &= \lim_{dx\to 0}\frac{\sin x\cos dx + \cos x\sin dx - \sin x}{dx} \\
&= \lim_{dx\to 0}\left(\sin x\frac{\cos dx - 1}{dx} + \cos x\frac{\sin dx}{dx}\right) \\
&= \cos x
\end{aligned}$$

- $\frac{d}{dx}\cos x = -\sin x.$ Proof:

$$\frac{d}{dx}\cos x = \frac{d}{dx}\sin\left(\frac{\pi}{2}-x\right) = -\cos\left(\frac{\pi}{2}-x\right) = -\sin x$$

- $\frac{d}{dx}\tan x = \frac{1}{\cos^2 x}$. The proof uses the quotient rule:

$$\frac{d}{dx}\tan x = \frac{d}{dx}\left(\frac{\sin x}{\cos x}\right) = \frac{\sin^2 x + \cos^2 x}{\cos^2 x} = \frac{1}{\cos^2 x}$$

- $\frac{d}{dx}\arcsin x = \frac{1}{\sqrt{1-x^2}}$. The proof uses the relation

$$y = \arcsin x \quad \Longleftrightarrow \quad x = \sin y$$

From $dx = \cos y\, dy$, it follows that $dy = \frac{1}{\cos y}\, dx$, so

$$d(\arcsin x) = dy = \frac{1}{\cos y}\, dx = \frac{1}{\sqrt{1-\sin^2 y}}\, dx = \frac{1}{\sqrt{1-x^2}}\, dx$$

- $\frac{d}{dx}\arccos x = \frac{-1}{\sqrt{1-x^2}}$. In a similar way, now using the relation

$$y = \arccos x \quad \Longleftrightarrow \quad x = \cos y$$

From $dx = -\sin y\, dy$, it follows that $dy = \frac{-1}{\sin y}\, dx$, so

$$d(\arccos x) = dy = \frac{-1}{\sin y}\, dx = \frac{-1}{\sqrt{1-\cos^2 y}}\, dx = \frac{-1}{\sqrt{1-x^2}}\, dx$$

- $\frac{d}{dx}\arctan x = \frac{1}{1+x^2}$. Here we use the relations

$$1 + \tan^2 y = \frac{1}{\cos^2 y} \quad \text{(see page 144, exercise 17.25)}$$

$$\text{and} \quad y = \arctan x \quad \Longleftrightarrow \quad x = \tan y$$

From $dx = \frac{1}{\cos^2 y}\, dy$, it follows that $dy = \cos^2 y\, dx$, so

$$d(\arctan x) = dy = \cos^2 y\, dx = \frac{1}{1+\tan^2 y}\, dx = \frac{1}{1+x^2}\, dx$$

# Answers

# Formula overview

# Index

# Answers

## I Numbers

### 1. Calculating with integers

1.1 a. 6321 b. 22700

1.2 a. 4815 b. 1298 c. 5635

1.3 a. 3026 b. 3082 c. 5673 d. 605 e. 2964

1.4 a. 29382 b. 36582 c. 36419 d. 66810 e. 70844

1.5 a. $(q, r) = (11, 11)$ b. $(16, 3)$ c. $(27, 10)$ d. $(27, 8)$ e. $(21, 3)$

1.6 a. $(44, 2)$ b. $(63, 100)$ c. $(130, 12)$ d. $(13, 315)$ e. $(86, 49)$

1.7 a. $(1405, 2)$ b. $(46, 72)$ c. $(2753, 2)$ d. $(315, 82)$ e. $(256, 28)$

1.8 a. $(1032, 22)$ b. $(133, 34)$ c. $(360, 1)$ d. $(75, 30)$ e. $(1110, 53)$

1.9 a. $2^3 \times 3$ b. $2^3 \times 3^2$ c. $2 \times 5^3$ d. $2^5 \times 3$ e. $2 \times 7^2$

1.10 a. $2^5 \times 3^2$ b. $2^{10}$ c. $3^2 \times 5 \times 7$ d. $2^2 \times 3^2 \times 11$ e. $3 \times 5^4$

1.11 a. $2^2 \times 3^5$ b. $2^2 \times 13^2$ c. $3^4 \times 5^2$ d. $2 \times 3 \times 11 \times 17$ e. $2^2 \times 5 \times 43$

1.12 a. $3 \times 5 \times 17$ b. $3^2 \times 7^2$ c. $2 \times 19^2$ d. $2^4 \times 3^3$ e. $5 \times 197$

1.13 a. $2^4 \times 5^3$ b. $3 \times 23 \times 29$ c. $2 \times 7 \times 11 \times 13$ d. 2003 (prime)
e. $2^2 \times 3 \times 167$

1.15 a. $\{1, 2, 3, 4, 6, 12\}$ b. $\{1, 2, 4, 5, 10, 20\}$ c. $\{1, 2, 4, 8, 16, 32\}$
d. $\{1, 2, 3, 4, 6, 9, 12, 18, 27, 36, 54, 108\}$ e. $\{1, 2, 3, 4, 6, 8, 9, 12, 16, 18, 24, 36, 48, 72, 144\}$

1.16 a. $\{1, 2, 3, 4, 6, 8, 9, 12, 18, 24, 36, 72\}$ b. $\{1, 2, 4, 5, 10, 20, 25, 50, 100\}$
c. $\{1, 7, 11, 13, 77, 91, 143, 1001\}$ d. $\{1, 3, 11, 17, 33, 51, 187, 561\}$
e. $\{1, 2, 4, 7, 14, 28, 49, 98, 196\}$

1.17 a. 6 b. 12 c. 9 d. 8 e. 17

1.18 a. 45 b. 72 c. 2 d. 27 e. 24

1.19 a. 32 b. 33 c. 125 d. 490 e. 128

1.20 a. 1 b. 1 c. 25 d. 1960 e. 8

1.21 a. 60 b. 135 c. 126 d. 80 e. 363

1.22 a. 156 b. 320 c. 720 d. 1690 e. 204

1.23 a. 250 b. 432 c. 1560 d. 4440 e. 1364

1.24 a. 720 b. 828 c. 3528 d. 945 e. 6851

1.25 a. (3,180) b. (6,360) c. (5,210) d. (9,378) e. (3,1512)

1.26 a. (7,980) b. (16,2240) c. (13,780) d. (12,1008) e. (63,3780)

## 2. Calculating with fractions

2.1 a. $\frac{3}{4}$ b. $\frac{2}{5}$ c. $\frac{3}{7}$ d. $\frac{1}{3}$ e. $\frac{1}{4}$

2.2 a. $\frac{5}{12}$ b. $\frac{2}{3}$ c. $\frac{5}{9}$ d. $\frac{27}{32}$ e. $\frac{7}{12}$

2.3 a. $\left(\frac{4}{12}, \frac{3}{12}\right)$ b. $\left(\frac{14}{35}, \frac{15}{35}\right)$ c. $\left(\frac{20}{45}, \frac{18}{45}\right)$ d. $\left(\frac{28}{44}, \frac{33}{44}\right)$ e. $\left(\frac{24}{156}, \frac{65}{156}\right)$

2.4 a. $\left(\frac{3}{18}, \frac{2}{18}\right)$ b. $\left(\frac{9}{30}, \frac{4}{30}\right)$ c. $\left(\frac{9}{24}, \frac{20}{24}\right)$ d. $\left(\frac{20}{36}, \frac{21}{36}\right)$ e. $\left(\frac{6}{40}, \frac{5}{40}\right)$

2.5 a. $\left(\frac{20}{60}, \frac{15}{60}, \frac{12}{60}\right)$ b. $\left(\frac{70}{105}, \frac{63}{105}, \frac{30}{105}\right)$ c. $\left(\frac{9}{36}, \frac{6}{36}, \frac{4}{36}\right)$ d. $\left(\frac{6}{30}, \frac{2}{30}, \frac{25}{30}\right)$ e. $\left(\frac{30}{72}, \frac{28}{72}, \frac{27}{72}\right)$

2.6 a. $\left(\frac{16}{216}, \frac{30}{216}, \frac{45}{216}\right)$ b. $\left(\frac{28}{60}, \frac{9}{60}, \frac{50}{60}\right)$ c. $\left(\frac{40}{210}, \frac{45}{210}, \frac{49}{210}\right)$ d. $\left(\frac{32}{504}, \frac{60}{504}, \frac{9}{504}\right)$
e. $\left(\frac{25}{390}, \frac{50}{390}, \frac{18}{390}\right)$

2.7 a. $\frac{6}{19}$ b. $\frac{7}{15}$ c. $\frac{11}{18}$ d. $\frac{11}{36}$ e. $\frac{25}{72}$

2.8 a. $\frac{2}{3}$ b. $\frac{14}{85}$ c. $\frac{39}{84}$ d. $\frac{31}{90}$ e. $\frac{29}{60}$

2.9 a. $\frac{7}{12}$ b. $\frac{1}{30}$ c. $\frac{16}{63}$ d. $\frac{2}{99}$ e. $\frac{17}{30}$

2.10 a. $\frac{17}{12}$ b. $\frac{1}{35}$ c. $\frac{29}{28}$ d. $\frac{5}{72}$ e. $\frac{119}{165}$

2.11 a. $\frac{5}{12}$ b. $-\frac{1}{45}$ c. $\frac{11}{24}$ d. $\frac{7}{6}$ e. $-\frac{1}{30}$

2.12 a. $\frac{29}{315}$ b. $\frac{17}{108}$ c. $\frac{67}{360}$ d. $\frac{1}{10}$ e. $\frac{13}{42}$

2.13 a. $\frac{47}{60}$ b. $\frac{13}{42}$ c. $\frac{29}{180}$ d. $\frac{1}{42}$ e. $\frac{31}{120}$

2.14 a. $\frac{7}{8}$ b. $\frac{3}{4}$ c. $-\frac{7}{24}$ d. $\frac{1}{12}$ e. $\frac{1}{5}$

2.15 a. $\frac{3}{8}$ b. $\frac{1}{4}$ c. $-\frac{13}{24}$ d. $-\frac{1}{36}$ e. $\frac{1}{3}$

2.16 a. $\frac{7}{27}$ b. $\frac{7}{15}$ c. $-\frac{59}{180}$ d. 1 e. $\frac{11}{30}$

2.17 a. $\frac{11}{70}$ b. $\frac{4}{3}$ c. $\frac{71}{84}$ d. $\frac{85}{286}$ e. $\frac{57}{170}$

2.18 a. $\frac{10}{21}$ b. $\frac{8}{45}$ c. $\frac{10}{91}$ d. $\frac{63}{26}$ e. $\frac{13}{300}$

2.19 a. 3 b. $\frac{2}{3}$ c. $\frac{4}{3}$ d. $\frac{15}{14}$ e. $\frac{16}{9}$

2.20 a. $\frac{14}{15}$ b. $\frac{14}{5}$ c. $\frac{27}{8}$ d. $\frac{15}{8}$ e. $\frac{4}{9}$

2.21 a. 3 b. 1 c. $\frac{20}{27}$ d. $\frac{10}{49}$ e. $\frac{2}{3}$

2.22 a. $\frac{14}{15}$ b. $\frac{2}{3}$ c. 30 d. $\frac{27}{25}$ e. $\frac{28}{25}$

2.23 a. $\frac{3}{2}$ b. $\frac{1}{2}$ c. 6 d. $\frac{14}{15}$ e. $\frac{2}{3}$

2.24 a. $\frac{8}{9}$ b. $\frac{4}{3}$ c. $\frac{8}{3}$

2.25 a. 2 b. $-\frac{154}{25}$ c. $\frac{7}{26}$

2.26 a. $\frac{470}{399}$ b. $\frac{105}{8}$ c. $-\frac{1463}{5220}$

## 3. Powers and roots

3.1 a. 8 b. 9 c. 1024 d. 625 e. 256

3.2 a. -8 b. 9 c. -1024 d. 625 e. 64

3.3 a. $\frac{1}{8}$ b. $\frac{1}{16}$ c. $\frac{1}{81}$ d. $\frac{1}{7}$ e. $\frac{1}{128}$

3.4 a. 1 b. $\frac{1}{9}$ c. $\frac{1}{121}$ d. $\frac{1}{729}$ e. $\frac{1}{10000}$

3.5 a. $-64$ b. $\frac{1}{243}$ c. $-\frac{1}{27}$ d. 16 e. $\frac{1}{16}$

3.6 a. 1 b. 0 c. $\frac{1}{12}$ d. 49 e. $-\frac{1}{128}$

3.7 a. $\frac{4}{9}$ b. $\frac{1}{16}$ c. $\frac{64}{125}$ d. $\frac{4}{49}$

3.8 a. $\frac{9}{4}$ b. 8 c. $\frac{9}{7}$ d. $\frac{16}{81}$

3.9 a. $\frac{9}{16}$ b. 16 c. $\frac{5}{4}$ d. $\frac{243}{32}$

3.10 a. 4 b. 1 c. $\frac{64}{27}$ d. $\frac{16}{625}$

3.11 a. $\frac{36}{49}$ b. 1 c. $\frac{49}{36}$ d. $\frac{8}{343}$

3.12 a. $\frac{64}{729}$ b. $\frac{27}{125}$ c. $\frac{25}{121}$ d. 32

3.13 a. 6 b. 9 c. 11 d. 8 e. 13

3.14 a. 15 b. 4 c. 14 d. 16 e. 21

3.15 a. $2\sqrt{2}$ b. $2\sqrt{3}$ c. $3\sqrt{2}$ d. $2\sqrt{6}$ e. $5\sqrt{2}$

3.16 a. $6\sqrt{2}$ b. $4\sqrt{2}$ c. $2\sqrt{5}$ d. $7\sqrt{2}$ e. $2\sqrt{10}$

3.17 a. $3\sqrt{6}$ b. $3\sqrt{11}$ c. $4\sqrt{5}$ d. $4\sqrt{6}$ e. $10\sqrt{2}$

3.18 a. $7\sqrt{3}$ b. $11\sqrt{2}$ c. $5\sqrt{5}$ d. $6\sqrt{6}$ e. $12\sqrt{2}$

3.19 a. $15\sqrt{3}$ b. $9\sqrt{5}$ c. $16\sqrt{2}$ d. $13\sqrt{2}$ e. $14\sqrt{3}$

3.20 a. $11\sqrt{11}$ b. $18\sqrt{3}$ c. 45 d. $19\sqrt{2}$ e. 26

3.21 a. $3\sqrt{2}$ b. $5\sqrt{6}$ c. $-42\sqrt{6}$ d. $-220\sqrt{6}$ e. $36\sqrt{35}$

3.22 a. $\sqrt{15}$ b. $-\sqrt{14}$ c. $\sqrt{30}$ d. $12\sqrt{21}$ e. $-240\sqrt{3}$

3.23 a. 720 b. $-1000$ c. $84\sqrt{15}$ d. $-90\sqrt{7}$ e. $3024\sqrt{5}$

3.24 a. $\frac{3}{4}$ b. $\frac{9}{2}$ c. $\frac{3}{2}$ d. $\frac{2}{27}\sqrt{2}$ e. $6\sqrt{6}$

3.25 a. $\frac{1}{4}\sqrt{2}$ b. $\frac{2}{9}\sqrt{6}$ c. $\frac{49}{64}$ d. $\frac{3}{4}\sqrt{6}$ e. $\frac{32}{27}\sqrt{3}$

3.26 a. $\frac{1}{3}\sqrt{6}$ b. $\frac{1}{2}\sqrt{6}$ c. $\frac{1}{5}\sqrt{30}$ d. $\frac{1}{2}\sqrt{14}$ e. $\frac{1}{7}\sqrt{14}$

3.27 a. $\frac{1}{6}\sqrt{15}$ b. $\frac{2}{9}\sqrt{3}$ c. $\frac{3}{10}\sqrt{5}$ d. $\frac{1}{5}\sqrt{10}$ e. $\frac{1}{8}\sqrt{14}$

3.28 a. $\frac{1}{2}\sqrt{6}$ b. $\frac{1}{3}\sqrt{15}$ c. $\frac{1}{11}\sqrt{77}$ d. $\frac{1}{5}\sqrt{55}$ e. $\frac{1}{11}\sqrt{22}$

3.29 a. $\frac{1}{2}\sqrt{30}$ b. $\frac{1}{5}\sqrt{30}$ c. $\frac{4}{5}\sqrt{15}$ d. $-\frac{1}{3}\sqrt{30}$ e. $2\sqrt{2}$

3.30 a. 2 b. 3 c. 5 d. 4 e. 6

3.31 a. $-3$ b. 2 c. 3 d. $-2$ e. 12

3.32 a. $2\sqrt[3]{2}$ b. $3\sqrt[4]{3}$ c. $5\sqrt[3]{3}$ d. $2\sqrt[5]{3}$ e. $3\sqrt[3]{2}$

3.33 a. $-2\sqrt[3]{5}$ b. $2\sqrt[4]{3}$ c. $2\sqrt[5]{10}$ d. $6\sqrt[3]{2}$ e. $2\sqrt[6]{3}$

3.34 a. $\sqrt[3]{35}$ b. $\sqrt[4]{56}$ c. $2\sqrt[3]{3}$ d. $3\sqrt[4]{10}$ e. $2\sqrt[5]{6}$

3.35 a. 6 b. $6\sqrt[3]{2}$ c. $3\sqrt[5]{5}$ d. $6\sqrt[6]{2}$ e. $10\sqrt[3]{7}$

3.36 a. $\frac{1}{7}$ b. doesn't exist c. $-\frac{2}{3}$ d. $\frac{6}{11}$ e. $\frac{6}{5}$

3.37 a. $\frac{2}{3}$ b. $\frac{5}{2}$ c. $\frac{2}{3}$ d. $\frac{3}{5}$ e. $\frac{12}{5}$

3.38 a. $\frac{1}{2}\sqrt[3]{2}$ b. $\frac{1}{3}\sqrt[4]{6}$ c. $\frac{1}{5}\sqrt[3]{15}$ d. $\frac{1}{3}\sqrt[3]{15}$ e. $\frac{1}{2}\sqrt[6]{24}$

3.39 a. $\frac{1}{6}\sqrt[3]{45}$ b. $\frac{1}{6}\sqrt[4]{126}$ c. $\frac{1}{6}\sqrt[5]{60}$ d. $\frac{1}{10}\sqrt[3]{90}$

3.40 a. $\frac{1}{3}\sqrt[3]{18}$ b. $\frac{1}{2}\sqrt[4]{6}$ c. $\frac{1}{2}\sqrt[5]{2}$ d. $\frac{1}{3}\sqrt[6]{54}$

3.41 a. $-\frac{1}{2}\sqrt[3]{12}$ b. $\frac{1}{2}\sqrt[4]{12}$ c. $-\frac{1}{3}\sqrt[5]{63}$ d. $\frac{1}{6}\sqrt[3]{210}$

Attention: for the next two exercises, the answers are often not given in the standard form.

3.42 a. $\sqrt{2}$ b. $\sqrt{27}$ c. $\sqrt[3]{49}$ d. $\sqrt[4]{3125}$ e. $\sqrt[3]{256}$

3.43 a. $\sqrt{\frac{1}{3}}$ b. $\sqrt{\frac{1}{343}}$ c. $\sqrt[3]{\frac{1}{4}}$ d. $\sqrt[5]{\frac{1}{81}}$ e. $\sqrt{\frac{1}{2}}$

3.44 a. $5^{\frac{1}{3}}$ b. $7^{\frac{1}{2}}$ c. $2^{\frac{1}{4}}$ d. $12^{\frac{1}{6}}$ e. $5^{\frac{1}{5}}$

3.45 a. $5^{-\frac{1}{2}}$ b. $6^{-\frac{1}{3}}$ c. $2^{-\frac{5}{4}}$ d. $3^{\frac{1}{2}}$ e. $7^{\frac{4}{5}}$

3.46 a. $2^{\frac{2}{3}}$ b. $2^{\frac{3}{2}}$ c. $2^{\frac{5}{4}}$ d. $2^{\frac{2}{3}}$ e. $2^{\frac{5}{3}}$

3.47 a. $2^{\frac{3}{2}}$ b. $2^{-\frac{3}{2}}$ c. $2^{\frac{7}{3}}$ d. $2^{\frac{1}{4}}$ e. $2^{-\frac{10}{3}}$

3.48 a. $\sqrt[6]{32}$ b. $\sqrt[6]{243}$ c. $4\sqrt[12]{2}$ d. $3\sqrt[15]{81}$ e. 4

3.49 a. $7\sqrt[6]{7}$ b. $3\sqrt[6]{3}$ c. $\sqrt[6]{3125}$ d. $3\sqrt[20]{3^{11}}$ e. 7

3.50 a. $\sqrt[6]{2}$ b. $\sqrt[6]{3}$ c. $\sqrt[4]{2}$ d. $\sqrt[15]{3}$ e. $\frac{1}{2}\sqrt[6]{32}$

# II Algebra

## 4. Calculating with letters

4.1 a. 18 b. $-6$ c. 102 d. $-108$ e. 9

4.2 a. 12 b. 6 c. 24 d. $-10$ e. $-20$

4.3 a. 40 b. $-32$ c. $-16$ d. $-8$ e. 64

4.4 a. $-15$ b. $-45$ c. $-45$ d. $-77$ e. $25$

4.5 a. $36$ b. $96$ c. $-192$ d. $36$ e. $-54$

4.6 a. $20$ b. $-36$ c. $72$ d. $42$ e. $29$

4.7 a. $320$ b. $-55$ c. $30$ d. $6075$ e. $8100$

4.8 a. $56$ b. $196$ c. $-56$ d. $15$ e. $16$

4.9 a. $a^8$ b. $b^5$ c. $a^{11}$ d. $b^4$ e. $a^{14}$

4.10 a. $a^6$ b. $b^{12}$ c. $a^{25}$ d. $b^8$ e. $a^{54}$

4.11 a. $a^4b^4$ b. $a^4b^6$ c. $a^{12}b^3$ d. $a^8b^{12}$ e. $a^{15}b^{20}$

4.12 a. $a^8$ b. $6a^{10}$ c. $60a^6$ d. $210a^8$ e. $6a^6$

4.13 a. $8a^6$ b. $81a^{12}b^{16}$ c. $16a^4b^4$ d. $125a^{15}b^9$ e. $16a^4b^{20}$

4.14 a. $15a^3b^5$ b. $24a^9b^6$ c. $6a^5b^5$ d. $35a^{12}b^8$ e. $144a^8b^{10}$

4.15 a. $24a^{10}$ b. $120a^{10}$ c. $40a^{11}$ d. $18a^{15}$ e. $-24a^7$

4.16 a. $-8a^6$ b. $9a^6$ c. $625a^{16}$ d. $-a^{10}b^{20}$ e. $-128a^{21}b^{35}$

4.17 a. $12a^8$ b. $72a^{12}$ c. $-135a^{18}$ d. $750a^{16}$ e. $320a^{12}$

4.18 a. $18a^7b^{10}$ b. $-72a^{10}b^{22}$ c. $-64a^{12}b^6$ d. $-324a^{16}b^{18}$ e. $1152a^{18}b^{13}$

4.19 a. $72a^7b^{12}c^{17}$ b. $-64a^{12}b^{21}c^{16}$ c. $-810a^{15}b^{10}c^{12}$ d. $1000000a^{24}b^{16}c^{22}$
e. $108a^{12}b^{12}c^{12}$

4.20 a. $a^{36}$ b. $16a^{24}$ c. $11664a^{26}b^{24}$ d. $-32a^{35}$ e. $-108a^{36}$

4.21 a. $6a+15$ b. $40a-16$ c. $-15a+10$ d. $-60a+12$ e. $-49a-42$

4.22 a. $2a^2-10a$ b. $14a^2+84a$ c. $-117a^2+65a$ d. $64a^2-120a$
e. $-63a^2-189a$

4.23 a. $2a^3+18a$ b. $12a^3-21a^2$ c. $-10a^4-20a^2$ d. $9a^4+18a^3$ e. $-3a^3+12a^2$

4.24 a. $12a^4+8a^3+12a^2$ b. $-6a^5-15a^4+3a^3$ c. $14a^5+21a^4-42a^3$
d. $-72a^5-24a^4+12a^3-12a^2$ e. $-15a^6-5a^4+10a^2$

4.25 a. $6a+8b$ b. $-10a+25b$ c. $2a^2+4ab$ d. $-64a^2+96ab$ e. $-176a^2+242ab$

4.26 a. $27a^2+15ab-36a$ b. $14a^3-12a^2b$ c. $-56a^3-32a^2b+8a^2$
d. $-12a^3+12a^2b+12a^2$ e. $-169a^3-156a^2b+182a^2$

4.27 a. $6a^4+4a^2b-6a^2$ b. $-10a^5-5a^4+10a^3b$ c. $6a^2b^2+4b^4$
d. $-8a^5+20a^3b^2-8a^3b$ e. $-196a^2b^3-28ab^3+70b^5$

4.28 a. $2a^4+6a^3b$ b. $-15a^4-10a^3b+15a^2b^2$ c. $6a^6+4a^5b^2-2a^3b^2$
d. $-6a^7-6a^6b^2-6a^5b^2$ e. $-49a^6+21a^5b-28a^4b^2$

4.29 a. $2a^3b+4a^2b^2-2ab^3$ b. $15a^3b^2-10a^2b^3+30ab^2$ c. $12a^3b^3-30a^2b^3-6ab^4$
d. $144a^4b^4-72a^3b^3+144a^2b^2$ e. $12a^3b^3+54a^2b^3-6a^2b^4$

4.30 a. $-5a^5b^5+2a^5b^4-a^4b^5$ b. $a^5b^5+a^4b^4+14a^2b^3$

c. $-15a^7b^7 - 90a^6b^6 + 15a^5b^7$ d. $-13a^9b^9 + 12a^7b^7 - 9a^6b^9$

e. $-49a^5b^2 - 49a^3b^4 - 7a^2b^2$

4.31 a. $2a^2 + 8a - 8$ b. $-12a^2 - 22a - 6$ c. $-9a$ d. $-8a^2 + 66a - 10$

e. $10a^2 - 15a - 5$ f. $-2a^2 - 3a + 1$

4.32 a. $3a^2 + 8ab - 2b$ b. $-a^2 + b$ c. $4a^2 + 4ab - 2b^2 - 2a + 2b$

d. $-2a^2 - 2b^2 + 6a - 3b$

4.33 a. $6(a + 2)$ b. $4(3a + 4)$ c. $3(3a - 4)$ d. $5(3a - 2)$ e. $27(a + 3)$

4.34 a. $3(a - 2b + 3)$ b. $4(3a + 2b - 4)$ c. $3(3a + 4b + 1)$ d. $6(5a - 4b + 10)$

e. $12(2a + 5b - 3)$

4.35 a. $-3(2a - 3b + 5)$ b. $-7(2a - 5b + 3)$ c. $-6(3a + 4b + 2c)$

d. $-14(2a + 5b - 3c)$ e. $-9(5a - 3b + 7c + 2)$

4.36 a. $a(a + 1)$ b. $a^2(a - 1)$ c. $a(a^2 - a + 1)$ d. $a^2(a^2 + a - 1)$ e. $a^3(a^3 - a + 1)$

4.37 a. $3a(a + 2)$ b. $3a(3a^2 + 2a - 1)$ c. $5a^2(3a^2 - 2a + 5)$ d. $9a^2(3a^4 - 2a^2 - 4)$

e. $12a(4a^3 - 2a^2 + 3a + 5)$

4.38 a. $3ab(a + 2)$ b. $9ab(a - b)$ c. $4ab(3b - 1)$ d. $7ab^2(2a - 3)$ e. $3a^2b(6b - 5)$

4.39 a. $3a^2b(ab + 2)$ b. $3a^2b(2a^2b^2 - 3ab + 4)$ c. $5abc(2a^2bc - ac - 3)$

d. $4a^3b^4c^3(2a^3bc - 3a + 5)$ e. $a^3b^3c(c^2 + c + 1)$

4.40 a. $2a^2bc^2(-2b^2 + b - 3)$ b. $a^3b^5c^3(a^3c - abc - b^2)$ c. $-2a^2c^2(ac^2 - b^2c + 2b)$

d. $-a^5b^6(a^2 - ab + 1)$ e. $-a^6b^6c^6(a^2b + ac - 1)$

4.41 a. $(a + 3)(b + 3)$ b. $(a - 2)(b - 1)$ c. $(2a + 7)(b + 4)$

d. $(a^2 + 2)(2b - 1)$ e. $(a - 1)(b - 2)$

4.42 a. $a(a - 1)(b + 1)$ b. $6(a + 2)(2b + 1)$ c. $-2(a - 2)(b - 1)$

d. $a^2(a - 1)(4b + 3)$ e. $-3a(2a + 3)(2b + 3)$

4.43 a. $(a + 4)(b + 1)$ b. $2b(2a - 1)$ c. $(3a + 2)(2b - 1)$

d. $(2a + 1)(a + 3)$ e. $(a + 2)(a + 1)$

4.44 a. $2(a + 5)(a + 3)$ b. $(a + 1)(a + 3)(b + 1)$ c. $-3(a - 1)(a + 2)$

d. $12(a - 1)(a + 2)(a - 2)$ e. $4(a + 1)(a + 4)^2$

4.45 a. $a^2 + 4a + 3$ b. $2a^2 + 9a + 9$ c. $3a^2 - 17a - 6$ d. $20a^2 - 9a - 20$

e. $6a^2 + 3a - 45$ f. $24a^2 + 12a - 120$

4.46 a. $-24a^2 + 73a - 24$ b. $56a^2 + 19a - 132$ c. $17a^2 - 288a - 17$

d. $6a^2 - 6a - 36$ e. $ab - 5a + 3b - 15$ f. $6ab + 10a + 24b + 40$

4.47 a. $-4ab + 4a + b - 1$ b. $-3ab + 9a + b - 3$ c. $156ab - 169a + 144b - 156$

d. $a^3 - 4a^2 + 4a - 16$ e. $a^3 - a^2 + 7a - 7$ f. $a^4 + 12a^2 + 27$

4.48 a. $2a^3+14a^2-7a-49$ b. $6a^4-13a^2+6$ c. $2a^4+3a^3-2a^2$
d. $-6a^4+23a^3-20a^2$ e. $-6a^4-6a^3+5a^2+5a$ f. $18a^4-49a^3-49a^2$

4.49 a. $-24a^4+55a^3+24a^2$ b. $-10a^5+13a^3-4a$ c. $-a^5+a^3$
d. $54a^7+18a^6-30a^5-10a^4$ e. $56a^6-35a^4-8a^3+5a$ f. $24a^8+38a^7+15a^6$

4.50 a. $6a^2b^2-2ab^2+3a^2b-ab$ b. $6a^3b^3-9a^3b^2+2a^2b^3-3a^2b^2$
c. $-4a^3b^4+10a^3b^3-6a^3b^2$ d. $-32a^5b^5-16a^4b^4+24a^3b^6+12a^2b^5$
e. $-a^8b^8+2a^6b^{10}-a^4b^{12}$ f. $2a^3+7a^2+2a-6$

4.51 a. $-12a^3+11a^2-5a+2$ b. $2a^2+3ab+8a+b^2+4b$
c. $-9a^2+18ab+9a-9b^2-9b$ d. $18a^2-81ab+13a-18b+2$
e. $a^4+a$ f. $6a^3+10a^2+a-2$

4.52 a. $2a^3+7a^2+11a+4$ b. $a^2-b^2-a-b$
c. $a^4+a^3b-ab^3-b^4$ d. $a^3+6a^2+11a+6$
e. $a^3-2a^2-5a+6$ f. $4a^3+4a^2-5a-3$

4.53 a. $4a^3-4a^2b-ab^2+b^3$ b. $60a^3-133a^2b+98ab^2-24b^3$
c. $-3a^4+6a^3-9a^2+18a$ d. $3a^3+5a^2-11a+3$
e. $2a^6+2a^4-4a^2$ f. $a^4b^3+a^3b^4+a^4b^2-a^2b^4-a^3b^2-a^2b^3$

4.54 a. $6a^5b+6a^4b^2-6a^3b^3-6a^2b^4$ b. $a^4+2a^3+a+2$
c. $a^4+a^3+a^2+3a+2$ d. $-6a^4+13a^3-a^2-5a-1$
e. $3a^5-6a^4+15a^3-6a^2+12a$ f. $10a^2+ab-21a-2b^2+12b-10$

## 5. Special products

5.1 a. $a^2+12a+36$ b. $a^2-4a+4$ c. $a^2+22a+121$ d. $a^2-18a+81$
e. $a^2+2a+1$

5.2 a. $b^2+10b+25$ b. $b^2-24b+144$ c. $b^2+26b+169$ d. $b^2-14b+49$
e. $b^2+16b+64$

5.3 a. $a^2+28a+196$ b. $b^2-10b+25$ c. $a^2-30a+225$ d. $b^2+4b+4$
e. $a^2-20a+100$

5.4 a. $4a^2+20a+25$ b. $9a^2-36a+36$ c. $121a^2+44a+4$ d. $16a^2-72a+81$
e. $169a^2+364a+196$

5.5 a. $25b^2+20b+4$ b. $4a^2-12a+9$ c. $81b^2+126b+49$ d. $16a^2-24a+9$
e. $64b^2+16b+1$

5.6 a. $4a^2+20ab+25b^2$ b. $9a^2-78ab+169b^2$ c. $a^2+4ab+4b^2$
d. $4a^2-4ab+b^2$ e. $36a^2+84ab+49b^2$

5.7 a. $144a^2-120ab+25b^2$ b. $4a^2-4ab+b^2$ c. $49a^2-70ab+25b^2$

d. $196a^2 - 84a + 9$ e. $a^2 + 22ab + 121b^2$

5.8 a. $a^4 + 10a^2 + 25$ b. $a^4 - 6a^2 + 9$ c. $b^4 - 2b^2 + 1$ d. $a^6 + 4a^3 + 4$

e. $b^8 - 14b^4 + 49$

5.9 a. $4a^2 + 28ab + 49b^2$ b. $9a^2 + 48ab + 64b^2$ c. $25a^2 - 90ab + 81b^2$

d. $49a^2 - 112ab + 64b^2$ e. $36a^2 - 132ab + 121b^2$

5.10 a. $a^4 + 6a^2 + 9$ b. $b^4 - 8b^2 + 16$ c. $4a^6 - 52a^3 + 169$ d. $25b^4 + 140b^2 + 196$

e. $144a^6 + 120a^3 + 25$

5.11 a. $4a^4 - 12a^2b + 9b^2$ b. $9a^4 + 12a^2b + 4b^2$ c. $81a^4 - 90a^2b^2 + 25b^4$

d. $144a^6 + 48a^3b^2 + 4b^4$ e. $400a^4 - 240a^2b^3 + 36b^6$

5.12 a. $5a^2 + 10a + 10$ b. $-18a + 9$ c. $5a^2 + 6a - 8$ d. $5a^2 + 8ab + 5b^2$

e. $-32a^4 + 32b^4$

5.13 a. $(a+4)(a-4)$ b. $(a+1)(a-1)$ c. $(a+12)(a-12)$

d. $(a+9)(a-9)$ e. $(a+11)(a-11)$

5.14 a. $(a+6)(a-6)$ b. $(a+2)(a-2)$ c. $(a+13)(a-13)$

d. $(a+16)(a-16)$ e. $(a+32)(a-32)$

5.15 a. $(2a+3)(2a-3)$ b. $(3a+1)(3a-1)$ c. $(4a+5)(4a-5)$

d. $(5a+9)(5a-9)$ e. $(12a+13)(12a-13)$

5.16 a. $(6a+7)(6a-7)$ b. $(8a+11)(8a-11)$ c. $(20a+21)(20a-21)$

d. $(14a+15)(14a-15)$ e. $(12a+7)(12a-7)$

5.17 a. $(a+b)(a-b)$ b. $(2a+5b)(2a-5b)$ c. $(3a+b)(3a-b)$

d. $(4a+9b)(4a-9b)$ e. $(14a+13b)(14a-13b)$

5.18 a. $(ab+2)(ab-2)$ b. $(ab+25)(ab-25)$ c. $(3ab+5c)(3ab-5c)$

d. $(5a+4bc)(5a-4bc)$ e. $(10ab+3c)(10ab-3c)$

5.19 a. $(a^2+b)(a^2-b)$ b. $(5a^2+4b)(5a^2-4b)$ c. $(4a^2+b^2)(2a+b)(2a-b)$

d. $(9a^2+4b^2)(3a+2b)(3a-2b)$ e. $(16a^2+25b^2)(4a+5b)(4a-5b)$

5.20 a. $(a^2b+1)(a^2b-1)$ b. $(ab^2+c)(ab^2-c)$ c. $(a^2+9b^2c^2)(a+3bc)(a-3bc)$

d. $(a^4+b^4)(a^2+b^2)(a+b)(a-b)$ e. $(16a^4+b^4)(4a^2+b^2)(2a+b)(2a-b)$

5.21 a. $a(a+1)(a-1)$ b. $2(2a+5)(2a-5)$ c. $3(3a+2b)(3a-2b)$

d. $5a(5a+3)(5a-3)$ e. $24a^3(5a+1)(5a-1)$

5.22 a. $3b(ab+3)(ab-3)$ b. $2ab(8ab+3)(8ab-3)$ c. $a^2b(a^2b+1)(a^2b-1)$

d. $5abc(5+ab)(5-ab)$ e. $3b(a+1)(a-1)$

5.23 a. $a(a^2+1)(a+1)(a-1)$ b. $2a(a^2+4)(a+2)(a-2)$

c. $ab(a^2b^2+9)(ab+3)(ab-3)$ d. $a(25-a^3)(25+a^3)$

e. $ab(a^4+16b^4)(a^2+4b^2)(a+2b)(a-2b)$

5.24 a. $2a+5$ b. $(3a+1)(a-3)$ c. $(3a+8)(2-a)$
d. $4a(2-2a)=8a(1-a)$ e. $(5a+3)(-a-1)$

5.25 a. $a^2-4$ b. $a^2-49$ c. $a^2-9$ d. $a^2-144$ e. $a^2-121$

5.26 a. $4a^2-25$ b. $9a^2-1$ c. $16a^2-9$ d. $81a^2-144$ e. $169a^2-196$

5.27 a. $36a^2-81$ b. $225a^2-1$ c. $49a^2-64$ d. $256a^2-25$ e. $441a^2-625$

5.28 a. $a^4-25$ b. $a^4-81$ c. $4a^4-9$ d. $36a^4-25$ e. $81a^4-121$

5.29 a. $a^6-16$ b. $a^{10}-100$ c. $81a^4-4$ d. $121a^8-9$ e. $144a^{12}-169$

5.30 a. $4a^2-9b^2$ b. $36a^2-100b^2$ c. $81a^2-4b^2$ d. $49a^2-25b^2$ e. $a^2-400b^2$

5.31 a. $a^4-b^2$ b. $4a^4-9b^2$ c. $25a^4-9b^4$ d. $36a^4-121b^4$ e. $169a^4-225b^4$

5.32 a. $a^6-4b^4$ b. $4a^4-81b^6$ c. $25a^8-9b^6$ d. $49a^4-361b^8$ e. $225a^{10}-64b^8$

5.33 a. $4a^2b^2-c^2$ b. $9a^4b^2-4c^2$ c. $25a^2b^4-c^4$ d. $81a^4b^4-16c^4$
e. $324a^6b^4-49c^6$

5.34 a. $4a^4-9b^2c^4$ b. $49a^6b^2-64c^6$ c. $169a^{10}b^6-196c^{10}$ d. $25a^2b^2c^2-1$
e. $81a^4b^2c^6-49$

5.35 a. $a^2+8a+16$ b. $a^2-16$ c. $a^2+7a+12$ d. $4a+12$ e. $a^2-a-12$

5.36 a. $a^2-a-42$ b. $a^2+14a+49$ c. $a^2-36$ d. $a^2-12a+36$
e. $2a^2-6a-36$

5.37 a. $a^2+26a+169$ b. $a^2-28a+196$ c. $a^2-a-182$ d. $3a^2-26a-169$
e. $182a^2-27a-182$

5.38 a. $4a^2+32a+64$ b. $a^2-10a+16$ c. $3a^2-18a$ d. $4a^2-64$ e. $2a^2+8a+8$

5.39 a. $a^2-13a-68$ b. $a^2-34a+289$ c. $a^2+13a-68$ d. $16a^2-289$
e. $68a^2+273a-68$

5.40 a. $a^2+42a+441$ b. $a^2+9a-252$ c. $441a^2-144$ d. $a^2-24a+144$
e. $12a^2+123a-252$

5.41 a. $a^4+2a^3-3a^2-8a-4$ b. $a^4+2a^3-3a^2-8a-4$ c. $a^4-2a^2+1$
d. $4a^4+24a^3+5a^2-24a-9$ e. $4a^4+12a^3+5a^2-12a-9$

5.42 a. $a^4-2a^2+1$ b. $a^4-2a^2+1$ c. $a^4-2a^2+1$ d. $16a^4-72a^2+81$
e. $a^4+4a^3+6a^2+4a+1$

5.43 a. $a^4-1$ b. $8a^3-18a$ c. $a^4-16$ d. $54a^6-24a^2$ e. $2a^5-1250a$

5.44 a. 391 b. 2475 c. 4899 d. 8091 e. 4884

5.45 a. $2a^2+12a+26$ b. $2a^2-2a-24$ c. $12a$ d. $6a^2-8a-6$ e. $10a-2$

5.46 a. $42a-98$ b. $-12a^2-28a$ c. $-a^4+81a^2+36a+8$ d. $9a^2+2$
e. $2a^4+2a^2$

5.47 a. $a^4 - 5a^2 + 4$ b. $a^4 - 41a^2 + 400$ c. $a^8 - 5a^4 + 4$ d. $a^3 + 4a^2 + 5a + 2$ e. $a^3 + 6a^2 + 12a + 8$

5.48 a. $-a^3 - 14a^2 - 25a$ b. $4a^2 + 8a$ c. $-a^3 + 25a^2 - 6a$ d. $5a^3 - 25a^2 + 125a - 625$ e. $2a^3 - a^2 + 3$

## 6. Fractions with letters

6.1 a. $\frac{a}{a-3} + \frac{3}{a-3}$ b. $\frac{2a}{a-b} + \frac{3b}{a-b}$ c. $\frac{a^2}{a^2-3} + \frac{3a}{a^2-3} + \frac{1}{a^2-3}$ d. $\frac{2a}{ab-3} - \frac{b}{ab-3} + \frac{3}{ab-3}$ e. $\frac{2}{b-a^3} - \frac{5a}{b-a^3}$

6.2 a. $\frac{a^2}{a^2-b^2} + \frac{b^2}{a^2-b^2}$ b. $\frac{ab}{a-2b} + \frac{bc}{a-2b} - \frac{ca}{a-2b}$ c. $\frac{b^2}{a^2-1} - \frac{1}{a^2-1}$ d. $\frac{4abc}{c-ab} + \frac{5}{c-ab}$ e. $\frac{5ab^2}{ab-c} - \frac{abc}{ab-c}$

6.3 a. $\frac{6}{a^2-9}$ b. $\frac{2a}{a^2-9}$ c. $\frac{a+9}{a^2-9}$ d. $\frac{a^2-2a+3}{a^2-9}$ e. $\frac{6a}{a^2-9}$

6.4 a. $\frac{7a+1}{a^2+a-6}$ b. $\frac{2a^2+2}{a^2-1}$ c. $\frac{-a}{a^2+7a+12}$ d. $\frac{5a^2-10a+7}{a^2-3a+2}$ e. $\frac{12a}{a^2+2a-8}$

6.5 a. $\frac{a^2-3ab+b^2}{a^2-3ab+2b^2}$ b. $\frac{2a}{a^2-b^2}$ c. $\frac{-2a^2+2a+2ab-4}{a^2-2a-ab+2b}$ d. $\frac{a^2-ab+2a+3b}{2a^2+ab-3b^2}$ e. $\frac{2ab+6a}{a^2-9}$

6.6 a. $\frac{2ab+2ac}{a^2-c^2}$ b. $\frac{3a^2-a+ab+3b}{a^2-b^2}$ c. $\frac{16a+4b-4a^2-ab-a^2b-4ab^2}{4a^2+17ab+4b^2}$ d. $\frac{a^2-ab-5ac+3bc+2b^2+3b-3c}{ab-b^2-ac+bc}$ e. $\frac{2a+2a^2+2b+8}{a^2-b^2-8b-16}$

6.7 a. $\frac{a+6}{3b-2}$ b. $a$ c. $\frac{2}{a}$ d. $\frac{1}{a-2b}$ e. $\frac{a+b^2}{b-3}$

6.8 a. $\frac{a+b}{3c}$ b. $\frac{a-4}{1+2a}$ c. $\frac{4b-3b^2}{a-bc}$ d. $\frac{a+b}{a-b}$ e. $a^2 + b$

6.9 a. $\frac{a+2}{a^2-9}$ b. $\frac{3}{a^2-9}$ c. $\frac{6a^2+2a}{a^2-9}$ d. $-1$ e. $\frac{2a}{a+1}$

6.10 a. $\frac{6ab-3b^2}{a^2-ab-2b^2}$ b. $\frac{a^2-ab+b}{a-b}$ c. $\frac{1}{a-2}$ d. $\frac{11a^2-5ab-3b^2}{3a^2-3ab}$ e. $\frac{4-3a}{2a}$

# III Sequences

## 7. Factorials and binomial coefficients

7.1 a. $a^3 + 3a^2 + 3a + 1$ b. $a^3 - 3a^2 + 3a - 1$ c. $8a^3 - 12a^2 + 6a - 1$ d. $a^3 + 6a^2 + 12a + 8$ e. $8a^3 - 36a^2 + 54a - 27$

7.2 a. $1 - 3a^2 + 3a^4 - a^6$ b. $a^3b^3 + 3a^2b^2 + 3ab + 1$ c. $a^3 + 6a^2b + 12ab^2 + 8b^3$ d. $a^6 - 3a^4b^2 + 3a^2b^4 - b^6$ e. $8a^3 - 60a^2b + 150ab^2 - 125b^3$

7.3 a. $9a^3 - 18a^2 + 18a - 9$ b. $a^3 - 6a^2b + 12ab^2 - 8b^3$ c. $a^3 + 9a^2b + 27ab^2 + 27b^3$ d. $125a^3 + 150a^2 + 60a + 8$ e. $2a^3 + 294a$

7.4 a. $a^6 - 3a^4b + 3a^2b^2 - b^3$ b. $a^{12} + 6a^8b^2 + 12a^4b^4 + 8b^6$

c. $2a^3 + 24ab^2$ d. $12a^2b + 16b^3$ e. $-7a^3 - 6a^2b + 6ab^2 + 7b^3$

7.5 a. $a^4 + 4a^3 + 6a^2 + 4a + 1$ b. $a^4 - 4a^3 + 6a^2 - 4a + 1$

c. $16a^4 - 32a^3 + 24a^2 - 8a + 1$ d. $a^4 + 8a^3 + 24a^2 + 32a + 16$

e. $16a^4 - 96a^3 + 216a^2 - 216a + 81$

7.6 a. $a^8 - 4a^6 + 6a^4 - 4a^2 + 1$ b. $a^4b^4 + 4a^3b^3 + 6a^2b^2 + 4ab + 1$

c. $a^4 + 8a^3b + 24a^2b^2 + 32ab^3 + 16b^4$ d. $a^8 - 4a^6b^2 + 6a^4b^4 - 4a^2b^6 + b^8$

e. $2a^4 + 12a^2b^2 + 2b^4$

7.7 1 8 28 56 70 56 28 8 1 $(n = 8)$
1 9 36 84 126 126 84 36 9 1 $(n = 9)$
1 10 45 120 210 252 210 120 45 10 1 $(n = 10)$

7.8 $a^8 + 8a^7 + 28a^6 + 56a^5 + 70a^4 + 56a^3 + 28a^2 + 8a + 1$,
$a^9 - 9a^8 + 36a^7 - 84a^6 + 126a^5 - 126a^4 + 84a^3 - 36a^2 + 9a - 1$,
$a^{10} - 10a^9b + 45a^8b^2 - 120a^7b^3 + 210a^6b^4 - 252a^5b^5 + 210a^4b^6 - 120a^3b^7 + 45a^2b^8 - 10ab^9 + b^{10}$

7.9 to 7.11: verify the answers yourself; for 7.12 to 7.15 see Pascal's triangle (completed up to $n = 10$).

7.16 a. 1 b. 15 c. 1287 d. 210 e. 3060

7.17 a. 792 b. 462 c. 1128 d. 18424 e. 1225

7.18 a. 680 b. 51 c. 220

7.19 a. 11480 b. 1716 c. 80730

7.20 a. 76076 b. 2002 c. 20475

7.21 a. $\sum_{k=0}^{7} \binom{7}{k} a^{7-k}$ b. $\sum_{k=0}^{12} \binom{12}{k} a^{12-k}(-1)^k$ c. $\sum_{k=0}^{12} \binom{12}{k} a^{12-k}10^k$

d. $\sum_{k=0}^{9} \binom{9}{k}(2a)^{9-k}(-1)^k$ e. $\sum_{k=0}^{10} \binom{10}{k}(2a)^{10-k}b^k$

7.22 a. $\sum_{k=0}^{7} \binom{7}{k} a^{7-k}5^k$ b. $\sum_{k=0}^{5} \binom{5}{k}(-a)^k$ c. $\sum_{k=0}^{18} \binom{18}{k}(ab)^{18-k}$

d. $\sum_{k=0}^{9} \binom{9}{k} a^{9-k}(2b)^k$ e. $\sum_{k=0}^{8} \binom{8}{k} a^{8-k}(-b)^k$

7.23 a. $2^8$ $(a = b = 1)$ b. $0$ $(a = 1, b = -1)$ c. $3^8$ $(a = 1, b = 2)$

7.24 a. $1$ $(a = 1, b = -2)$ b. $2^n$ $(a = 1, b = 1)$ c. $0$ $(a = 1, b = -1)$

7.25 a. 91 b. 0 c. 70

7.26 a. 25 b. $\frac{47}{6}$ c. 978

## 8. Sequences and limits

8.1 a. 2007006 b. 494550 c. 750000 d. 49800 e. 9899100 f. 9902700

8.2 a. 670 b. 16958 c. 3892 d. 570 e. 25250

8.3 49 (the sum of all even serial numbers is $\frac{1}{2} \times 48 \times 98$, the sum of all odd serial numbers is $\frac{1}{2} \times 49 \times 98$)

8.4 a. 510 b. $\frac{511}{256}$ c. 2186 d. $\frac{1330}{729}$ e. 0.3333333

8.5 a. 8 b. 3 c. $\frac{8}{15}$ d. $\frac{70}{9}$ e. $\frac{10}{19}$

8.6 a. $a = 0.1, r = -0.1$ so $a/(1-r) = \frac{1}{11}$ b. $a = 0.3, r = 0.1$ so $a/(1-r) = \frac{1}{3}$
c. $a = 0.9, r = 0.1$ so $a/(1-r) = 1$ d. $a = 0.12, r = 0.01$ so $a/(1-r) = \frac{4}{33}$
e. $a = 0.98, r = -0.01$ so $a/(1-r) = \frac{98}{101}$

8.7 a. $0.3333\ldots = 0.3 + 0.03 + 0.003 + 0.0003 + \cdots = \frac{1}{3}$
b. $0.9999\ldots = 0.9 + 0.09 + 0.009 + 0.0009 + \cdots = 1$
c. $0.12121212\ldots = 0.12 + 0.0012 + 0.000012 + 0.00000012 + \cdots = \frac{4}{33}$

d. See the hint e. See the hint

8.8 a. $0.2^{100} \approx 0.12677 \times 10^{-69}, 0.2^{1000} \approx 0.10715 \times 10^{-698}$
b. $0.5^{100} \approx 0.78886 \times 10^{-30}, 0.5^{1000} \approx 0.93326 \times 10^{-301}$
c. $0.7^{100} \approx 0.32345 \times 10^{-15}, 0.7^{1000} \approx 0.12533 \times 10^{-154}$
d. $0.9^{100} \approx 0.26561 \times 10^{-4}, 0.9^{1000} \approx 0.17479 \times 10^{-45}$
e. $0.99^{100} \approx 0.36603, 0.99^{1000} \approx 0.43171 \times 10^{-4}$

8.9 a. $\frac{2}{9}$ b. $\frac{31}{99}$ c. 2 d. $\frac{41}{333}$ e. $\frac{37}{300}$

8.10 a. $\frac{10}{99}$ b. $\frac{110}{333}$ c. $\frac{1210}{999}$ d. $\frac{1}{9000}$ e. $\frac{3061}{990}$

8.11 a. $\frac{1001}{45}$ b. $\frac{700}{999}$ c. $\frac{233}{333}$ d. $\frac{1828}{225}$ e. $\frac{112}{99}$

8.12 a. $\frac{11111}{100000}$ b. $\frac{181}{495}$ c. $\frac{31412}{9999}$ d. $\frac{271801}{99990}$ e. $\frac{1}{11}$

8.13 a. $1.02^{101} \approx 7.38954, 1.02^{1001} \approx 0.40623 \times 10^{9}$
b. $(-2)^{101} \approx -0.25353 \times 10^{31}, (-2)^{1001} \approx -0.21430 \times 10^{302}$
c. $10.1^{101} \approx 0.27319 \times 10^{102}, 10.1^{1001} \approx 0.21169 \times 10^{1006}$
d. $(-0.999)^{101} \approx -0.90389, (-0.999)^{1001} \approx -0.36733$
e. $9.99^{101} \approx 0.90389 \times 10^{101}, 9.99^{1001} \approx 0.36733 \times 10^{1001}$

8.14 a. $a_{100} = 0.1 \times 10^{5}, a_{1000} = 0.1 \times 10^{7}$

b. $a_{100} = 0.1 \times 10^{-5}, a_{1000} = 0.1 \times 10^{-8}$

c. $a_{100} \approx 0.63096 \times 10^{-2}, a_{1000} \approx 0.50119 \times 10^{-3}$

d. $a_{100} \approx 0.15849 \times 10^{2001}, a_{1000} \approx 0.19953 \times 10^{3001}$

e. $a_{100} = 0.1 \times 10^{2} (= 10), a_{1000} \approx 0.31623 \times 10^{2}$

8.15 a. $a_{100} \approx 0.21577, a_{1000} \approx 0.10023$, b. $a_{100} \approx 0.10233 \times 10, a_{1000} \approx 0.10023 \times 10$
c. $a_{100} \approx 0.10715 \times 10, a_{1000} \approx 0.10069 \times 10$, d. $a_{100} \approx 0.95499, a_{1000} \approx 0.99541$
e. $a_{100} \approx 0.99895, a_{1000} \approx 0.99989$

8.16 a. 1 b. 2 c. 1 d. $\frac{1}{3}$ e. 0 f. $\frac{2}{5}$

8.17 a. 2 b. 0 c. $\infty$ d. 1 e. 1 f. 0

8.18 a. 1 b. $\infty$ c. $\frac{4}{3}$ d. 1 e. $\frac{4}{5}$

8.19 a. 0 b. this limit doesn't exist c. 1 d. 1 e. $\infty$

8.20 a. 0 b. 0 c. $\infty$ d. $\infty$ e. 1

8.21 a. 0 b. 1 c. 2 d. 1 e. 0 (write $2^{3n} = 8^n$ and $3^{2n} = 9^n$)

8.22 a. $-1$ b. $\infty$ c. $-1$

d. 0 (since $(3n)! = n! \times (n+1)(n+2)\cdots 3n > n! \times n^n$ et cetera) e. 0

8.23 a. 0 b. $\infty$ c. $\infty$ d. 0 e. $\infty$

8.24 a. 0 b. 0 c. $\infty$ d. 0 e. $\infty$

# IV Equations

## 9. Linear equations

9.1 a. $x = 3$ b. $x = 16$ c. $x = -13$ d. $x = 3$ e. $x = -8$

9.2 a. $x = 9$ b. $x = -17$ c. $x = 27$ d. $x = 1$ e. $x = -19$

9.3 a. $x = 1$ b. $x = 5$ c. $x = 2$ d. $x = 3$ e. $x = -1$

9.4 a. $x = -2$ b. $x = -9$ c. $x = -3$ d. $x = -6$ e. $x = -3$

9.5 a. $x = \frac{3}{2}$ b. $x = 7$ c. $x = \frac{7}{2}$ d. $x = \frac{29}{5}$ e. $x = \frac{3}{2}$

9.6 a. $x = -21$ b. $x = \frac{1}{3}$ c. $x = -\frac{5}{6}$ d. $x = -\frac{4}{3}$ e. $x = \frac{19}{5}$

9.7 a. $x = 1$ b. $x = -4$ c. $x = -13$ d. $x = 3$ e. $x = -3$

9.8 a. $x = -3$ b. $x = \frac{3}{4}$ c. $x = -14$ d. $x = \frac{1}{2}$ e. $x = \frac{19}{2}$

9.9 a. $x = 3$ b. $x = \frac{13}{5}$ c. no solution d. $x = \frac{5}{11}$ e. $x = \frac{3}{2}$

9.10 a. $x = \frac{1}{4}$ b. $x = \frac{1}{5}$ c. $x = -2$ d. $x = \frac{3}{2}$ e. $x = \frac{1}{7}$

9.11 a. $x = -3$ b. $x = \frac{1}{8}$ c. $x = -\frac{75}{32}$ d. $x = \frac{45}{2}$ e. $x = -\frac{2}{21}$

9.12 a. $x = -\frac{28}{5}$ b. $x = \frac{26}{5}$ c. $x = -3$

9.13 a. $x = 3$ b. no solution c. $x = \frac{13}{6}$

9.14 a. $x < 2$ b. $x > 14$ c. $x \leq -2$ d. $x \geq -2$ e. $x > 1$

9.15 a. $x > -2$ b. $x < -5$ c. $x \geq 3$ d. $x \leq 1$ e. $x < \frac{1}{2}$

9.16 a. $x < -14$ b. $x > 3$ c. $x \leq -\frac{1}{2}$ d. $x \geq -2$ e. $x > \frac{1}{2}$

9.17 a. $x > -1$ b. $x < -7$ c. $x \geq 8$ d. $x \leq -2$ e. $x < 2$

9.18 a. $x < \frac{6}{5}$ b. $x > \frac{9}{2}$ c. $x \leq -3$ d. $x \leq -\frac{1}{3}$ e. $x < -21$

9.19 a. $x > -\frac{12}{5}$ b. $x < -\frac{15}{2}$ c. $x \geq \frac{14}{15}$ d. $x \leq -\frac{1}{3}$ e. $x < -29$

9.20 a. $-4 < x < 3$ b. $-1 < x < 1$ c. $-2 \leq x < 1$ d. $-1 < x \leq \frac{3}{2}$ e. $0 \leq x \leq \frac{1}{2}$

9.21 a. $-1 < x < 4$ b. $3 < x < 4$ c. $1 < x \leq 3$ d. $-\frac{1}{2} \leq x < 2$ e. $\frac{1}{2} \leq x \leq 1$

9.22 a. $x = -\frac{4}{5}$ b. $x = 8$ c. $x = -\frac{1}{5}$ d. $x = \frac{7}{6}$ e. $x = \frac{1}{4}$

9.23 a. $x = \frac{4}{5}$ b. no solution c. no solution (!!) d. $x = \frac{3}{22}$

e. no solution

9.24 a. $x = 0, x = -2$ b. $x = 1, x = 7$ c. $x = -4, x = 6$ d. $x = \frac{1}{2}, x = -\frac{3}{2}$
e. $x = -1, x = \frac{5}{3}$

9.25 a. $x = -2 \pm \sqrt{3}$ b. $x = 1 \pm \sqrt{2}$ c. $x = 3 \pm \sqrt{5}$ d. $x = -\frac{1}{2} \pm \frac{1}{2}\sqrt{6}$
e. $x = 3 \pm \sqrt{2}$

9.26 a. $x = 2$ b. $x = -6$ c. $x = 0$ d. $x = 2$ e. $x = -\frac{5}{4}$

9.27 a. $x = 1, x = 3$ b. $x = -3, x = 1$ c. $x = \frac{3}{2} \pm \frac{1}{2}\sqrt{2}$ d. $x = -3, x = 0$
e. $x = -\frac{1}{3}, x = 3$

9.28 a. $x = 0, x = 2$ b. $x = 0, x = \frac{1}{2}$ c. $x = 0, x = 2$ d. $x = -\frac{2}{3}, x = -8$
e. $x = -1, x = -\frac{3}{5}$

9.29 a. $x = 2, x = -\frac{2}{3}$ b. $x = -\frac{3}{4}$ c. $x = -4, x = -1$ d. $x = \frac{3}{7}, x = \frac{7}{3}$
e. $x = -\frac{1}{10}, x = -\frac{5}{2}$

## 10. Quadratic equations

10.1 a. $x = \pm 3$ b. $x = \pm 2$ c. $x = \pm 2$ d. $x = \pm 3$ e. $x = \pm 7$

10.2 a. $x = \pm\sqrt{2}$ b. no solutions c. $x = \pm 2\sqrt{2}$ d. $x = 0$ e. $x = \pm\sqrt{2}$

10.3 a. $x = \pm 2$ b. $x = \pm\frac{1}{2}\sqrt{3}$ c. $x = \pm\frac{2}{3}$ d. $x = \pm\frac{5}{4}$ e. $x = \pm\frac{3}{4}\sqrt{2}$

10.4 a. $x = \pm\frac{1}{3}\sqrt{3}$ b. $x = \pm\frac{3}{2}$ c. no solutions d. $x = \pm\sqrt{14}$ e. $x = \pm\frac{1}{2}\sqrt{5}$

10.5 a. $x = 0, x = -3$ b. $x = -1, x = 5$ c. $x = 1, x = -1$ d. $x = -7, x = 2$
e. $x = 3, x = -9$

10.6 a. $x = 0, x = \frac{1}{2}$ b. $x = -\frac{1}{2}, x = 3$ c. $x = -\frac{2}{3}, x = \frac{3}{2}$ d. $x = -\frac{3}{5}, x = \frac{5}{3}$
e. $x = \frac{2}{3}$

10.7 a. $x = 1, x = -3$ b. $x = 1, x = -5$ c. $x = -\frac{1}{2}, x = \frac{4}{3}$ d. $x = -\frac{2}{3}, x = -\frac{1}{2}$
e. $x = \frac{2}{3}, x = -\frac{2}{3}$

10.8 a. $x = -6, x = \frac{2}{3}$ b. $x = \frac{6}{5}, x = \frac{6}{7}$ c. $x = \frac{16}{9}, x = \frac{3}{2}$

10.9 a. $x = -2 \pm \sqrt{3}$ b. $x = -3 \pm \sqrt{11}$ c. $x = -4 \pm \sqrt{13}$ d. $x = 1 \pm \sqrt{2}$
e. $x = -5 \pm 2\sqrt{5}$

10.10 a. $x = 6 \pm \sqrt{30}$ b. $x = \frac{13}{2} \pm \frac{1}{2}\sqrt{197}$ c. $x = 6, x = -7$ d. $x = 3, x = 9$
e. $x = -3 \pm \sqrt{21}$

10.11 a. $x = -\frac{7}{2} \pm \frac{1}{2}\sqrt{53}$ b. $x = 1, x = -4$ c. $x = -2$ d. $x = 2 \pm 2\sqrt{2}$
e. $x = \frac{11}{2} \pm \frac{1}{2}\sqrt{93}$

10.12 a. $x = -10 \pm 2\sqrt{10}$ b. $x = 9 \pm \sqrt{161}$ c. $x = -\frac{13}{2} \pm \frac{1}{2}\sqrt{337}$ d. $x = 7, x = 8$
e. $x = -20, x = -40$

10.13 a. $x = -\frac{1}{4} \pm \frac{1}{4}\sqrt{13}$ b. $x = \frac{1}{3}, -\frac{5}{3}$ c. $x = \frac{1}{6} \pm \frac{1}{6}\sqrt{5}$ d. $-\frac{3}{4} \pm \frac{1}{4}\sqrt{19}$
e. $\frac{1}{5} \pm \frac{1}{5}\sqrt{6}$

10.14 a. $x = -\frac{3}{8} \pm \frac{1}{8}\sqrt{33}$ b. $x = -1, -\frac{3}{2}$ c. $x = \frac{1}{3}$ d. $\frac{3}{4} \pm \frac{1}{4}\sqrt{21}$
e. $-\frac{2}{5} \pm \frac{2}{5}\sqrt{6}$

10.15 a. $x = \pm 1$ b. $x = \pm\sqrt{7}$ c. no solutions d. $x = \pm\sqrt{2}$
e. $x = -1, x = \sqrt[3]{12}$

10.16 a. $x = 9$ b. $x = 1, x = 289$ c. $x = 9$ d. $x = 4, x = 169$ e. $x = 1$

10.17 a. $x = \frac{-5 \pm \sqrt{21}}{2}$ b. $x = 1, x = 2$ c. $x = \frac{-7 \pm \sqrt{37}}{2}$ d. no solutions
e. $x = \frac{-11 \pm \sqrt{77}}{2}$

10.18 a. $x = \frac{-3 \pm \sqrt{5}}{2}$ b. $x = 1, x = 3$ c. $x = \frac{-9 \pm \sqrt{89}}{2}$ d. $x = 6 \pm \sqrt{33}$
e. $x = \frac{5 \pm \sqrt{21}}{2}$

10.19 a. no solutions b. $x = \frac{6 \pm 3\sqrt{2}}{2}$ c. $x = \frac{-6 \pm 2\sqrt{15}}{3}$ d. $x = \frac{-3 \pm 2\sqrt{2}}{2}$
e. $x = \frac{6 \pm \sqrt{42}}{6}$

10.20 a. $x = \frac{1}{2}, x = -1$ b. no solutions c. $x = -2 \pm \sqrt{5}$ d. $x = \frac{-9 \pm \sqrt{39}}{6}$
e. $x = \frac{2 \pm \sqrt{3}}{2}$

10.21 a. $x = 1 \pm \sqrt{2}$ b. $x = \frac{4 \pm \sqrt{10}}{2}$ c. $x = \frac{9 \pm \sqrt{69}}{6}$ d. $x = \frac{-3 \pm 3\sqrt{2}}{2}$
e. $x = \frac{1 \pm \sqrt{5}}{2}$

10.22 a. no solutions b. $x = -\frac{1}{2}, x = 2$ c. $x = \frac{3 \pm \sqrt{29}}{4}$ d. $x = \frac{-9 \pm \sqrt{87}}{6}$
e. no solutions

10.23 a. $x = -1 \pm \sqrt{3}$ b. $x = \frac{-3 \pm 3\sqrt{3}}{2}$ c. $x = 1 \pm \sqrt{3}$ d. $x = \frac{-15 \pm \sqrt{385}}{8}$
e. $x = \frac{-5 \pm 3\sqrt{5}}{5}$

10.24 a. $x = \frac{-3 \pm \sqrt{11}}{2}$ b. no solutions c. $\frac{-1 \pm \sqrt{17}}{4}$ d. $x = \frac{-3 \pm \sqrt{59}}{4}$
e. no solutions

10.25 a. $x = -1, x = 2$ b. $x = -\frac{5}{3} \pm \frac{1}{3}\sqrt{19}$ c. $x = 1, x = \frac{4}{3}$ d. $x = \pm 2\sqrt{5}$
e. $x = 1 \pm \sqrt{2}$

10.26 a. $x = \pm\frac{1}{2}\sqrt{10 + 2\sqrt{29}}$ b. $x = \pm\frac{1}{2}\sqrt[4]{8}$ c. $x = 6 \pm 4\sqrt{2}$ d. $x = 3 - 2\sqrt{2}$
e. $x = \sqrt[3]{2 \pm \sqrt{2}}$

## 11. Systems of linear equations

11.1 a. $x = 2, y = 0$ b. $x = 1, y = 1$ c. $x = 1, y = -1$ d. $x = 2, y = 1$

e. $x = \frac{1}{2}, y = -\frac{1}{2}$

11.2 a. $x = \frac{4}{33}, y = \frac{5}{33}$ b. $x = -2, y = -3$ c. $x = -2, y = -3$

d. $x = 1, y = -2$ e. $x = 3, y = -4$

11.3 a. $x = -2, y = 2$ b. $x = -4, y = -3$ c. $x = 17, y = -10$

d. $x = 27, y = 7$ e. $x = -12, y = -8$

11.4 a. $x = \frac{21}{19}, y = -\frac{11}{19}$ b. $x = -\frac{22}{9}, y = -\frac{4}{3}$ c. $x = -38, y = 9$

d. $x = -4, y = 5$ e. $x = 29, y = 46$

11.5 a. $x = -1, y = 0, z = 2$ b. $x = 1, y = 1, z = 1$ c. $x = 2, y = -1, z = 1$

d. $x = 1, y = 0, z = -2$ e. $x = -1, y = 3, z = 1$

11.6 a. $x = -1, y = 0, z = -1$ b. $x = -1, y = -1, z = 1$ c. $x = -2, y = 1, z = 0$

d. $x = 0, y = 2, z = -1$ e. $x = 2, y = 1, z = -1$

11.7 a. Eliminating $x$ yields two equations in $y$ and $z$ that are the same (up to a factor), namely $y - 4z = 4$. Every chosen value of $z$ by means of this equation yields a corresponding value of $y$, namely $y = 4z + 4$ and via substitution also a value of $x$, namely $x = 2y - z = 2(4z + 4) - z = 7z + 8$. For instance, choosing $z = 1$ yields $y = 8$ and $x = 15$. Since $z$ may be chosen randomly, there are *infinitely many* solutions. On page 117, this will be treated in a geometric manner.
b. This time, elimination of $x$ yields a system of two equations in $y$ and $z$ that is *contradictory*. This means that there are no solutions.
c. Infinitely many solutions. For instance, choosing $x$ randomly, yields $y = 5 + 2x$ and $z = -1 - x + 3y = -1 - x + 3(5 + 2x) = 14 + 5x$.

11.8 a. No solutions b. No solutions c. Infinitely many solutions.

# V Geometry

## 12. Lines in the plane

For 12.1, 12.2 and 12.3, we only give the intersections of the line with the coordinate axes. You may then verify your drawings yourself.

12.1 a. $(1,0), (0,1)$ b. $(0,0)$ c. $(1,0), (0,2)$ d. $(2,0), (0,-1)$ e. $(4,0), (0,\frac{4}{3})$

12.2 a. $(-3,0), (0,\frac{3}{4})$ b. $(-5,0), (0,-\frac{5}{4})$ c. $(0,0)$ d. $(-2,0), (0,7)$

e. $(-\frac{4}{5},0), (0,-2)$

12.3 a. $(0,0)$ (the line is the $y$-axis) b. $(-3,0)$ (vertical line) c. $(0,0)$ d. $(0,-1)$ (horizontal line) e. $(\frac{1}{3},0), (0,-\frac{1}{2})$

12.4 a. Half plane to the left of the $y$-axis b. Half plane to the right of the vertical line through $(-3,0)$ c. Half plane to the right of the diagonal line $y = x$ d. Half plane below the horizontal line through $(0,-2)$ e. Half plane to the left of the line $y = 3x$

For the exercises 12.5 and 12.6, we don't give a drawing, but only the intersections of the bounding line with the coordinate axes, together with an interior point of the half plane. This way, you can easily verify your own drawings.

12.5 a. $(2,0)$, $(0,2)$, $(0,0)$ b. Half plane to the right of the line $y = 2x$ c. $(1,0)$, $(0,2)$, $(0,0)$

d. $(1,0)$, $(0,-\frac{2}{3})$, $(2,0)$ e. $(\frac{4}{3},0)$, $(0,\frac{4}{3})$, $(2,0)$

12.6 a. $(\frac{3}{5},0)$, $(0,-\frac{3}{4})$, $(1,0)$ b. $(\frac{9}{2},0)$, $(0,-\frac{9}{7})$, $(5,0)$ c. $(-\frac{2}{3},0)$, $(0,-2)$, $(-1,0)$

d. $(-\frac{2}{7},0)$, $(0,2)$, $(-1,0)$ e. $(-5,0)$, $(0,-\frac{5}{2})$, $(0,0)$

12.7

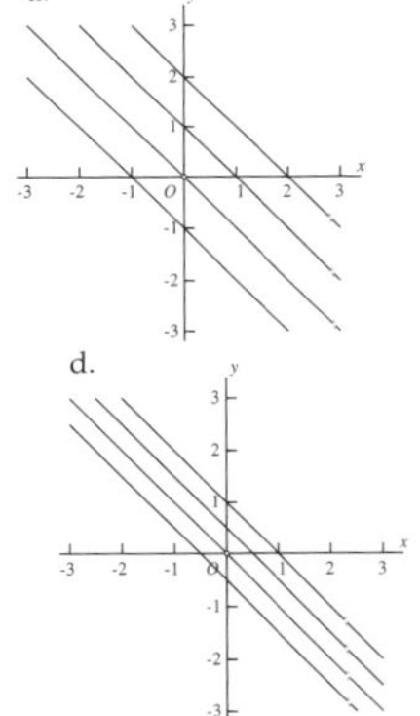

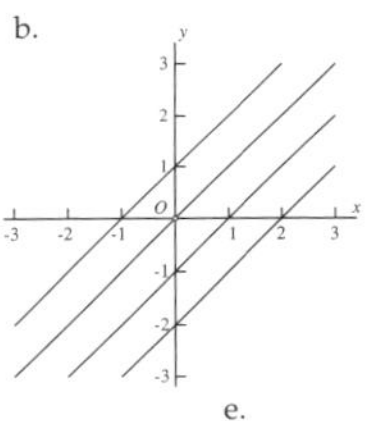

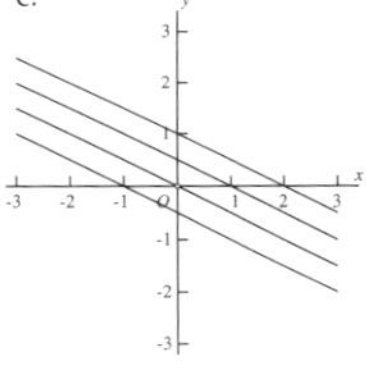

12.8 a. $x + y = 3$ b. $y = 0$ c. $-5x + y = 5$ d. $-5x + 2y = 10$ e. $x = -2$

12.9 a. $2x - 3y = 6$ b. $x = 3$ c. $5x + 2y = 10$ d. $x + y = 0$ e. $x - y = 2$

12.10 a. $x + y = 3$ b. $-2x + 4y = 4$ c. $4x - 2y = -6$ d. $6x - 2y = -16$

e. $x - 5y = 9$

12.11 a. $7x - 2y = 11$ b. $x - y = 6$ c. $4x - 5y = -9$ d. $4x - y = 14$

e. $2x - 5y = 13$

12.12 a. $x + 4y = 0$ b. $3x - 2y = 0$ c. $5x + 2y = -5$ d. $x + y = 1$
e. $2x + y = -4$

12.13 a. $x + y = 10$ b. $y = -1$ c. $5x + 4y = 17$ d. $3x - 5y = 34$ e. $8x - y = 9$

12.14 a. yes b. yes c. yes d. yes e. no

12.15 a. yes b. yes c. no d. yes e. no

12.16 a. $(\frac{3}{2},\frac{1}{2})$ b. $(3,0)$ c. $(-\frac{12}{13},-\frac{4}{13})$ d. parallel lines e. $(\frac{61}{56},-\frac{4}{7})$

12.17 a. $(-\frac{27}{7},-\frac{29}{14})$ b. $(-\frac{16}{13},\frac{1}{13})$ c. $(\frac{14}{5},\frac{7}{10})$ d. $(-\frac{17}{4},\frac{11}{9})$ e. $(\frac{11}{98},-\frac{1}{49})$

12.18 a. $(4, -1)$ b. $(4, -5)$ c. $(-3, -\frac{5}{2})$ d. $(\frac{23}{17}, \frac{18}{17})$ e. parallel lines

12.19 a. $(\frac{17}{3}, \frac{22}{3})$ b. $(-\frac{43}{47}, -\frac{161}{47})$ c. $(\frac{52}{33}, \frac{101}{132})$ d. coinciding lines e. $(\frac{1}{47}, \frac{81}{47})$

12.20 a. $x + y = 0$ b. $2x - y = 2$ c. $-x + 4y = 12$ d. $-5x + 2y = -7$
e. $8x + 7y = 19$

## 13. Distances and angles

13.1 a. $3$ b. $4$ c. $\sqrt{26}$ d. $2\sqrt{2}$ e. $2\sqrt{10}$

13.2 a. $4$ b. $\sqrt{2}$ c. $4\sqrt{2}$ d. $\sqrt{5}$ e. $\sqrt{10}$

13.3 a. $3\sqrt{2}$ b. $\sqrt{2}$ c. $\sqrt{29}$ d. $\sqrt{41}$ e. $\sqrt{5}$

13.4 a. $2\sqrt{5}$ b. $\sqrt{5}$ c. $\sqrt{29}$ d. $5$ e. $\sqrt{10}$

13.5 a. $x = y$ b. $4x + 2y = 5$ c. $x = -1$ d. $x + y = 3$ e. $4x + 2y = -15$

13.6 a. $x + 2y = 2$ b. $2x - 4y = 7$ c. $10x + 4y = 21$ d. $6x - 8y = 3$
e. $x + 3y = 3$

13.7 a. $y = 0,\ x = a$ b. $x = y,\ x + y = a + b$ c. $ax + by = 0,\ bx - ay = 0$
d. $ax + by = a + b$ e. $ax + by = a^2 + b^2$

13.8 a. $x = 0$ b. $x + 2y = 10$ c. $x - y = -3$ d. $4x - 3y = 20$
e. $7x - 6y = 19$

13.9 a. $x + y = -1$ b. $-x + 3y = 17$ c. $5x - 8y = -39$ d. $-2x + 9y = 32$
e. $5x - 7y = -16$

13.10 a. $x + y = 4$ b. $y = 2$ c. $2x - y = -6$ d. $5x - y = -27$
e. $2x - 3y = 7$

13.11 a. $x - 2y = -4$ b. $-3x + 4y = 41$ c. $-x + 3y = 25$ d. $2x + 11y = 73$
e. $7x + 5y = 20$

13.12 a. $3x + 2y = 6$ b. $5x - 4y = 23$ c. $7x + y = -6$ d. $3x - 4y = 48$
e. $x + y = -1$

13.13 a. $9x + 4y = 0$ b. $7x - 2y = 6$ c. $5x + y = -9$ d. $5x - 4y = -4$
e. $7x + 2y = -26$

13.14 a. $(-\frac{3}{13}, -\frac{2}{13})$ b. $(-\frac{1}{2}, -\frac{1}{2})$ c. $(\frac{4}{5}, \frac{3}{5})$ d. $(-\frac{1}{5}, -\frac{8}{5})$ e. $(-\frac{9}{10}, -\frac{13}{10})$

13.15 a. $(\frac{21}{17}, \frac{1}{17})$ b. $(\frac{8}{5}, -\frac{9}{5})$ c. $(-1, 2)$ d. $(-\frac{1}{2}, \frac{5}{2})$ e. $(\frac{16}{5}, \frac{2}{5})$

13.16 a. $-\frac{1}{2}\sqrt{2}, 135°$ b. $0, 90°$ c. $\frac{11}{130}\sqrt{130}, 15°$ d. $-\frac{1}{5}\sqrt{5}, 117°$ e. $-1, 180°$

13.17 a. $-\frac{1}{10}\sqrt{10}, 108°$ b. $-\frac{1}{2}\sqrt{2}, 135°$ c. $\frac{2}{13}\sqrt{13}, 56°$ d. $0, 90°$ e. $-\frac{1}{170}\sqrt{170}$,
$94°$

13.18 a. $79°$ b. $65°$ c. $90°$ d. $18°$ e. $45°$

13.19 a. $7°$ b. $27°$ c. $32°$ d. $74°$ e. $15°$

## 14. Circles

14.1 a. $x^2+y^2-4=0$ b. $x^2+y^2-4x=0$ c. $x^2+y^2+6y-16=0$
d. $x^2+y^2-2x-4y-11=0$ e. $x^2+y^2+4x-4y=0$

14.2 a. $x^2+y^2-8x+15=0$ b. $x^2+y^2-6x+4y=0$
c. $x^2+y^2-4x+2y-20=0$ d. $x^2+y^2-2x-14y+1=0$
e. $x^2+y^2+10x-24y=0$

14.3 a. $M=(-2,1), r=2$ b. $M=(-\frac{1}{2},\frac{1}{2}), r=\frac{1}{2}\sqrt{6}$ c. $M=(-1,-1), r=\sqrt{2}$
d. $M=(4,0), r=2$ e. no circle

14.4 a. no circle b. $M=(2,0), r=3$ c. $M=(0,2), r=0$ d. $M=(0,\frac{1}{3}), r=\frac{1}{3}$
e. $M=(2,1), r=\frac{1}{2}$

14.5 a. $x^2+y^2-2x-2y=0$ b. $x^2+y^2-2x-4y=0$ c. $x^2+y^2-6x-8y=0$
d. $x^2+y^2-4x=0$ e. $x^2+y^2-3x-4y=0$

14.6 a. $x^2+y^2-6x-6y+10=0$ b. $x^2+y^2-2y-4=0$ c. $x^2+y^2+2y-1=0$
d. $x^2+y^2-4x-4y+6=0$ e. $x^2+y^2+2x+2y+1=0$

14.7 a. $(-2\pm\sqrt{3},0),(0,1)$ b. $(\frac{-1\pm\sqrt{5}}{2},0),(0,\frac{1\pm\sqrt{5}}{2})$ c. $(0,0),(-2,0),(0,-2)$
d. $(2,0),(6,0)$ e. $(\frac{-3\pm\sqrt{5}}{2},0),(0,2\pm\sqrt{3})$

14.8 a. $(-1,0),(5,0),(0,\pm\sqrt{5})$ b. $(0,1),(0,5)$ c. $(0,2\pm\sqrt{2})$ d. $(0,0),(0,\frac{2}{3})$
e. no intersection points

14.9 a. $(2,\pm\sqrt{5})$ b. $(\frac{6}{5}\sqrt{5},\frac{3}{5}\sqrt{5}),(-\frac{6}{5}\sqrt{5},-\frac{3}{5}\sqrt{5})$ c. $(3,0),(0,3)$
d. $(-3,0),(\frac{9}{5},-\frac{12}{5})$ e. $(\frac{3}{2}\sqrt{2},-\frac{3}{2}\sqrt{2})$

14.10 a. $(\pm2\sqrt{3},-2)$ b. $(\frac{16}{5},\frac{12}{5}),(-\frac{16}{5},-\frac{12}{5})$ c. $(-4,0),(0,-4)$
d. $(4,0),(-\frac{12}{5},-\frac{16}{5})$ e. $(2\sqrt{3},2),(-2\sqrt{3},-2)$

14.11 a. $(0,0),(-1,-1)$ b. $(3+\sqrt{2},4-\sqrt{2}),(3-\sqrt{2},4+\sqrt{2})$
c. $(-1,0),(-6,-5)$ d. $(0,2),(\frac{8}{5},-\frac{14}{5})$ e. $(1,5),(-\frac{1}{5},\frac{13}{5})$

14.12 a. $(1,4),(1,-2)$ b. no solutions c. $(4,-3),(-3,4)$
d. $(-1,1),(-\frac{38}{5},-\frac{6}{5})$ e. $(1,1),(-\frac{3}{5},-\frac{11}{5})$

14.13 a. $(1,\pm\sqrt{3})$ b. $(0,-3),(\frac{12}{5},\frac{9}{5})$ c. $(-3,-4),(-\frac{24}{5},\frac{7}{5})$ d. $(2,0),(0,2)$
e. no solutions

14.14 a. $(0,0),(\frac{2}{5},-\frac{6}{5})$ b. $(0,0),(3,1)$ c. no solutions d. $(1,1),(\frac{17}{5},\frac{11}{5})$
e. no solutions

14.15 a. $(1,2),(-3,-2)$ b. $(1,1),(\frac{161}{101},\frac{107}{101})$ c. $(-1,0),(-\frac{34}{73},-\frac{104}{73})$
d. $(-1,1),(\frac{1}{5},\frac{23}{5})$ e. $(3,0),(\frac{45}{13},-\frac{4}{13})$

14.16 a. $(1,0)$ b. no solutions c. no solutions d. $(1,-1)$
e. $(2,-3), (\frac{2}{5}, \frac{1}{5})$

14.17 a. $x^2+y^2=16$ b. $(x-2)^2+y^2=2$ c. $x^2+(y-2)^2=\frac{36}{25}$
d. $(x+1)^2+(y+1)^2=\frac{9}{5}$ e. $(x-1)^2+(y-2)^2=8$

14.18 a. $x+2y=5$ b. $x-y=2$ c. $y=1$ d. $x+y=2$ e. $4x-y=5$

14.19 a. $(0,5), (0,-1), (5,0), (-1,0)$
b. $-2x+3y=15, 2x+3y=-3, 3x-2y=15, 3x+2y=-3$

14.20 a. $x=-1\pm\sqrt{3}, y=\pm\sqrt{3}$ b. $x=-2\pm\sqrt{33}, y=3\pm\sqrt{33}$ c. $x=1\pm\sqrt{17}$, $y=2\pm\sqrt{17}$ d. $x=-1\pm\sqrt{17}, y=-4\pm\sqrt{17}$ e. $x=1\pm2\sqrt{3}, y=3\pm2\sqrt{3}$

## 15. Geometry in space

15.1 a. $\sqrt{10}$ b. $3\sqrt{2}$ c. $3\sqrt{3}$ d. $3$ e. $\sqrt{33}$

15.2 a. $\sqrt{3}$ b. $\sqrt{11}$ c. $\sqrt{26}$ d. $\sqrt{6}$ e. $\sqrt{11}$

15.3 N.B.: for typographical reasons, we list the coordinates of the vectors horizontally between brackets and not vertically.
a. $(-3,-1,3)$ b. $(-2,-1,1)$ c. $(-1,6,-1)$ d. $(3,2,4)$ e. $(-1,-3,1)$

15.4 a. $(1,-2,-4)$ b. $(-3,2,-2)$ c. $(3,0,6)$ d. $(1,0,-4)$ e. $(1,1,3)$

15.5 a. $\frac{2}{13}\sqrt{13}, 56°$ b. $-\frac{1}{11}\sqrt{55}, 132°$ c. $-\frac{1}{11}\sqrt{11}, 108°$ d. $-\frac{3}{10}\sqrt{2}, 115°$
e. $-1, 180°$

15.6 a. $\frac{1}{15}\sqrt{15}, 75°$ b. $0, 90°$ c. $-\frac{3}{35}\sqrt{10}, 106°$ d. $0, 90°$ e. $\frac{1}{30}\sqrt{30}, 79°$

15.7 For instance, it is the angle between the vectors $(1,1,1)$ and $(1,0,0)$. Its cosine is $\frac{1}{3}\sqrt{3}$, so the angle, rounded off to degrees, is 55°.

15.8 a. $(1,0,0), (0,3,0), (0,0,-3)$ b. $(\frac{1}{4},0,0), (0,\frac{1}{2},0), (0,0,\frac{1}{3})$
c. $(\frac{1}{a},0,0), (0,\frac{1}{b},0), (0,0,\frac{1}{c})$

15.9 a. $x+y+z=1$ b. $\frac{x}{2}+\frac{y}{3}+\frac{z}{4}=1$ c. $x-y-\frac{z}{3}=1$ d. $z=0$
e. $x=1$

15.10 a. $x=z$ b. $x+y-4z=-6$ c. $x-y+4z=-4$ d. $x+y+z=\frac{3}{2}$ e. $x=z$

15.11 a. $4y-2z+3=0$ b. $-x+2y+z=4$ c. $4x+2y-8z=13$
d. $6x-2y+2z=13$ e. $x=z$

15.12 a. $x=1$ b. $x+z=2$ c. $-2x+3y+z=-2$ d. $x+y-z=0$
e. $y+2z=7$

15.13 a. $3x-2y=-6$ b. $3y-z=11$ c. $3x-y+z=12$ d. $x=6$ e. $y=0$

15.15 a. $3x+2y-4z=16$ b. $2x-2y-3z=4$ c. $-2x+3y-z=5$
d. $5x-y+7z=-16$ e. $x+2z=3$ f. $x=4$

15.16 a. $(1,-1,1)$ b. $(2,0,1)$ c. $(2,-1,1)$ d. $(0,2,1)$ e. $(-3,1,1)$ f. $(4,-4,1)$ g. $(0,1,1)$ h. No intersections with $z=1$. The whole line is in the plane $z=-1$.

15.17 a. $(1,0,1)$ b. $(0,1,-2)$ c. $(-1,1,0)$ d. The planes $\alpha$ and $\gamma$ are parallel. e. The planes intersect in a line. Points on trhis line are, e.g., $(-1,-1,0)$ and $(-9,-3,1)$. f. $(1,-1,1)$ g. The planes $\alpha$ and $\gamma$ are equal; on the intersection line of $\alpha=\gamma$ and $\beta$ are, e.g., the points $(0,-\frac{14}{3},-5)$ and $(1,-3,-3)$ h. $(0,0,-1)$

15.18 For 11.5 and 11.6, each time we have three planes intersecting in one point. Exercise 11.7 a. Three planes through one line. b. The planes intersect pairwise in three parallel lines. c. Three planes through one line.
Exercise 11.8 a. Three parallel intersection lines. b. Three parallel intersection lines. c. Three planes through one line.

15.19 a. $x^2+y^2+z^2-2z-3=0$ b. $x^2+y^2+z^2-4x-4y+4=0$

c. $x^2+y^2+z^2-2x+6z-15=0$ d. $x^2+y^2+z^2-2x-4y+4z=0$

e. $x^2+y^2+z^2+4x-4y-41=0$

15.20 a. $x^2+y^2+z^2-8x-2z+16=0$ b. $x^2+y^2+z^2-6x-2y+4z+1=0$

c. $x^2+y^2+z^2-4x+2z-20=0$ d. $x^2+y^2+z^2-2x-14y+4z+5=0$

e. $x^2+y^2+z^2+10x-4y-2z+21=0$

15.21 a. $M=(-2,1,-1), r=\sqrt{6}$ b. $M=(-\frac{1}{2},\frac{1}{2},0), r=\frac{1}{2}\sqrt{6}$

c. $M=(-1,0,-2), r=\sqrt{5}$ d. $M=(0,0,4), r=2$ e. no sphere

15.22 a. no sphere b. $M=(0,0,2), r=3$ c. $M=(0,2,2), r=2$

d. $M=(0,\frac{1}{3},0), r=\frac{1}{3}$ e. no sphere

15.23 a. $x+2y+2z=9$ b. $x-z=2$ c. $-y+z=0$ d. $3x+3y-2z=8$

e. $4x-y-2z=5$

15.24 a. $y^2+z^2-4y-4z-7=0, x^2+z^2-2x-4z-7=0$

b. $M=(0,2,2), r=\sqrt{15}$, and $M=(1,0,2), r=2\sqrt{3}$, respectively

c. $(1\pm2\sqrt{2},0,0)$, $(0,2\pm\sqrt{11},0)$, $(0,0,2\pm\sqrt{11})$

d. $\sqrt{2}x-y-z=4+\sqrt{2}$, $\sqrt{2}x+y+z=-4+\sqrt{2}$,
$x-\sqrt{11}y+2z=-11-2\sqrt{11}$, $x+\sqrt{11}y+2z=-11+2\sqrt{11}$,
$x+2y-\sqrt{11}z=-11-2\sqrt{11}$, $x+2y+\sqrt{11}z=-11+2\sqrt{11}$

# VI Functions

## 16. Functions and graphs

16.1 a. $-\frac{3}{5}$ b. 2 c. 2 d. 0 e. $-\frac{1}{5}$

16.2 a. $\frac{2}{7}$ b. $\frac{1}{3}$ c. $\frac{5}{2}$ d. $\frac{2}{11}$ e. $\frac{1}{2}$

16.3 a. $18°$ b. $72°$ c. $-53°$ d. $82°$ e. $14°$

16.4 a. $68°$ b. $76°$ c. $45°$ d. $-47°$ e. $-72°$

16.5 a. 0.32 b. 1.25 c. −0.93 d. 1.43 e. 0.24

16.6 a. 1.19 b. 1.33 c. 0.79 d. −0.83 e. −1.25

16.7 a. $y = 3x$ b. $y = -2x + 1$ c. $y = 0.13x + 1.87$ d. $y = -x$ e. $y = 4x - 11$

16.8 a. $y = -4x + 16$ b. $y = 2.22x - 10.66$ c. $y = -3$ d. $y = -1.5x + 0.5$
e. $y = 0.4x - 1.6$

16.9 a. $(0, -1)$ b. $(0, 7)$ c. $(-1, 0)$ d. $(2, 1)$ e. $(-1, -1)$

16.10 a. $(-3, 4)$ b. $(2, -8)$ c. $(\frac{7}{6}, \frac{73}{12})$ d. $(-3, -23)$ e. $(-2, -26)$

16.11 a. $(-1, -4)$ b. $(1, -4)$ c. $(-1, -9)$ d. $(-\frac{1}{4}, -\frac{9}{8})$ e. $(\frac{1}{6}, -\frac{25}{12})$

16.12 a. $(-1, 4)$ b. $(2, 1)$ c. $(-\frac{1}{2}, \frac{9}{4})$ d. $(\frac{3}{4}, -\frac{25}{8})$ e. $(-\frac{1}{3}, -\frac{4}{3})$

16.13 a. $y = 2x^2$ b. $y = -2x^2$ c. $y = \frac{1}{4}x^2$ d. $y = -\frac{1}{2}x^2$ e. $y = -5x^2$

16.14 a. $y = x^2 + 1$ b. $y = -\frac{1}{4}x^2 - 1$ c. $y = -3x^2 - 2$ d. $y = -\frac{1}{8}x^2 + \frac{3}{4}x - \frac{9}{8}$
e. $y = \frac{1}{16}x^2 + \frac{1}{4}x + \frac{1}{4}$

16.15 a. $y = x^2 - 2x + 3$ b. $y = x^2 + 2x + 3$ c. $y = 2x^2 - 8x + 7$ d. $y = x^2 + 3$
e. $y = 3x^2 + 18x + 27$

16.16 a. $y = \frac{2}{3}x^2$ b. $y = x^2 - x - \frac{1}{4}$ c. $y = 11x^2 - \frac{22}{3}x + \frac{2}{9}$ d. $y = 2x^2 + \frac{3}{2}$
e. $y = 2x^2 + 3x + \frac{15}{8}$

16.17 Only the intersections are given.
a. $(2, 4), (-2, 0)$ b. $(-2, -1)$ c. $(0, -3), (-1, -2)$ d. $(\frac{1}{2}, 0), (1, 1)$
e. $(-\frac{1}{3}, -4), (-1, -2)$

16.18 Only the intersections are given.
a. $(1, 2), (-1, 2)$ b. $(1, 0)$ c. $(1, 0), (2, 5)$ d. $(1, 3), (-\frac{1}{3}, \frac{7}{9})$ e. $(1, -4)$

16.19 a. $x \leq -2$ or $x \geq 1$ b. $x \leq -1$ or $x \geq 1$ c. $-1 \leq x \leq 2$ d. $x \leq \frac{3-\sqrt{33}}{4}$ of $x \geq \frac{3+\sqrt{33}}{4}$ e. $0 \leq x \leq \frac{2}{3}$

16.20 a. $x \leq 0$ of $x \geq 1$ b. $1 - \sqrt{3} \leq x \leq 1 + \sqrt{3}$ c. $1 \leq x \leq \frac{3}{2}$ d. $x \leq -1$ of $x \geq \frac{5}{3}$ e. $-\frac{4}{5} \leq x \leq 2$

16.21 a. $-2 < p < 2$ b. $p > \frac{1}{4}$ c. $p < -1$ d. $0 < p < 4$ e. $-6 < p < 2$

16.22 a. for all $p$ b. $p > -\frac{5}{4}$ c. $p > -\frac{1}{3}$ and $p \neq 0$ d. $p < -2, p > 6$ e. $p \neq -2$ and $p \neq -1$

16.23 We only give the two asymptotes and the intersection points with the coordinate axes (if any).
a. $x = 1, y = 0, (0, -1)$ b. $x = 0, y = 1, (-1, 0)$ c. $x = 2, y = 0, (0, -\frac{3}{4})$ d. $x = 5$, $y = 2, (0, 0)$ e. $x = 2, y = 1, (0, -1), (-2, 0)$

16.24 a. $x = \frac{3}{2}, y = -\frac{1}{2}, (0,0)$ b. $x = \frac{1}{3}, y = \frac{1}{3}, (-2,0), (0,-2)$ c. $x = \frac{1}{5}, y = -\frac{2}{5}, (2,0), (0,-4)$ d. $x = -\frac{4}{7}, y = \frac{1}{7}, (-3,0), (0, \frac{3}{4})$ e. $x = \frac{1}{2}, y = -\frac{1}{2}, (\frac{3}{2},0), (0, -\frac{3}{2})$

16.25 a. $x \leq -4$ or $x \geq -2$ b. $0 \leq x \leq 2$ c. $x \leq 0$ or $x \geq 10$ d. $x \leq \frac{3}{4}$

e. $-\frac{1}{5} \leq x \leq 3$

16.26 a. $x \leq -1$ or $x \geq 1$ b. $x \leq -6$ or $x \geq -\frac{2}{5}$ c. $x \leq -3$ or $x \geq 2$

d. $x \leq \frac{2}{7}$ or $x \geq 2$ e. $x \leq -7$ or $x \geq -\frac{1}{3}$

16.27 a. $(-4,-8), (1,2)$ b. $(2,0), (-3,5)$ c. $(-3,1), (4,8)$ d. $(\frac{5}{2},-2), (1,1)$

16.28 a. $(-\frac{2}{3},-7), (2,1)$ b. $(-4,-2), (-1,1)$ c. $(-2,4), (0,2)$ d. $(1,-2), (3,4)$

16.29 Just do it!

16.30 Just do it!

16.31 Just do it!

16.32 Just do it!

16.33 a. $0 \leq x \leq 1$ b. $x = 0, x = 1, x = -1$ c. $x \leq -1, x \geq 1, x = 0$ d. $x = 0, x \geq 1$ e. $-1 \leq x \leq 1$

16.34 a. $x = -\frac{1}{2}, x = -\frac{5}{2}$ b. no solution c. $x = \frac{3}{2}$ d. $-\frac{1}{2} \leq x \leq \frac{3}{2}$

e. $1 - \sqrt{2} < x < 1,\ \ 1 < x < 1 + \sqrt{2}$

16.35 a. $x = 0, x = 1$ b. $x = \frac{5 \pm \sqrt{5}}{2}$ c. $x \geq 3$ d. $x \leq \frac{-1+\sqrt{5}}{2}$ e. $x \geq -\frac{1}{2}$

16.36 a. $x(x-1)(x-2)(x-3)(x-4)$ b. $x^2(x-1)(x-2)(x-3)$ c. $x^3(x-1)(x-2)$ d. $x^4(x-1)$ e. $x^5$ f. $x^6+1$

16.37 a. $x+1$ b. $2x+2$ c. $x^2-x+1$ d. $x^5+x^4+x^3+x^2+x+1$ e. $2x^2-4x+4$ f. $x^3+x^2-x-1$ g. $-x^2-5x+2$

16.38 a. $2x^3+2x^2+2x+2$ b. $x^2-x+2$ c. $x^2-2x+4$ d. $x^3+2x^2+4x+8$ e. $x^2-2x$ f. $2x^2-4x+4$ g. $x^3-10x^2+4x-4$

16.39 a. VIII b. V c. II d. III e. I f. IV g. VII h. IX i. VI

## 17. Trigonometry

17.1 a. $\frac{\pi}{6}$ b. $\frac{\pi}{4}$ c. $\frac{\pi}{3}$ d. $\frac{7\pi}{18}$ e. $\frac{\pi}{12}$

17.2 a. $\frac{\pi}{9}$ b. $\frac{5\pi}{18}$ c. $\frac{4\pi}{9}$ d. $\frac{5\pi}{9}$ e. $\frac{5\pi}{6}$

17.3 a. $\frac{13\pi}{18}$ b. $\frac{3\pi}{4}$ c. $\frac{10\pi}{9}$ d. $\frac{4\pi}{3}$ e. $\frac{11\pi}{6}$

17.4 a. $30°$ b. $210°$ c. $60°$ d. $120°$ e. $45°$

17.5 a. $225°$ b. $75°$ c. $82.5°$ d. $337.5°$ e. $345°$

17.6 a. $177.5°$ b. $307.5°$ c. $250°$ d. $97.5°$ e. $155°$

17.7 a. $330°$ b. $85°$ c. $200°$ d. $340°$ e. $155°$

17.8 a. $140°$ b. $70°$ c. $110°$ d. $280°$ e. $110°$

17.9 a. 290° b. 215° c. 120° d. 120° e. 10°

17.10 a. $\frac{9}{2}$ b. $\frac{\pi}{8}$ c. $2\sqrt{\pi}$

17.11 a. $z^2 + z^2 = 1$ so $z = \sqrt{\frac{1}{2}} = \frac{1}{2}\sqrt{2}$
b. $z^2 = 1^2 - \frac{1}{2}^2 = \frac{3}{4}$ so $z = \sqrt{\frac{3}{4}} = \frac{1}{2}\sqrt{3}$
c. You can read off the answers directly from the figures.

17.12 a. $\frac{1}{2}\sqrt{3}$ b. $-\frac{1}{2}\sqrt{2}$ c. $\frac{1}{2}\sqrt{3}$ d. 1 e. $\frac{1}{2}$

17.13 a. 0 b. 0 c. $-1$ d. 0 e. 0

17.14 a. $-\frac{1}{2}\sqrt{3}$ b. $-1$ c. $-\frac{1}{2}\sqrt{3}$ d. $\sqrt{3}$ e. $-\frac{1}{2}\sqrt{2}$

17.15 a. $\sqrt{3}$ b. $-\frac{1}{2}\sqrt{2}$ c. $\frac{1}{2}$ d. 1 e. $\frac{1}{2}\sqrt{3}$

17.16 a. $-1$ b. 0 c. 0 d. 0 e. 1

17.17 a. $-\frac{1}{2}$ b. $-1$ c. $-\frac{1}{2}\sqrt{3}$ d. $-\frac{1}{3}\sqrt{3}$ e. $-\frac{1}{2}\sqrt{2}$

17.18 If $\alpha$ runs from $-\frac{1}{2}\pi$ to $\frac{1}{2}\pi$, the tangent runs from $-\infty$ to $\infty$.

17.19 a. $a = 1.5898, b = 2.5441$ b. $a = 1.7820, b = 0.9080$ c. $b = 1.9314, c = 2.7803$
d. $a = 6.9734, b = 0.6101$ e. $a = 0.6377, c = 3.0670$

17.20 a. $b = 7.0676, c = 7.6779$ b. $a = 2.3007, c = 3.0485$ c. $a = 7.7624, b = 1.9354$
d. $a = 0.7677, c = 2.1423$ e. $a = 1.2229, c = 4.1828$

17.21 a. $a = 2.6736, b = 1.3608$ b. $a = 1.9177, b = 3.5103$
c. $b = 9.8663, c = 10.0670$ d. $a = 18.0051, c = 19.3179$ e. $b = 3.5617, c = 4.6568$

17.22 a. $\frac{2}{5}\sqrt{6}, \frac{1}{12}\sqrt{6}$ b. $\frac{3}{7}\sqrt{5}, \frac{3}{2}\sqrt{5}$ c. $\frac{1}{8}\sqrt{55}, \frac{3}{55}\sqrt{55}$ d. $\frac{1}{5}\sqrt{21}, \frac{1}{2}\sqrt{21}$
e. $\frac{2}{7}\sqrt{6}, \frac{2}{5}\sqrt{6}$

17.23 a. $\frac{1}{4}\sqrt{7}, \frac{3}{7}\sqrt{7}$ b. $\frac{1}{6}\sqrt{35}, \sqrt{35}$ c. $\frac{3}{8}\sqrt{7}, \frac{1}{21}\sqrt{7}$ d. $\frac{1}{8}\sqrt{39}, \frac{1}{5}\sqrt{39}$ e. $\frac{12}{13}, \frac{12}{5}$

17.24 a. $\frac{5}{31}\sqrt{37}, \frac{6}{185}\sqrt{37}$ b. $\frac{3}{23}\sqrt{57}, \frac{3}{4}\sqrt{57}$ c. $\frac{2}{3}, \frac{1}{2}\sqrt{5}$ d. $\frac{1}{3}\sqrt{2}, \frac{1}{7}\sqrt{14}$
e. $\frac{1}{4}\sqrt{6}, \frac{1}{5}\sqrt{15}$

17.25 We give the elaboration of part (a.). The other parts are left to the reader.
a. $\cos 2\alpha = \cos^2\alpha - \sin^2\alpha = \cos^2\alpha - (1 - \cos^2\alpha) = 2\cos^2\alpha - 1$
$= 2(1 - \sin^2\alpha) - 1 = 1 - 2\sin^2\alpha$

17.26 $\cos\left(\frac{\pi}{2} - x\right) = \cos\frac{\pi}{2}\cos x + \sin\frac{\pi}{2}\sin x = 0 \times \cos x + 1 \times \sin x = \sin x$.
$\sin\left(\frac{\pi}{2} - x\right) = \sin\frac{\pi}{2}\cos x - \cos\frac{\pi}{2}\sin x = 1 \times \cos x - 0 \times \sin x = \cos x$.

17.27 Expand and simplify.

17.28 a. Use that $\sin^2\frac{\pi}{8} = (1 - \cos\frac{\pi}{4})/2 = \frac{2-\sqrt{2}}{4}$. It follows that:
$\sin\frac{\pi}{8} = \frac{1}{2}\sqrt{2 - \sqrt{2}}$

b. $\frac{1}{2}\sqrt{2 + \sqrt{2}}$ c. $\frac{\sqrt{2-\sqrt{2}}}{\sqrt{2+\sqrt{2}}}$ d. $\frac{1}{2}\sqrt{2 - \sqrt{3}}$ e. $\frac{1}{2}\sqrt{2 + \sqrt{3}}$

17.29 a. $\sin\frac{3}{8}\pi = \cos\frac{1}{8}\pi = \frac{1}{2}\sqrt{2 + \sqrt{2}}$ b. $-\frac{1}{2}\sqrt{2 + \sqrt{2}}$ c. $-\frac{\sqrt{2+\sqrt{2}}}{\sqrt{2-\sqrt{2}}}$

d. $\frac{1}{2}\sqrt{2+\sqrt{3}}$ e. $-\frac{\sqrt{2+\sqrt{3}}}{\sqrt{2-\sqrt{3}}}$

17.30 a. $-\frac{1}{2}\sqrt{2+\sqrt{2}}$ b. $-\frac{1}{2}\sqrt{2-\sqrt{3}}$ c. $-\frac{\sqrt{2-\sqrt{2}}}{\sqrt{2+\sqrt{2}}}$ d. $\frac{\sqrt{2-\sqrt{3}}}{\sqrt{2+\sqrt{3}}}$ e. $\frac{1}{2}\sqrt{2-\sqrt{2}}$

17.31 a. $x = -\frac{\pi}{6} + 2k\pi$ or $x = \frac{7\pi}{6} + 2k\pi$ b. $x = \frac{\pi}{3} + 2k\pi$ or $x = -\frac{\pi}{3} + 2k\pi$
c. $x = \frac{3\pi}{4} + k\pi$

17.32 a. $x = \frac{\pi}{4} + 2k\pi$ or $x = \frac{3\pi}{4} + 2k\pi$ b. $x = \frac{5\pi}{6} + 2k\pi$ or $x = \frac{7\pi}{6} + 2k\pi$
c. $x = \frac{2\pi}{3} + k\pi$

17.33 a. $x = \frac{\pi}{6} + k\pi$ b. $x = \frac{3\pi}{4} + 2k\pi$ or $x = \frac{5\pi}{4} + 2k\pi$ c. $x = \frac{\pi}{2} + k\pi$

In exercises 17.34 to 17.42 we only give the period length and the zeroes. In each case, $k$ is an arbitrary integer. Then, you can verify your graphs yourself.

17.34 a. $2\pi, x = k\pi$ b. $2\pi, x = \frac{\pi}{2} + k\pi$ c. $\pi, x = k\pi$ d. $2\pi, x = k\pi$
e. $2\pi, x = \frac{\pi}{2} + k\pi$

17.35 a. $\pi, x = \frac{\pi}{2} + k\pi$ b. $2\pi, x = \frac{2\pi}{3} + k\pi$ c. $2\pi, x = \frac{\pi}{3} + k\pi$
d. $2\pi, x = \frac{3\pi}{4} + k\pi$ e. $\pi, x = \frac{\pi}{3} + k\pi$

17.36 a. $\pi, x = \frac{5\pi}{6} + k\pi$ b. $2\pi, x = \frac{\pi}{2} + k\pi$ c. $2\pi, x = \frac{\pi}{3} + k\pi$
d. $2\pi, x = \frac{\pi}{4} + k\pi$ e. $\pi, x = \frac{5\pi}{6} + k\pi$

17.37 a. $2\pi, x = \frac{\pi}{2} + k\pi$ b. $2\pi, x = \frac{\pi}{4} + k\pi$ c. $\pi, x = \frac{2\pi}{3} + k\pi$ d. $2\pi$, $x = \frac{5\pi}{6} + k\pi$ e. $2\pi, x = \frac{\pi}{3} + k\pi$

17.38 a. $\frac{\pi}{2}, x = k\frac{\pi}{2}$ b. $\frac{2\pi}{3}, x = \frac{\pi}{6} + \frac{k\pi}{3}$ c. $3\pi, x = \frac{3k\pi}{2}$ d. $\frac{8\pi}{5}, x = \frac{2\pi}{5} + \frac{4k\pi}{5}$
e. $\frac{\pi}{8}, x = \frac{k\pi}{8}$

17.39 a. $\frac{\pi}{3}, x = \frac{5\pi}{18} + k\frac{\pi}{3}$ b. $\pi, x = \frac{\pi}{4} + \frac{k\pi}{2}$ c. $\pi, x = \frac{\pi}{6} + \frac{k\pi}{2}$ d. $4\pi$, $x = \frac{\pi}{2} + 2k\pi$ e. $3\pi, x = \frac{5\pi}{2} + 3k\pi$

17.40 a. $1, x = \frac{k}{2}$ b. $\frac{2}{3}, x = \frac{1}{6} + \frac{k}{3}$ c. $1, x = k$ d. $1, x = \frac{k}{2}$ e. $1, x = \frac{1}{6} + \frac{k}{2}$

17.41 a. $\frac{1}{6}, x = \frac{k}{6}$ b. $\frac{1}{2}, x = \frac{1}{8} + \frac{k}{4}$ c. $1, x = \frac{1}{6} + \frac{k}{2}$ d. $\frac{8}{5}, x = \frac{2}{5} + \frac{4k}{5}$ e. $3, x = 3k$

17.42 a. $1, x = \frac{5}{6} + k$ b. $4, x = 1 + 2k$ c. $\frac{2}{7}, x = \frac{1}{21} + \frac{k}{7}$ d. $\frac{2}{5}, x = \frac{3}{20} + \frac{k}{5}$
e. $\frac{4}{3}, x = \frac{2}{9} + \frac{4k}{3}$

17.43 a. $-\frac{\pi}{2}$ b. $\frac{\pi}{2}$ c. $-\frac{\pi}{4}$ d. $\frac{\pi}{4}$ e. $\frac{5\pi}{6}$

17.44 a. $-\frac{\pi}{6}$ b. $\frac{3\pi}{4}$ c. $\frac{\pi}{3}$ d. $-\frac{\pi}{3}$ e. $\pi$

17.45 a. $0$ b. $\pi$ c. $0$ d. $\frac{\pi}{3}$ e. $\frac{\pi}{4}$

17.46 a. $\frac{2}{3}\sqrt{2}$ b. $\frac{1}{4}\sqrt{2}$ c. $\frac{4}{9}\sqrt{2}$ d. $\frac{2}{3} - \frac{1}{6}\sqrt{2}$ e. $\frac{1}{6}\sqrt{18+12\sqrt{2}}$

17.47 a. $-\frac{5}{7}$ b. $1$ c. $\frac{3}{4}$ d. $\frac{1}{2}\sqrt{2}$ e. $-\frac{1}{2}\sqrt{2}$

17.48 a. $\frac{4}{5}$ b. $\frac{1}{3}\sqrt{5}$ c. $\frac{4}{5}$ d. $\frac{5}{12}\sqrt{6}$ e. $-\frac{4}{17}\sqrt{17}$

17.49 a. $\frac{\pi}{2} - \frac{\pi}{5} = \frac{3\pi}{10}$ b. $\frac{\pi}{14}$ c. $-\frac{\pi}{6}$ d. $-\frac{\pi}{10}$ e. $-\frac{\pi}{5}$

17.50 Instead of a drawing, we give some characteristics of the graph, so you can check your answer yourself.
a. domain $[-\frac{1}{2}, \frac{1}{2}]$, range $[-\frac{\pi}{2}, \frac{\pi}{2}]$, zero $x = 0$
b. domain $[-\frac{1}{2}, \frac{1}{2}]$, range $[0, \pi]$, zero $x = \frac{1}{2}$
c. domain $\langle -\infty, \infty \rangle$, range $\langle -\frac{\pi}{2}, \frac{\pi}{2} \rangle$, zero $x = 0$, horizontal asymptotes: $y = \pm\frac{\pi}{2}$, decreasing function
d. reflection of (a) in the $y$-axis
e. domain $[-3, 3]$, range $[0, \pi]$, zero $x = -3$

17.51 a. domain $[-3, 3]$, range $[-\frac{\pi}{2}, \frac{\pi}{2}]$, zero $x = 0$
b. domain $[-2, 2]$, range $[0, \pi]$, zero $x = -2$
c. domain $\langle -\infty, \infty \rangle$, range $\langle -\frac{\pi}{2}, \frac{\pi}{2} \rangle$, zero $x = 0$, horizontal asymptotes: $y = \pm\frac{\pi}{2}$
d. domain $[0, 2]$, range $[-\frac{\pi}{2}, \frac{\pi}{2}]$, zero $x = 1$
e. domain $[-2, 0]$, range $[0, \pi]$, zero $x = 0$

17.52 a. $\frac{\pi}{2}$ b. $-\frac{\pi}{2}$ c. $-\frac{\pi}{2}$ d. $-\frac{\pi}{2}$ e. $\frac{\pi}{2}$

17.53 a. domain $\langle -\infty, \infty \rangle$, range $\langle -\frac{\pi}{2}, \frac{\pi}{2} \rangle$, zero $x = \frac{1}{3}$, horizontal asymptotes: $y = \pm\frac{\pi}{2}$
b. domain $[0, 1]$, range $[-\frac{\pi}{2}, \frac{\pi}{2}]$, zero $x = \frac{1}{2}$
c. domain $[-1, 1]$, range $[0, \pi]$, zero $x = -1$
d. domain $\langle -\infty, \infty \rangle$, range $\langle -\frac{\pi}{2}, \frac{\pi}{4} ]$, zeroes $x = 1$ and $x = -1$, horizontal asymptote: $y = -\frac{\pi}{2}$
e. domain $\langle -\infty, 0 \rangle$ en $\langle 0, \infty \rangle$, range $\langle -\frac{\pi}{2}, 0 \rangle$ en $\langle 0, \frac{\pi}{2} \rangle$, no zeroes, horizontal asymptote: $y = 0$, $\lim_{x \downarrow 0} f(x) = \frac{\pi}{2}$, $\lim_{x \uparrow 0} f(x) = -\frac{\pi}{2}$

17.54 a. $\frac{1}{2}$ b. 1 c. $\frac{7}{3}$ d. 4 e. $\frac{1}{3}$

17.55 a. 1 b. 16 c. $\frac{2}{3}$ d. 0 e. $\frac{1}{3}$

17.56 a. Put $y = x - \pi$, then $x = y + \pi$ and $y \to 0$ if $x \to \pi$, so

$$\lim_{x \to \pi} \frac{\sin x}{x - \pi} = \lim y \to 0 \frac{\sin(y + \pi)}{y} = \lim y \to 0 \frac{-\sin y}{y} = -1$$

b. $-\frac{1}{2}$ c. $\frac{1}{3}$ d. 1 (put $y = \frac{\pi}{2} - \arccos x$, then $x = \cos(\frac{\pi}{2} - y) = \sin y$) e. $\sqrt{2}$

17.57 a. 1 b. $\frac{2}{3}$ c. 1 d. 1 e. 1

17.58 a. $b = 4.3560, O = 5.3524$ b. $a = 2.6994, O = 1.0980$ c. $c = 7.9806, O = 15.9394$ d. $\gamma = 0.6214, O = 6.9731$ e. $\alpha = 0.2983, O = 3.1983$ f. $\gamma = 0.6029, O = 3.4024$ g. $\gamma = 1.4455, O = 9.9216$ h. $\alpha = 0.9273, O = 12.0000$ i. $a = 8.5965, O = 14.6492$ j. $a = 3.0155, O = 2.8366$ k. $c = 6.8946, O = 14.6175$

## 18. Exponentials and logarithms

In the next two exercises, sometimes several 'simple' solutions exist.

18.1 a. $2^{243}$ b. $2^{30}$ c. $2^{118}$ d. $2 \cdot 6^x$ e. $2^{3x}$ (or $8^x$)

18.2 a. $2^{12-8x}$ (or $16^{3-2x}$) b. $3^{x-1}$ c. $10^{4x+2}$ (or $100^{2x+1}$) d. $2^{2-4x}$ (or $4^{1-2x}$) e. $10000^{-x}$

In the next four exercises, each time we give the horizontal asymptote, the intersection of the graph with the $y$-axis and the increasing or decreasing behaviour of the graph. This suffices to check your graphs.

18.3 a. $y = 0$, $(0, \frac{1}{2})$, increasing b. $y = 0$, $(0, 2)$, decreasing c. $y = 0$, $(0, 1)$, increasing

d. $y = 0$, $(0, 1.21)$, increasing e. $y = 0$, $(0, 1)$, increasing

18.4 a. $y = 0$, $(0, \frac{1}{3})$, increasing b. $y = 0$, $(0, \frac{1}{27})$, increasing c. $y = 0$, $(0, \frac{1}{10})$, decreasing

d. $y = 0$, $(0, \frac{100}{81})$, decreasing e. $y = 0$, $(0, \frac{4}{9})$, decreasing

18.5 a. $y = -\frac{1}{2}$, $(0, 0)$, increasing b. $y = -8$, $(0, -6)$, decreasing c. $y = -100$, $(0, -99)$, increasing d. $y = 1$, $(0, 2.21)$, increasing e. $y = -9$, $(0, -8)$, increasing

18.6 a. $y = -4$, $(0, -\frac{7}{2})$, increasing b. $y = -49$, $(0, -42)$, decreasing c. $y = 2$, $(0, \frac{19}{9})$, decreasing d. $y = 1$, $(0, \frac{7}{3})$, increasing e. $y = -13$, $(0, -12)$, increasing

18.7

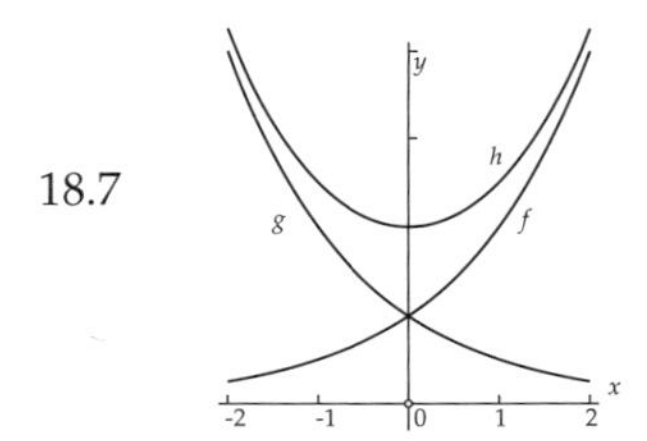


18.8

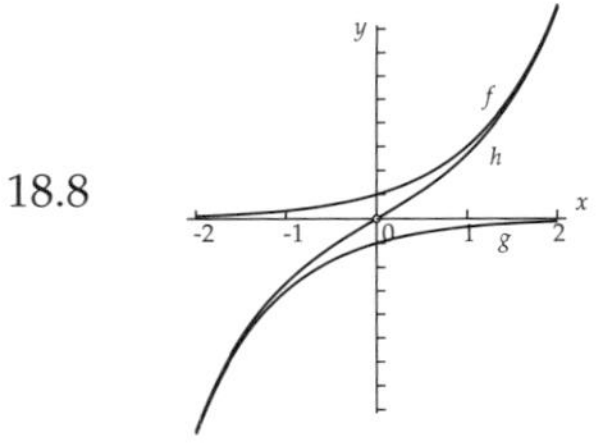


18.9 a. $-2$ b. $\frac{3}{2}$ c. $\log_2 15$ d. $1 - \log_3 4$ (or $\log_3 \frac{3}{4}$) e. $7 \log_{10} 2$

18.10 a. $5 \log_5 2$ b. $\log_5 2$ c. $6(\log_5 2)^2$ d. $\frac{3}{2}$ e. $0$

In the next four exercises, we give the vertical asymptote and the intersection(s) with the horizontal axis.

18.11 a. $x = 1$, $(2, 0)$ b. $x = 1$, $(0, 0)$ c. $x = 0$, $(\frac{1}{2}, 0)$ d. $x = -2$, $(-1, 0)$

e. $x = 0$, $(\frac{1}{3}, 0)$

18.12 a. $x = 2$, $(3, 0)$ b. $x = 0$, $(\frac{1}{4}, 0)$ c. $x = -\frac{10}{3}$, $(-3, 0)$ d. $x = \frac{2}{3}$, $(1, 0)$

e. $x = 0$, $(\frac{1}{32}, 0)$

18.13 a. $x = 0$, $(1, 0)$, $(-1, 0)$ b. $x = 0$, $(\frac{1}{4}, 0)$, $(-\frac{1}{4}, 0)$ c. $x = 1$, $(2, 0)$, $(0, 0)$

d. $x = 0$, $(1, 0)$ e. $x = 0$, $(1, 0)$, $(-1, 0)$

18.14 a. $x = 0$, $(2, 0)$ b. $x = 0$, $(\frac{1}{2}\sqrt{6}, 0)$, $(-\frac{1}{2}\sqrt{6}, 0)$ c. $x = \frac{10}{3}$, $(3, 0)$, $(\frac{11}{3}, 0)$

d. $x = 0$, $(\frac{1}{4}\sqrt[5]{16}, 0)$ e. $x = 0$, $(10\sqrt{10}, 0)$, $(-10\sqrt{10}, 0)$

18.15 Note that $g(x) = f(x) + 1$ and $h(x) = f(x) + 2$.

18.16 Note that $g(x) = f(-x)$ and that $h(x) = f(x) + g(x)$ on the domain $D_h = \{-1 < x < 1\}$.

18.17 a. $x = \log_2 5$ b. $x = (\log_5 2) - 5$ c. $x = 3$ d. $x = \log_{10} 5$ e. $x = \frac{4}{5}$

18.18 a. $x = 100$ b. $x = \sqrt[4]{10}$ c. $x = 13$ d. $x = \pm 2\sqrt{2}$ e. $x = 81$

18.19

a. 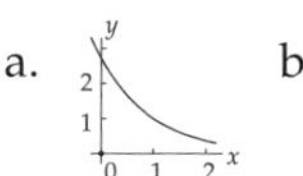 b. 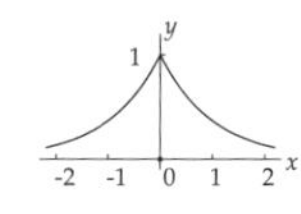 c. 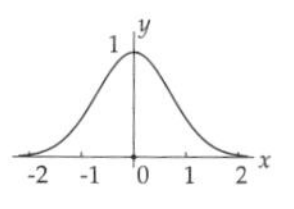 d. 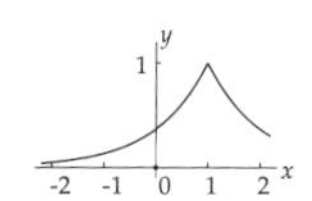 e. 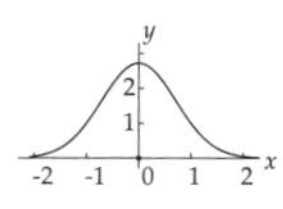

18.20

a. 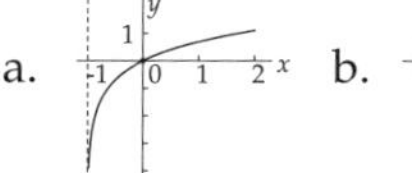 b. c. 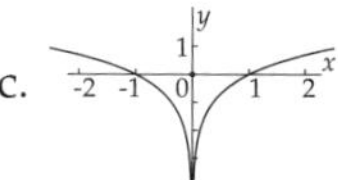 d. 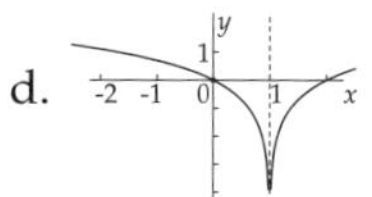 e. 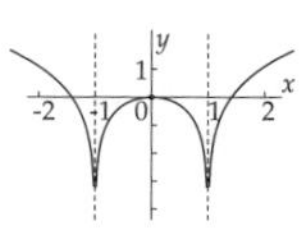

18.21

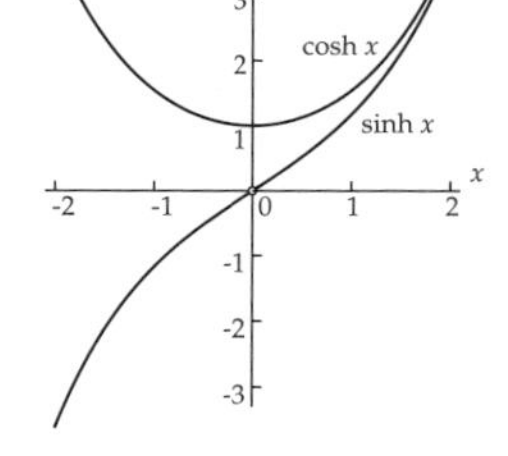


18.22 a. $-1$ b. $2$ c. $-3$ d. $a$ e. $0$

18.23 a. e (stel $y = x - 1$) b. $\mathrm{e}^2$ c. $\mathrm{e}^a$ d. $1$ e. $1$

In the next two exercises, we give the domain $D$ and, if necessary, a simplified function rule.

18.24 a. $D = \{x > 4\}$ b. $D = \{x > 0\}$ c. $D = \{x < 4\}$ d. $D = \{x > 1\}$ e. $D = \{x > \frac{3}{2}\}$ f. $D = \{x < \frac{2}{3}\}$ g. $D = \{x \neq 3\}$ h. $D = \{x > 0\}$ $f(x) = -\ln x$ i. $D = \{x > 1\}$ $f(x) = -\ln(x-1)$ j. $D = \{x > 2\}$ $f(x) = \ln 2 - \ln(x-2)$

18.25 a. $D = \{x \neq 0\}, f(x) = -2\ln|x|$ b. $D = \{x < \frac{1}{2}\}, f(x) = \ln 2 - \ln(1-2x)$ c. $D = \{x > 0\}, f(x) = -\frac{1}{2}\ln x$ d. $D = \{x \neq 0\}, f(x) = -\ln|x|$ e. $D = \{x \neq 2\}$, $f(x) = \ln 2 - \ln|x-2|$ f. $D = \{x > 0\}, f(x) = \ln 3 - 3\ln x$ g. $D = \{x \neq 1, -1\}$, $f(x) = \ln|x-1| - \ln|x+1|$

18.26 $\log_a b = \dfrac{\log_b b}{\log_b a} = \dfrac{1}{\log_b a}$

18.27 a. $-1$ b. $2$ c. $-\frac{3}{2}$ d. $0$ e. $-1$

18.28 a. $\frac{1}{\ln 2}$ b. $\frac{-1}{\ln 3}$ c. $\frac{1}{2}$ d. $\frac{1}{3}$ e. $\frac{1}{a}$

18.29 a. 0 b. 0 c. 0 d. 0 e. 0

18.30 a. 0 b. 0 c. 0 (put $y = -x$) d. 0 e. 0

18.31 a. 0 (put $y = -x$) b. 0 c. $-\infty$ d. 0 (put $y = -x$ and note that $2^3 = 8 < 3^2 = 9$) e. $-\infty$

18.32 a. 0 b. 0 c. 0 d. 0 e. 0

18.33 a. 0 b. 0 c. 0 d. 0 e. 0

## 19. Parameterized curves

19.1 clockwise tracing, $P_0 = (0, 2)$

19.2 $(-3\cos t, 2\sin t)$

19.3 $(4\cos t, 5\sin t)$

19.4 a. $(2\cos t, 2\sin t)$ b. $(-1 + 3\cos t, 3 + 3\sin t)$ c. $(2 + 5\cos t, -3 + 5\sin t)$
d. $(t^2, t)$ e. $(t, \frac{1}{t})$

19.5 a. Left-hand branche: $t < 0$, right-hand branche: $t > 0$.
Asymptotes are obtained by letting $x$ or $y$ go to infinity, which occurs if $t \to \infty$, $t \to -\infty$, $t \downarrow 0$ or $t \uparrow 0$. If $t \to \pm\infty$, then $x - y = \frac{2}{t} \to 0$, while both $x$ and $y$ go to infinity. Therefore, the line $x - y = 0$ is an asymptote.
If $t \downarrow 0$ or $t \uparrow 0$, then $x + y = 2t \to 0$, while both $x$ and $y$ go to infinity (with opposite signs). Therefore, the line $x + y = 0$ is also an asymptote.

19.6 a. V b. VIII c. VII d. III e. I f. II g. VI h. IV

19.7

a. 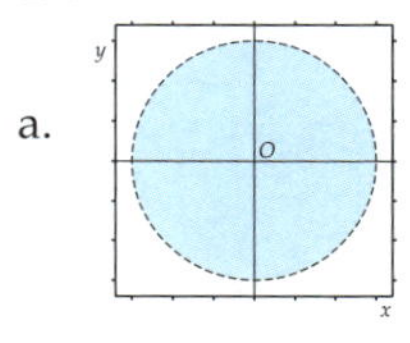


b. 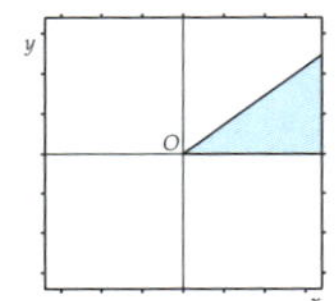


c. 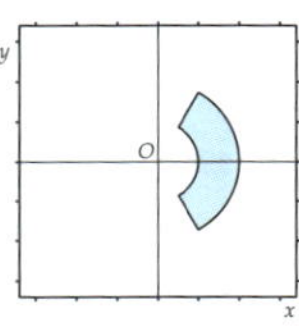


d. 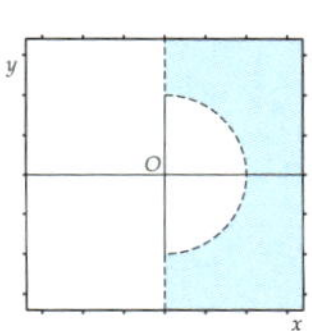


e. 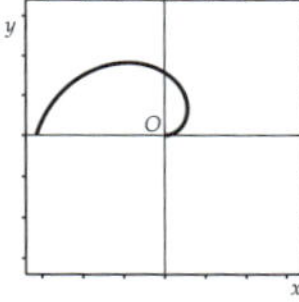


f. 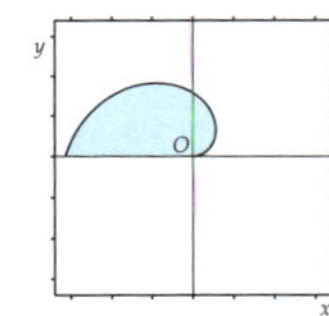


g. 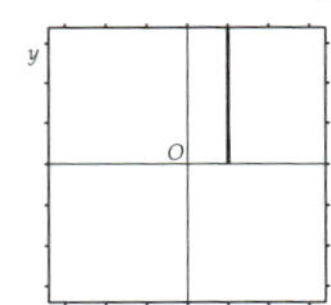


19.8 $d^2 = (r_1\cos\varphi - r_2)^2 + (r_1\sin\varphi)^2 = r_1^2(\cos^2\varphi + \sin^2\varphi) + r_2^2 - 2r_1r_2\cos\varphi = r_1^2 + r_2^2 - 2r_1r_2\cos\varphi$

19.9 a. VI b. I c. II d. III e. VII f. VIII g. IV h. V

19.10 $(\cos 8\pi t, -\sin 8\pi t, -t)$

19.11 All curves are drawn inside the cube with vertices $(\pm1, \pm1, \pm1)$.

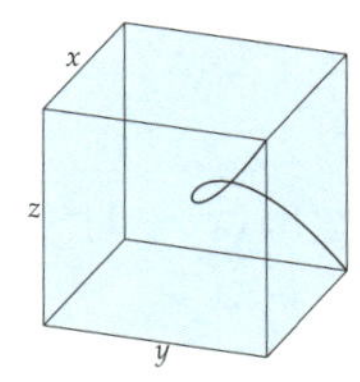


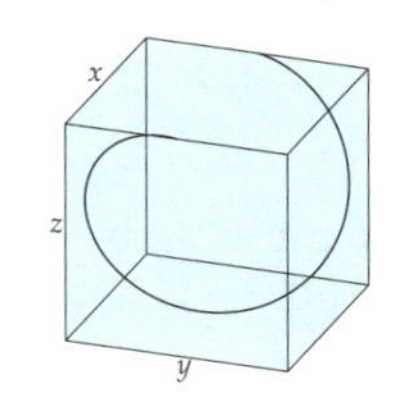


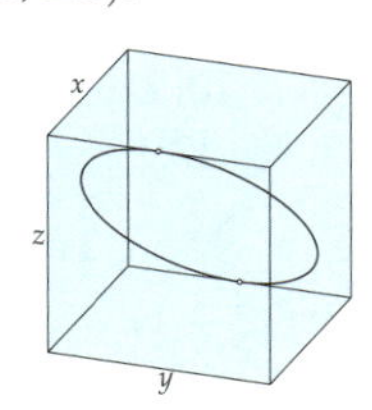


N.B.: the last curve is the intersection of the plane $x = z$ and the cylinder $x^2 + y^2 = 1$. It is an ellipse in space. The points $P_0 = (1, 0, 1)$ and $P_\pi = (-1, 0, -1)$ have been marked.

19.12 a. III b. V c. VI d. I e. IV f. VIII g. VII h. II

19.13 a. $(-1+2t, 1-3t)$ b. $(1-t, 2t)$ c. $(-1+2t, 2)$ d. $(t, 1-t)$
e. $(4t, -\frac{1}{2}+3t)$ f. $(7t, -\frac{2}{7}-5t)$ g. $(1, t)$ h. $(t, -3)$

19.14 a. $2x-3y=-5$ b. $x-y=-1$ c. $3x-y=22$ d. $y=3$ e. $x=0$
f. $x+2y=0$

19.15 a. $(-t, 1, 1+t)$ b. $(1+t, -1+t, 1-t)$ c. $(3-4t, -t, 1-t)$
d. $(1-3t, 4t, -1+2t)$ e. $(2-2t, -1+t, -1-t)$ f. $(\frac{1}{3}-\frac{1}{3}t, \frac{1}{3}+\frac{5}{3}t, t)$
g. $(-1-5t, t, -8t)$ h. $(2-\frac{3}{5}t, 1-\frac{9}{5}t, t)$ i. $(t, 2-6t, 1-4t)$ j. $(-2+3t, 5-2t, t)$

19.16 a. $(t,0,0)$, $(0,t,0)$, $(0,0,t)$ b. $(1, t, -1)$ c. $(t, t, t)$

# VII Calculus

## 20. Derivatives

20.1 a. $2$ b. $0$ c. $8x$ d. $70x^6$ e. $4+3x^2$

20.2 a. $3x^2$ b. $2x-2$ c. $4x^3-9x^2$ d. $64x^7$ e. $6x^5-24x^3$

20.3 a. $16x^3-6x$ b. $4000000x^{1999}$ c. $49x^6-36x^5$ d. $3x^2+49x^6$
e. $2x-15x^2+1$

20.4 a. $\frac{1}{2}x^{-\frac{1}{2}}$ b. $\frac{3}{2}x^{\frac{1}{2}}$ c. $\frac{3}{2}x^{\frac{1}{2}}$ d. $\frac{5}{2}x^{\frac{3}{2}}$ e. $\frac{1}{2}\sqrt{2}\,x^{-\frac{1}{2}}$

20.5 a. $\frac{1}{3}x^{-\frac{2}{3}}$ b. $\frac{2}{3}x^{-\frac{1}{3}}$ c. $\frac{1}{4}x^{-\frac{3}{4}}$ d. $\frac{5}{4}x^{\frac{1}{4}}$ e. $\frac{12}{5}x^{\frac{7}{5}}$

20.6 a. $\frac{2}{7}x^{-\frac{5}{7}}$ b. $\frac{3}{2}\sqrt{3}\,x^{\frac{1}{2}}$ c. $\frac{5}{3}\sqrt[3]{2}\,x^{\frac{2}{3}}$ d. $\frac{5}{4}x^{\frac{1}{4}}$ e. $\frac{7}{2}x^{\frac{5}{2}}$

20.7 a. $-x^{-2}$ b. $-4x^{-3}$ c. $-9x^{-4}$ d. $-\frac{1}{2}x^{-\frac{3}{2}}$ e. $-\frac{2}{3}x^{-\frac{5}{3}}$

20.8 a. $2.2x^{1.2}$ b. $4.7x^{3.7}$ c. $-1.6x^{-2.6}$ d. $0.333x^{-0.667}$ e. $-0.123x^{-1.123}$

20.9 a. $-x^{-2}$ b. $-\frac{3}{2}x^{-2}$ c. $-25x^{-6}$ d. $-\frac{1}{2}x^{-\frac{3}{2}}$ e. $-\frac{4}{3}x^{-\frac{7}{3}}$

20.10 a. $9(2+3x)^2$ b. $-35(3-5x)^6$ c. $6x(1-3x^2)^{-2}$ d. $-2x^{-\frac{1}{2}}(1-\sqrt{x})^3$
e. $-2(1-4x^3)(x-x^4)^{-3}$

20.11 a. $10(2x-3)^4$ b. $-2x(x^2+5)^{-2}$ c. $\frac{3}{2}(3x-4)^{-\frac{1}{2}}$ d. $\frac{1}{2}(2x+1)(x^2+x)^{-\frac{1}{2}}$
e. $-3(1+12x^2)(x+4x^3)^{-4}$

20.12 a. $\frac{1}{2}(1+2x)(1+x+x^2)^{-\frac{1}{2}}$ b. $\frac{1}{3}(1+2x)(1+x+x^2)^{-\frac{2}{3}}$ c. $8x(x^2-1)^3$
d. $\frac{3}{2}x^2(x^3+1)^{-\frac{1}{2}}$ e. $\frac{3}{2}(2x+1)(x^2+x)^{\frac{1}{2}}$

20.13 a. $\sin x + x\cos x$ b. $\cos 2x - 2x\sin 2x$ c. $2x\ln x + x$ d. $\tan x + \frac{x+1}{\cos^2 x}$
e. $2\ln x + \frac{2x+1}{x}$

20.14 a. $\frac{\sqrt{x+1}}{x} + \frac{\ln x}{2\sqrt{x+1}}$ b. $\cos x \ln x^2 + \frac{2\sin x}{x}$ c. $\ln\sqrt[3]{x} + \frac{1}{3}$ d. $\ln\sin x + \frac{x\cos x}{\sin x}$
e. $\frac{\ln(1-x^2)}{2\sqrt{x}} - \frac{2x^{3/2}}{1-x^2}$

20.15 a. $\frac{1+\ln x}{\ln 2}$ b. $\frac{\ln x^3}{2\sqrt{x}\ln 5}+\frac{3}{\sqrt{x}\ln 5}$ c. $\frac{\ln x}{\ln 2}+\frac{x-1}{x\ln 2}$ d. $\mathrm{e}^{-x}-x\mathrm{e}^{-x}$
e. $2x\mathrm{e}^{-x^2}-2x^3\mathrm{e}^{-x^2}$

20.16 a. $\frac{1}{(x+1)^2}$ b. $\frac{2}{(x+1)^2}$ c. $\frac{x^2+2x}{(x+1)^2}$ d. $\frac{1-x^2}{(x^2+1)^2}$ e. $\frac{-x^2+2x+1}{(x^2+x)^2}$

20.17 a. $\frac{-x-1}{2\sqrt{x}(x-1)^2}$ b. $\frac{x^2+4x+1}{(x+2)^2}$ c. $\frac{4x}{(x^2+1)^2}$ d. $\frac{14}{(4x+1)^2}$ e. $\frac{-1}{(2-x)^2}$

20.18 a. $\frac{1}{1+\cos x}$ b. $\frac{-x\sin x-\sin x-\cos x}{(1+x)^2}$ c. $\frac{x+1-\sqrt{1-x^2}\arcsin x}{\sqrt{1-x^2}(x+1)^2}$ d. $\frac{\sin x-x\cos x\ln x}{x\sin^2 x}$
e. $\frac{\mathrm{e}^x}{(1+\mathrm{e}^x)^2}$

20.19 a. $\cos(x-3)$ b. $-2\sin(2x+5)$ c. $3\cos(3x-4)$ d. $-2x\sin(x^2)$ e. $\frac{\cos\sqrt{x}}{2\sqrt{x}}$

20.20 a. $\frac{1}{\cos^2(x+2)}$ b. $\frac{2}{\cos^2(2x-4)}$ c. $2x\cos(x^2-1)$ d. $\frac{\sin(1/x)}{x^2}$ e. $\frac{1}{3\sqrt[3]{x^2}\cos^2\sqrt[3]{x}}$

20.21 a. $\frac{2}{\sqrt{1-4x^2}}$ b. $\frac{1}{\sqrt{-x^2-4x-3}}$ c. $\frac{-2x}{\sqrt{1-x^4}}$ d. $\frac{1}{2\sqrt{x}(1+x)}$ e. $-\tan x$

20.22 a. $2\mathrm{e}^{2x+1}$ b. $-\mathrm{e}^{1-x}$ c. $-2\mathrm{e}^{-x}$ d. $-3\mathrm{e}^{1-x}$ e. $2x\mathrm{e}^{x^2}$

20.23 a. $(2x-1)\mathrm{e}^{x^2-x+1}$ b. $-2x\mathrm{e}^{1-x^2}$ c. $-3\mathrm{e}^{3-x}$ d. $\frac{1}{\sqrt{x}}\mathrm{e}^{\sqrt{x}}$ e. $\frac{1}{2\sqrt{x}}\mathrm{e}^{1+\sqrt{x}}$

20.24 a. $(\ln 2)\,2^{x+2}$ b. $(-\ln 3)\,3^{1-x}$ c. $(-3\ln 2)\,2^{2-3x}$ d. $(2x\ln 5)\,5^{x^2}$
e. $(\frac{1}{3}x^{-\frac{2}{3}}\ln 3)\,3^{\sqrt[3]{x}}$

20.25 a. $\frac{-2}{1-2x}$ b. $\frac{6x}{3x^2-8}$ c. $\frac{3-8x}{3x-4x^2}$ d. $\frac{3x^2+6x^5}{x^3+x^6}$ e. $\frac{2x}{x^2+1}$

20.26 a. $\frac{1}{2x+2}$ b. $\frac{2}{x}$ c. $\frac{1}{3x}$ d. $\frac{1}{3x-3}$ e. $\frac{2}{x-4}$

20.27 a. $\frac{1}{x\ln 2}$ b. $\frac{3}{x\ln 3}$ c. $\frac{1}{(x+1)\ln 10}$ d. $\frac{1}{(2x+2)\ln 10}$ e. $\frac{2x+1}{(x^2+x+1)\ln 2}$

20.28 a. $x=1$ b. $x=1, x=-1$ c. $x=0$ d. $x=2$ e. $x=0$

20.29 a. $x=0$ b. no value of $x$ c. $x=k\pi$ (integer $k$) d. $x=0$ e. $x=0$

20.30 a. $y=-5+4x$ b. $y=10+11x$ c. $y=-3+2x$ d. $y=384-192x$
e. $y=4+11x$

20.31 a. $y=-7+5x$ b. $y=-22+\frac{29}{4}x$ c. $y=-3+x$ d. $y=3-4x$
e. $y=18+20x$

20.32 a. $-\frac{1}{4}(x+1)^{-3/2}$ b. $\frac{-4}{(x+1)^3}$ c. $\frac{2(1-x^2)}{(x^2+1)^2}$ d. $\frac{1}{x}$ e. $2\cos x-x\sin x$
f. $2\cos 2x-8x\sin 2x-4x^2\cos 2x$

20.33 a. $-\frac{\sqrt{x}\sin\sqrt{x}+\cos\sqrt{x}}{4x\sqrt{x}}$ b. $\frac{2\sin x}{\cos^3 x}$ c. $\frac{-2x}{(1+x^2)^2}$ d. $\frac{3x-4}{4(x-1)\sqrt{x-1}}$
e. $\frac{(2-x^2)\sin x-2x\cos x}{x^3}$ f. $2\cos^2 x-2\sin^2 x$

20.34 a. $0$ b. $10!$ c. $11!x$ d. $\mathrm{e}^{-x}$ e. $2^{10}\mathrm{e}^{2x}$ f. $\mathrm{e}^{x+1}$

20.35 a. $10!(x+1)^{-11}$ b. $-9!x^{-10}$ c. $-2^{10}\sin 2x$ d. $-\sin(x+\frac{\pi}{4})$ e. $(10+x)\mathrm{e}^x$
f. $(-10+x)\mathrm{e}^{-x}$

20.36 a. $x=0$, monotonically increasing on $\infty<x<\infty$

b. $x = 0$, $x = 3$, monotonically decreasing on $x \leq 3$, monotonically increasing on $x \geq 3$
c. $x = 0$, monotonically decreasing on $x \leq 0$, monotonically increasing on $x \geq 0$

20.37 a. no zeroes of the derivative, monotonically increasing on $\infty < x < \infty$
b. $x = 1$, monotonically decreasing on $x \leq 1$, monotonically increasing on $x \geq 1$
c. no zeroes of the drivative, monotonically increasing on $x < 0$, monotonically decreasing on $x > 0$

20.38 a. $x = -1 - \sqrt{2}$, $x = -1 + \sqrt{2}$, monotonically increasing on $x \leq -1 - \sqrt{2}$, monotonically decreasing on $-1 - \sqrt{2} \leq x < -1$, monotonically decreasing on $-1 < x \leq -1 + \sqrt{2}$, monotonically increasing on $x \geq -1 + \sqrt{2}$
b. no zeroes of tthe derivative, function is monotonically increasing on $\infty < x < \infty$
c. $x = 0$, monotonically decreasing on $x \leq 0$, monotonically increasing on $x \geq 0$

20.39 a. true. b. false; counter example: any constant function. c. true. d. false; counter example: any constant function. e. false; counter example: $f(x) = x^3$ on $\mathbb{R}$.

20.40 a. false; counter example $f(x) = x$ on $\mathbb{R}$. b. true. c. true.

20.41 a. true. b. false; counter example $f(x) = g(x) = x$.

20.42 a. $x = -\frac{1}{3}\sqrt{3}$ (local maximum) $x = \frac{1}{3}\sqrt{3}$ (local minimum) b. $x = -1$ (g. min.) $x = 0$ (l. max.) $x = 1$ (g. min.) c. $x = -\sqrt{3}$ (g. min.) $x = 0$ (l. max.) $x = \sqrt{3}$ (g. min.) d. $x = 1$ (g. min.) e. $x = -1$ (g. min.) $x = 0$ (l. max.) $x = 1$ (g. min.)

20.43 a. $x = \frac{1}{2}\pi + 2k\pi$ (g. max.), $x = \frac{3}{2}\pi + 2k\pi$ (g. min.) (integer $k$)
b. $x = \pm\sqrt{\frac{1}{2}\pi + 2k\pi}$ (g. max.), $x = \pm\sqrt{\frac{3}{2}\pi + 2k\pi}$ (g. min.) ($k \geq 0$ and integer), $x = 0$ (l. min.)
c. $x = (\frac{1}{2}\pi + 2k\pi)^2$ (g. max.), $x = (\frac{3}{2}\pi + 2k\pi)^2$ (g. min.) ($k \geq 0$ and integer), $x = 0$ (local boundary minimum)
d. $x = \pm\frac{1}{2}\pi + 2k\pi$ (g. max.), $x = \pm\frac{3}{2}\pi + 2k\pi$ (g. min.) ($k \geq 0$ and integer), $x = 0$ (l. min.)
e. $x = \frac{1}{2}\pi + k\pi$ (g. max.), $x = k\pi$ (g. min.) (integer $k$)

20.44 a. $x = \frac{1}{e}$ (g. min), $x = 0$ (l. boundary max.) b. $x = 1$ (g. min.) c. $x = -1$ (g. min.), $x = 1$ (g. max.) d. $x = 2k\pi$ (g. max.) (integer $k$) e. $x = k\pi$ (g. max.) (integer $k$)

20.45 a. $x = -1$ (g. min.) b. $x = 0$ (g. max.) c. $x = -\frac{1}{2}\sqrt{2}$ (g. min.), $x = \frac{1}{2}\sqrt{2}$ (g. max.) d. $x = \frac{1}{2}\pi + 2k\pi$ (g. max.), $x = \frac{3}{2}\pi + 2k\pi$ (g. min.) (integer $k$) e. $x = 0$ (g. max.)

20.46 a. $x = \frac{1}{k}$ (integer $k$, nonzero)
b. global maxima: $x = \frac{2}{1+4k}$, globale minima: $x = \frac{2}{3+4k}$ (integer $k$)
c. both limits are 0
d. This limit doesn't exist: each neighbourhood of 0 contains points where $f(x) = 1$ and points where $f(x) = -1$.

20.47 a. $x = 0$ is the only stationary point; it is also the only inflection point
b. st.p.: $x = \pm\frac{1}{3}\sqrt{3}$, i.p.: $x = 0$ c. st.p.: $x = 1$, i.p.: $x = \pm\frac{1}{6}\sqrt{6}$ d. st.p.: $x = 0$, $x = -\sqrt[3]{4}$, i.p.: $x = -1$ e. st.p.: $x = 0$, i.p.: $x = \pm\frac{1}{3}\sqrt{3}$

20.48 a. st.p.: $x = \frac{\pi}{2} + k\pi$, i.p.: $x = k\pi$ (integer $k$) b. no st.p., i.p.: $x = 0$ c. st.p.: $x = \frac{1}{\sqrt{e}}$, i.p.: $x = \frac{1}{e\sqrt{e}}$ d. st.p.: $x = 1$, i.p.: $x = 2$ e. st.p.: $x = 0$, i.p.: $x = \pm\frac{1}{2}\sqrt{2}$

20.49 a. $-1 \leq \sin\frac{\pi}{x} \leq 1$ for all $x \neq 0$, so $-x^2 \leq f(x) \leq x^2$ (also for $x = 0$). Equallity if the sine is $\pm 1$, so $f(x) = -x^2$ if $x = \frac{2}{3+4k}$ and $f(x) = x^2$ if $x = \frac{2}{1+4k}$ (integer $k$), and, of course, also if $x = 0$.
b. $f'(x) = 2x\sin\frac{\pi}{x} - \pi\cos\frac{\pi}{x}$.
c. $\lim_{x\to 0}\frac{f(x)}{x} = \lim_{x\to 0} x\sin\frac{\pi}{x} = 0$, since the absolute value of the sine is always less than or equal to 1.
d. $f'(\frac{1}{2k}) = -\pi$ and $f'(\frac{1}{2k+1}) = \pi$.
e. From the previous item, it follows that each neighbourhood of 0 contains points where $f(x) = \pi$ and where $f(x) = -\pi$. Therefore, the limit doesn't exist.
f. no g. no h. no

20.50 No, there are no more zeroes. The derivative is a fourth degree polynomial, so it has at most four zeroes. These, indeed, are present, as the drawing shows. The derivative is positive for $x$ less than its leftmost zero, and also for $x$ greater than its rightmost zero. So on the part of the graph not shown, the function is monotonically increasing, hence there are no more zeroes.

20.51 $f(x)$ V, $f'(x)$ VI, $f''(x)$ II, $g(x)$ III, $g'(x)$ I, $g''(x)$ IV (or $f$ and $g$ interchanged)

20.52 $f(x)$ II, $f'(x)$ I, $f''(x)$ V, $g(x)$ III, $g'(x)$ VI, $g''(x)$ IV (or $f$ and $g$ interchanged)

## 21. Differentials and integrals

21.1 a. $(6x+2)\,dx$ b. $(1+2\cos 2x)\,dx$ c. $(8x\sin(x+1)+4x^2\cos(x+1))\,dx$
d. $\left(3x^2\sqrt{x^3+1}+\frac{3x^5}{2\sqrt{x^3+1}}\right)dx$ e. $-2x\sin(x^2)\,dx$ f. $-2\,dx$

21.2 a. $dx$ b. $\frac{2x}{x^2+1}\,dx$ c. $2xe^{-x^2}\,dx$ d. $-\sin x e^{\cos x}\,dx$ e. $(1+\frac{1}{x^2})\,dx$

21.3 a. $(15x^2 - \frac{6x}{(x^2+1)^2})\,dx$ b. $4(x+4)^3\,dx$ c. $(4x^3\sin 2x + 2(x^4-1)\cos 2x)\,dx$
d. $\frac{1}{4}(x+1)^{-3/4}\,dx$ e. $\frac{1}{\cos^2(x+5)}\,dx$

21.4 a. $(\frac{2}{3}x^{-1/3} - \frac{2}{3}x^{-5/3})\,dx$ b. $(1-\frac{2x}{x^2+1})\,dx$ c. $-2\cos 2xe^{-\sin 2x}\,dx$
d. $\frac{4x}{(1-x^2)^2}\,dx$

21.5 a. $d(\frac{1}{3}x^3+x^2+2x)$ b. $d(\frac{1}{4}x^4-2x^2)$ c. $d(\frac{1}{5}x^5-2x^2+5x)$ d. $d(\frac{2}{3}x^{\frac{3}{2}})$
e. $d(\frac{-4}{x})$

21.6 a. $d(\frac{2}{5}x^{\frac{5}{2}})$ b. $d(\sqrt{x})$ c. $d(\frac{1}{5}(x+1)^5)$ d. $d(-\cos x)$ e. $d(-\frac{1}{5}\cos 5x)$

21.7 a. $d(x^3+x^2+2x)$ b. $d(\frac{1}{2}x^2-\frac{2}{3}x^{3/2})$ c. $d(\frac{1}{5}x^5-x^4+x^2-5x)$
d. $d(\frac{2}{3}(x+1)^{3/2})$ e. $d(\ln x)$

21.8 a. $d(\frac{3}{4}x^{4/3})$ b. $d(3x+\frac{1}{2}x^2-\frac{1}{2}\cos 2x)$ c. $d(-\cos(x+1))$
d. $d(\frac{1}{2}\sin(2x+1))$
e. $d(\ln(-x))$

21.9 a. $f(x_m) = 5.511376, k = 0.043$ b. $f(x_m) = 1.04511376, k = 0.00043$
c. $f(x_m) = -0.648656, k = 0.0078$ d. $f(x_m) = -163.8656, k = 0.78$
e. $f(x_m) = 0.8688021, k = 0.0050$ f. $f(x_m) = 4.333383, k = 0.20$
g. $f(x_m) = -0.8223451, k = 0.0023$ h. $f(x_m) = 1.480239974, k = 0.0023$
i. $f(x_m) = 3.782825067, k = 0.0023$ j. $f(x_m) = 0.3351206434, k = 0.0012$

21.10 Just do it!.

21.11 We use the inequality $|a \pm b| \leq |a| + |b|$, which holds for all real numbers $a$ and $b$.

a. $|(x_m + y_m) - (x_w + y_w)| = |(x_m - x_w) + (y_m - y_w)| \leq |x_m - x_w| + |y_m - y_w| \leq h_x + h_y$

b. $|(x_m - y_m) - (x_w - y_w)| = |(x_m - x_w) - (y_m - y_w)| \leq |x_m - x_w| + |y_m - y_w| \leq h_x + h_y$

c. $\frac{|x_m y_m - x_w y_w|}{|x_m y_m|} = \frac{|(x_m - x_w)y_m + x_w(y_m - y_w)|}{|x_m y_m|} \leq \frac{h_x}{|x_m|} + \frac{|x_w|}{|x_m|}\frac{h_y}{|y_m|} \approx q_x + q_y$ if we suppose that $x_m$ and $x_w$ are almost equal, in the sense that $x_m / x_w \approx 1$.

d. $\frac{|x_m/y_m - x_w/y_w|}{|x_m/y_m|} = \frac{|x_m y_w - x_w y_m|}{|x_m y_w|} = \frac{|(x_m - x_w)y_w + x_w(y_w - y_m)|}{|x_m y_w|} \leq \frac{h_x}{|x_m|} + \frac{|x_w|}{|x_m|}\frac{|y_m|}{|y_w|}\frac{h_y}{|y_m|} \approx q_x + q_y$ if we suppose that $x_m$ and $x_w$, or $y_m$ and $y_w$, respectively, are almost equal in the sense that $x_m / x_w \approx 1$, or $y_w / y_m \approx 1$, respectively.

21.12

a.

| $dx$ | $df = f'(x)\,dx$ | $\Delta f$ | $\Delta f - df$ | $\frac{1}{2}f''(x)\,(dx)^2$ |
|---|---|---|---|---|
| 0.1 | 0.4 | 0.41 | 0.01 | 0.01 |
| 0.01 | 0.04 | 0.0401 | 0.0001 | 0.0001 |
| 0.001 | 0.004 | 0.004001 | 0.000001 | 0.000001 |
| 0.0001 | 0.0004 | 0.00040001 | 0.00000001 | 0.00000001 |

b.

| $dx$ | $df = f'(x)\,dx$ | $\Delta f$ | $\Delta f - df$ | $\frac{1}{2}f''(x)\,(dx)^2$ |
|---|---|---|---|---|
| 0.1 | 0.1 | 0.09531 | −0.00468982 | −0.005000 |
| 0.01 | 0.01 | 0.00995033 | −0.000049669 | −0.00005000 |
| 0.001 | 0.001 | 0.0009950033 | −0.0000004996669 | −0.0000005000 |
| 0.0001 | 0.0001 | 0.0000999950 | −0.00000000499967 | −0.000000005000 |

c.

| $dx$ | $df = f'(x)\,dx$ | $\Delta f$ | $\Delta f - df$ | $\frac{1}{2}f''(x)\,(dx)^2$ |
|---|---|---|---|---|
| 0.1 | 0.2 | 0.22304888 | 0.02304888 | 0.02000 |
| 0.01 | 0.02 | 0.020202701 | 0.000202701 | 0.0002000 |
| 0.001 | 0.002 | 0.002002003 | 0.000002003 | 0.000002000 |
| 0.0001 | 0.0002 | 0.000200020 | 0.00000002000 | 0.00000002000 |

d.

| $dx$ | $df = f'(x)\,dx$ | $\Delta f$ | $\Delta f - df$ | $\frac{1}{2}f''(x)\,(dx)^2$ |
|---|---|---|---|---|
| 0.1 | 0.02 | 0.01922839 | −0.0007716010 | −0.0008000 |
| 0.01 | 0.002 | 0.001992029 | −0.000007971000 | −0.000008000 |
| 0.001 | 0.0002 | 0.000199920 | −0.000000079970 | −0.00000008000 |
| 0.0001 | 0.00002 | 0.0000199999 | −0.00000000079997 | −0.0000000008000 |

e.

| $dx$ | $df = f'(x)\,dx$ | $\Delta f$ | $\Delta f - df$ | $\frac{1}{2}f''(x)\,(dx)^2$ |
|---|---|---|---|---|
| 0.1 | 0 | −0.00499583 | −0.00499583 | −0.005000 |
| 0.01 | 0 | −0.0000499583 | −0.0000499583 | −0.00005000 |
| 0.001 | 0 | −0.000000499999958 | −0.000000499999958 | −0.0000005000 |
| 0.0001 | 0 | −0.000000005000000 | −0.000000005000000 | −0.0000000050000 |

f.

| $dx$ | $df = f'(x)\,dx$ | $\Delta f$ | $\Delta f - df$ | $\frac{1}{2}f''(x)\,(dx)^2$ |
|---|---|---|---|---|
| 0.1 | 0.1 | 0.0998334 | $-0.0001665$ | 0 |
| 0.01 | 0.01 | 0.0099998333 | $-0.1667 \times 10^{-6}$ | 0 |
| 0.001 | 0.001 | 0.0009999998333 | $-0.1667 \times 10^{-9}$ | 0 |
| 0.0001 | 0.0001 | 0.0000999999998333 | $-0.1667 \times 10^{-12}$ | 0 |

Since here we have $f''(0) = 0$, the linear approximation is even better: if $dx$ gets ten times smaller, then $\Delta f - df$ gets approximately *a thousand* times smaller!

21.13 $\frac{1}{4}$

21.14 $\frac{1}{20}$

21.15 2

21.16 2

21.17 $e - \frac{1}{e}$

21.18 a. $\frac{8}{3}$ b. $\frac{20}{3}$ c. $\frac{5}{3}$ d. $2\pi$ e. $e - \frac{1}{e}$ f. $\frac{62}{5}$ g. 1 h. $\frac{\pi}{2}$

21.19 a. 10 b. $\frac{73}{6}$ c. 5 d. $2\pi$

21.20 a. $\frac{1}{2}(1 - e^{-4})$ b. $e - \frac{1}{e}$ c. $\ln 2$ d. $\frac{\pi}{6}$

21.21 a. $-\frac{78}{5}$ b. $2 - \frac{4}{3}\sqrt{2}$ c. $\frac{15}{8} + 2\ln 2$ d. 0 e. $-1$

21.22 a. 16 b. $\frac{5}{6}(2\sqrt[5]{2} - 1)$ c. $\frac{1}{e} - e$ d. $-2 - \frac{\pi^2}{2}$ e. $\sqrt{2} - 1$

21.23 a. $\frac{3}{\ln 2}$ b. $e - 1$ c. 0 d. $-1$ e. 0

21.24 a. $-\frac{\pi}{4}$ b. $\ln 2$ c. $-\frac{2}{3}$ d. $\frac{\pi}{3}$ e. $-\sqrt{3}$

21.25 a. $x^2$ b. $x^2$ c. $-x^2$ d. $2x^2$

21.26 a. $-\sin x$ b. $-\sin x$ c. $2\cos 2x$ d. $2\cos x$

21.27 a. $\frac{3}{2}$ b. $\frac{99}{\ln 10}$ c. 12 d. $\frac{1}{2}\sqrt{2}$ e. $-\ln 2$

21.28 a. $\frac{\pi}{2}$ b. $\ln 3 - \ln 4$ c. $\frac{5}{16}$ d. $\frac{\pi}{2}$ e. $4 - 2\sqrt{2}$

21.29 d. $\int_0^{\pi} \sin^2 x\,dx = \int_0^{\pi} \cos^2 x\,dx = \frac{\pi}{2}$ e. $\int_0^{\pi/2} \sin^2 x\,dx = \int_0^{\pi/2} \cos^2 x\,dx = \frac{\pi}{4}$

21.30 a. $\frac{2}{3}\sqrt{2}$ b. $\frac{9}{2}$ c. $\frac{5}{12}$ d. $\frac{12+4\pi}{3\pi}$ e. 1

21.31 $Q = (-2, -8)$, area: $\frac{27}{4}$

21.32 a. $x^4 - \frac{2}{3}x^3 + \frac{1}{2}x^2 + x + c$ b. $3x - \frac{1}{2}x^4 + c$ c. $-\frac{1}{3}\cos 3x + c$
d. $\frac{2}{3}x\sqrt{x} + \frac{2}{x} + c$

21.33 a. $\frac{1}{2}x - \frac{1}{4}\sin 2x + c$ b. $\frac{1}{2}x + \frac{1}{12}\sin 6x + c$ c. $\frac{1}{2}x - \frac{1}{20}\sin 10x + c$
d. $\frac{1}{2}x + \frac{1}{2}\sin x + c$

21.34 a. $\frac{1}{8}\cos 4x - \frac{1}{12}\cos 6x + c$ b. $\frac{1}{2}\sin x + \frac{1}{10}\sin 5x + c$ c. $\frac{1}{4}\sin 2x - \frac{1}{12}\sin 6x + c$

21.35 We only give a proof of the first formula; the other proofs are similar.
$\int_0^{2\pi} \sin mx \cos nx\,dx = \frac{1}{2}\int_0^{2\pi} (\sin(n+m)x + \sin(n-m)x)\,dx =$
$\frac{1}{2}\left[-\frac{1}{n+m}\cos(n+m)x - \frac{1}{n-m}\cos(n-m)x\right]_0^{2\pi} = 0$ since $\cos(n+m)(2\pi) = \cos 0 = 1$ and $\cos(n-m)(2\pi) = \cos 0 = 1$.

21.36 a. $F(x) = \ln(x-1) + c_1$ if $x > 1$, $F(x) = \ln(1-x) + c_2$ if $x < 1$.
b. $F(x) = -\ln(x-2) + c_1$ if $x > 2$, $F(x) = -\ln(2-x) + c_2$ if $x < 2$.
c. $F(x) = \frac{3}{2}\ln(x-\frac{1}{2}) + c_1$ if $x > \frac{1}{2}$, $F(x) = \frac{3}{2}\ln(\frac{1}{2}-x) + c_2$ if $x < \frac{1}{2}$.
d. $F(x) = -\frac{4}{3}\ln(x-\frac{2}{3}) + c_1$ if $x > \frac{2}{3}$, $F(x) = -\frac{4}{3}\ln(\frac{2}{3}-x) + c_2$ if $x < \frac{2}{3}$.
e. $F(x) = -\frac{1}{x} + c_1$ if $x > 0$, $F(x) = -\frac{1}{x} + c_2$ if $x < 0$.
f. $F(x) = -\frac{1}{2(x-1)^2} + c_1$ if $x > 1$, $F(x) = -\frac{1}{2(x-1)^2} + c_2$ if $x < 1$.
21.37 a. $F(x) = 2\sqrt{x} + c_1$ if $x > 0$, $F(x) = -2\sqrt{-x} + c_2$ if $x < 0$.
b. $F(x) = \frac{3}{2}\sqrt[3]{x^2} + c_1$ if $x > 0$, $F(x) = \frac{3}{2}\sqrt[3]{x^2} + c_2$ if $x < 0$.
c. $F(x) = \frac{5}{4}\sqrt[5]{(x-1)^4} + c_1$ if $x > 1$, $F(x) = \frac{5}{4}\sqrt[5]{(x-1)^4} + c_2$ if $x < 1$.
d. $F(x) = 2\sqrt{x-2} + c_1$ if $x > 2$, $F(x) = -2\sqrt{2-x} + c_2$ if $x < 2$.
e. On each interval $\langle -\frac{1}{2}\pi + k\pi, \frac{1}{2}\pi + k\pi \rangle$ (integer $k$) we have $F(x) = \tan x + c_k$, where $c_k$ depends on the interval.
f. On each interval $\langle -\frac{1}{2} + k, \frac{1}{2} + k \rangle$ (integer $k$) we have $F(x) = \frac{1}{\pi}\tan \pi x + c_k$, where $c_k$ depends on the interval.

## 22. Integration techniques

22.1 a. $\frac{1023}{10}$ b. 868 c. $\frac{2186}{7}$ d. 8502 e. $\frac{1}{2}$ f. $\ln 3 - \ln 2$

22.2 a. $\frac{e^3}{2} - \frac{1}{2e}$ b. $\frac{1}{2}(e-1)$ c. 0 d. $\frac{1}{3}(e-1)$ e. $\arctan e - \frac{\pi}{4}$
f. $2\sqrt{2} - 2\sqrt{1+\frac{1}{e}}$

22.3 a. $\frac{3\sqrt{3}}{2}$ b. $-\frac{2}{\pi}$ c. 0 d. $\frac{1}{4}$ e. $\frac{3}{8}$ f. $\frac{11}{24}$

22.4 a. $\frac{16}{15}$ b. $\frac{1}{2}\sqrt{6} - \frac{2}{3}$ c. $-\ln 2$ d. $2\sin\sqrt{\pi}$ e. $1 - \cos(\frac{1}{2}\sqrt(2))$ f. $\ln 3 - \ln 2$

22.5 a. $\frac{9217}{110}$ b. $\frac{1}{30}$ c. $-\frac{181}{14}$ d. 0 e. $\frac{511}{18}$ f. $\frac{81}{8}$

22.6 a. $1 - \ln 2$ b. $2 - 3\ln 3$ c. $-\frac{1}{8} + \frac{1}{4}\ln 2$ d. $-\frac{1}{2}\ln 2$ e. $-\frac{1}{2}\ln 3 + \ln 2$ f. $\frac{1}{8}\ln 2$

22.7 a. $\frac{506}{15}$ b. 0 c. $\frac{186}{5}$ d. $\frac{10\sqrt{2}-8}{3}$ e. $\frac{2\sqrt{2}-4}{3}$

22.8 a. $\frac{33}{28}$ b. $2 - \frac{\pi}{2}$ c. $\frac{5}{3} - 2\ln 2$ d. $2\ln 2 - 1$ e. $\frac{\pi}{3}$

22.9 a. 2 b. $\frac{1}{4}(e^2+1)$ c. 1 d. $\frac{\pi}{4} - \frac{1}{2}$ e. $\pi^2 - 4$ f. $\frac{\pi}{12} + \frac{1}{2}\sqrt{3} - 1$

22.10 a. $\frac{1}{9}(2e^3+1)$ b. $2 - \frac{5}{e}$ c. $\frac{5}{e^2} - 1$ d. $\frac{1}{2} - \frac{1}{e}$ e. $\frac{2}{5}(e^{-2\pi} - e^{2\pi})$
f. $\frac{2}{9}(2 + e\sqrt{e})$

22.11 a. $2\pi$ b. $2e^2$ c. $\frac{1}{25}(6e^{5/3}+9)$ d. $\frac{\pi}{2}$ e. $\frac{3}{4}$ f. $\frac{\pi}{8} - \frac{1}{4}\ln 2$

22.12 a. $\frac{5}{2}$ b. $\frac{3}{4}$ c. $\frac{8}{3} - \sqrt{3}$ d. $(e-1)\ln 2 + 1$ e. $\pi$ f. $2 - 2\ln 2$

22.13 a. $\frac{1}{2}$ b. $\frac{1}{2}(1 - 5e^{-4})$ c. $\frac{2}{3}(2\sqrt{2}-1)$ d. 4 e. $\sqrt{e}$ f. $\frac{1}{8}(-2 + 5\arctan 2)$

22.14 a. $\frac{2}{3}\ln 3$ b. $\frac{1}{462}$ c. 0 d. $\frac{3}{8}\sqrt{3}$ e. $(e+\frac{1}{3})\ln(1+3e) - e$ f. $\frac{1}{2}(1-\frac{1}{e})$

22.15 a. 1 b. $\frac{1}{8}$ c. $\frac{1}{9}$ d. $\frac{1}{(p-1)2^{p-1}}$ e. $-\infty$

22.16 a. $\infty$ b. $\infty$ c. $\frac{\pi}{2}$ d. $\infty$ e. $\infty$

22.17 a. 1 b. $\frac{1}{3}$ c. $\frac{2}{e}$ d. 2 e. 2

22.18 a. 1 b. $\infty$ c. $-\frac{\pi^2}{8}$ d. doesn't exist e. $\frac{1}{2}$

22.19 a. 2 b. $\frac{3}{2}\sqrt[3]{4}$ c. $\frac{10}{9}$ d. $\frac{2^{1-p}}{1-p}$ e. $\infty$

22.20 a. $\infty$ b. 2 c. $-\frac{3}{2}$ d. 4 e. $2+2\sqrt{2}$

22.21 a. $-1$ b. $\infty$ c. $2\ln 2-2$ d. $\ln 3-1$ e. $8(\ln 2-1)$

22.22 a. $\infty$ b. $\infty$ c. $\pi$ d. 2 e. $-\frac{1}{4}$

22.23

a.

| $N$ | $\sum f(x_i)\,dx$ | $\sum f(x_i)\,dx - \int f(x)\,dx$ |
|---|---|---|
| 10 | 1.320000 | $-0.013333$ |
| 100 | 1.33320000 | $-0.00013333$ |
| 1000 | 1.3333320000 | $-0.0000013333$ |

b.

| $N$ | $\sum f(x_i)\,dx$ | $\sum f(x_i)\,dx - \int f(x)\,dx$ |
|---|---|---|
| 10 | 1.3932726 | $-0.0494224$ |
| 100 | 1.4377008 | $-0.0049942$ |
| 1000 | 1.4426950 | $-0.000499942$ |

c.

| $N$ | $\sum f(x_i)\,dx$ | $\sum f(x_i)\,dx - \int f(x)\,dx$ |
|---|---|---|
| 10 | 5.61563 | $-0.4757$ |
| 100 | 6.0460859 | $-0.045263755$ |
| 1000 | 6.0868470 | $-0.004502638$ |

22.24 a. $M=2$, $\frac{1}{2}M(b-a)\,dx=0.4$, resp. 0.04, resp. 0.004
b. $M=1.4$, $\frac{1}{2}M(b-a)\,dx=0.07$, resp. 0.007, resp. 0.0007
c. $M=0.44$, $\frac{1}{2}M(b-a)\,dx=1.76$, resp. 0.176, resp. 0.0176

22.25

a.

| $n$ | $M(n)$ | $T(n)$ | $S(n)$ |
|---|---|---|---|
| 8 | 0.8862269182191298 | 0.8862268965093698 | 0.8862269109825432 |
| 16 | 0.8862269139051738 | 0.8862269073642503 | 0.8862269117248660 |
| 32 | 0.8862269123605085 | 0.8862269106347116 | 0.8862269117852428 |
| 64 | 0.8862269119351357 | 0.8862269114976102 | 0.8862269117892939 |

b.

| $n$ | $M(n)$ | $T(n)$ | $S(n)$ |
|---|---|---|---|
| 8 | 0.8769251489660240 | 0.8710235901875318 | 0.8749579627065266 |
| 16 | 0.8754486864404032 | 0.8739743695767780 | 0.8749572474858615 |
| 32 | 0.8750800387547874 | 0.8747115280085906 | 0.8749572018393885 |
| 64 | 0.8749879067668622 | 0.8748957833816889 | 0.8749571989718044 |

c.

| $n$ | $M(n)$ | $T(n)$ | $S(n)$ |
|---|---|---|---|
| 8 | 0.9022135976794877 | 0.8801802371138100 | 0.8948691441575951 |
| 16 | 0.8966522494458846 | 0.8911969173966484 | 0.8948338054294725 |
| 32 | 0.8952851310507982 | 0.8939245834212664 | 0.8948316151742876 |
| 64 | 0.8949447892593938 | 0.8946048572360328 | 0.8948314785849401 |

22.26 a. $\int_x^\infty \varphi(t)\,dt = \int_{-\infty}^\infty \varphi(t)\,dt - \int_{-\infty}^x \varphi(t)\,dt = 1-\Phi(x)$

b. substitute $t=-u$: $\Phi(-x)=\frac{1}{\sqrt{2\pi}}\int_{-\infty}^{-x}\mathrm{e}^{-\frac{1}{2}t^2}dt=\frac{1}{\sqrt{2\pi}}\int_{u=\infty}^{u=x}\mathrm{e}^{-\frac{1}{2}(-u)^2}d(-u)=\frac{1}{\sqrt{2\pi}}\int_{u=x}^{u=\infty}\mathrm{e}^{-\frac{1}{2}u^2}du=1-\Phi(x)$ c. $\frac{1}{2}$ d. $\sqrt{\pi}$

22.27 a. $0,1,-1$ b. similar to the graph of $\Phi(x)$, but this graph goes through the origin and has asymptotes $y=\pm 1$. c. $\mathrm{Erf}(x)=2\Phi(x\sqrt{2})-1$

22.28 a. substitute $t=-u$: $\mathrm{Si}(-x)=\int_0^{-x}\frac{\sin t}{t}\,dt=-\int_{u=0}^{u=x}\frac{\sin u}{u}\,du=-\mathrm{Si}(x)$
b. local maxima at $x=-\pi+2k\pi$ and $x=-2k\pi$ ($k>0$ and integer), local minima at $x=2k\pi$ and $x=\pi-2k\pi$ ($k>0$ and integer) c. $\pi$ d. $\mathrm{Si}(mb)-\mathrm{Si}(ma)$

## 23. Applications

To save space, we write the coordinates of vectors horizontally and not vertically.

23.1 a. $(-3\sin 3t, 2\cos 2t)$
b. $(-2\sin 2t, 3\cos 3t)$, zero vector if $t=\pm\frac{\pi}{2}+2k\pi$
c. $(-3\cos^2 t\sin t, 3\sin^2 t\cos t)$, zero vector if $x=\frac{1}{2}k\pi$
d. $(-3\cos^2 t\sin t, \cos t)$, zero vector if$t=\frac{1}{2}\pi+k\pi$
e. $(-3\cos^2 t\sin t, 2\cos 2t)$
f. $(-\frac{1}{2}\sin\frac{1}{2}t, 3\sin^2 t\cos t)$, zero vector if $t=2k\pi$
g. $(-\frac{1}{3}(\cos t)^{-2/3}\sin t, \frac{1}{3}(\sin t)^{-2/3}\cos t)$
h. $(-\frac{1}{3}(\cos t)^{-2/3}\sin t, 3\sin^2 t\cos t)$, zero vector if $t=k\pi$

23.2 a. $(-\sin 2\varphi, \cos 2\varphi)$
b. $(-2\sin 2\varphi\cos\varphi-\cos 2\varphi\sin\varphi, -2\sin 2\varphi\sin\varphi+\cos 2\varphi\cos\varphi)$
c. $(-3\sin 3\varphi\cos\varphi-\cos 3\varphi\sin\varphi, -3\sin 3\varphi\sin\varphi+\cos 3\varphi\cos\varphi)$
d. $(\frac{1}{2}\cos\frac{1}{2}\varphi\cos\varphi-\sin\frac{1}{2}\varphi\sin\varphi, \frac{1}{2}\cos\frac{1}{2}\varphi\sin\varphi+\sin\frac{1}{2}\varphi\cos\varphi)$
e. $(-\frac{3}{2}\sin\frac{3}{2}\varphi\cos\varphi-\cos\frac{3}{2}\varphi\sin\varphi, -\frac{3}{2}\sin\frac{3}{2}\varphi\sin\varphi+\cos\frac{3}{2}\varphi\cos\varphi)$
f. $\left(-\frac{\sin 3\varphi}{\sqrt{\cos 2\varphi}}, \frac{\cos 3\varphi}{\sqrt{\cos 2\varphi}}\right)$
g. $(-\sin\varphi-\sin 2\varphi, \cos\varphi+\cos 2\varphi)$ (note: this is the zero vector if $\varphi=\pi+2k\pi$)
h. $(-21\sin 7\varphi\cos\varphi-(1+3\cos 7\varphi)\sin\varphi, -21\sin 7\varphi\sin\varphi+(1+3\cos 7\varphi)\cos\varphi)$

23.3 radius vector: $(\mathrm{e}^{c\varphi}\cos\varphi, \mathrm{e}^{c\varphi}\sin\varphi)$ with length $\mathrm{e}^{c\varphi}$.
tangent vector: $(c\mathrm{e}^{c\varphi}\cos\varphi-\mathrm{e}^{c\varphi}\sin\varphi, c\mathrm{e}^{c\varphi}\sin\varphi+\mathrm{e}^{c\varphi}\cos\varphi)$ with length $\mathrm{e}^{c\varphi}\sqrt{c^2+1}$.
The inner product is $c\mathrm{e}^{2c\varphi}$, so the cosine of the included angle is $\dfrac{c}{\sqrt{c^2+1}}$.
For $c=1$ the angle is 45°. Note that the angle is 90° if $c=0$ (the spiral then is a circle), and note that the angle descends to zero if $c\to\infty$.

23.4 a. $(1, 4t, 3t^2)$ b. $(\cos t, 2\cos 2t, -\sin t)$ c. $(\cos t, 2\cos 2t, -3\sin 3t)$
d. $(2\pi\cos 2\pi t, 1, -2\pi\sin 2\pi t)$ e. $(2\pi\cos 2\pi t, 2t, 3t^2)$ f. $(-\sin t, \cos t, -12\sin 12t)$

23.5 a. $2\pi$ b. $2\pi R$ c. $\sqrt{4\pi^2R^2+a^2}$ d. $\frac{\sqrt{c^2+1}}{c}(\mathrm{e}^{2\pi c}-1)$

23.6 a. $P_{t=0}=(0,0)$, $P_{t=\frac{\pi}{2}}=(\frac{\pi}{2}-1,1)$, $P_{t=\pi}=(\pi,2)$, $P_{t=\frac{3\pi}{2}}=(\frac{3\pi}{2}+1,1)$, $P_{t=2\pi}=(2\pi,0)$
b. The central angle at $M$ equals $t$ radians, so the length of the circular arc $PQ$ is $t$, which is also the distance $OQ$.
c. velocity vector: $(1-\cos t, \sin t)$, scalar velocity

$\sqrt{2-2\cos t} = \sqrt{4\sin^2\frac{t}{2}} = 2|\sin\frac{t}{2}|$. The velocity is zero af $t = 2k\pi$ and maximum (namely 2) if $t = \pi + 2k\pi$. The velocity vector then is $(2,0)$.
d. 8

23.7 $I = \int_0^h \pi\left(\frac{r}{h}y\right)^2 dy = \frac{1}{3}\pi r^2 h$

23.8 $\frac{\pi^2}{2}$

23.9 $\frac{\pi}{2}$

23.10 $\frac{2\pi}{3}$

23.11 $\pi$

23.12 $\infty$

23.13 $\frac{\pi}{2}$

23.14 $\frac{\pi}{2}$

23.15 $\frac{3\pi}{10}$

23.16 It is the solid of revolution resulting from rotating the graph of the function $y = \sqrt{R^2 - z^2}$ between the bounds $z = h$ and $z = R$ around the $z$-axis. Its volume is $\int_h^R \pi(R^2 - z^2)\,dz = \pi(\frac{2}{3}R^3 - R^2 h + \frac{1}{3}h^3)$

23.17 $O = \displaystyle\int_0^h 2\pi\frac{r}{h}y\sqrt{1+\frac{r^2}{h^2}}dy = \pi r\sqrt{h^2+r^2}$

23.18 Take $f(y) = \sqrt{R^2 - y^2}$ between the bounds $y = -R$ and $y = R$, then the surface area formula yields $O = \displaystyle\int_{-R}^{R} 2\pi\sqrt{R^2-y^2}\sqrt{1+\frac{y^2}{R^2-y^2}}\,dy = 4\pi R^2$

23.19 $2\pi R(R-h)$

23.20 $\frac{\pi}{6}(5\sqrt{5}-1)$

23.21 $O = \displaystyle\int_1^\infty 2\pi\frac{1}{x}\sqrt{1+\frac{1}{x^4}}\,dx > \int_1^\infty 2\pi\frac{1}{x}\,dx = \infty$

23.22 1.436

23.23 Simplifying $P(t+t_d) = 2P(t)$ yields the equation $\mathrm{e}^{\lambda t_d} = 2$ (independent of $t$), so $t_d = \frac{\ln 2}{\lambda}$.

23.24 Rough estimate: $2^{13} < 1000000/100 < 2^{14}$, so the answer will be between $13 \times 5$ and $14 \times 5$. A calculation yields $t = 66.43856$.

23.25 Halving time $t_h$ is the time it takes to reach $\mathrm{e}^{\lambda t_h} = \frac{1}{2}$. For $\lambda = -0.2$ this yields $t_h = 3.4657359$

23.26 Rough estimate: $2^{-17} < 0.001/100 < 2^{-16}$, so the answer will be between $16 \times 3$ and $17 \times 3$. A calcultation yields $t = 49.82892$.

23.27 a. Just expand!
b. Expand; let the 'integration constant' be called $c = a\lambda T$, in other words, take $T = \frac{c}{a\lambda}$.

c. $T = \frac{P_0^{-a}}{a\lambda}$. If $t \uparrow T$ then $a\lambda(T-t) \downarrow 0$ and so $(a\lambda(T-t))^{-1/a} \to \infty$.
d. $T = 39.810717$ and $P(t) = \frac{1}{(0.39810717 - 0.01t)^5}$.

23.28 a. $dP/dt = \mu(M-P)P$ only depends on $P$ and not on $t$. b. 1.6 c. Both are equal to 1.2 d. decreasing line elements. The solution functions are decreasing for $t \to \infty$ with the line $P = M$ as horizontal asymptote. e. Here also decreasing line elements. Now $P = 0$ is a horizontal asymptote for $t \to -\infty$.

23.29 a. IV b. III c. VIII d. VI e. II f. I g. VII h. V

23.30 In this exercise $t_0 = 0$.
a. $t = \frac{\ln 3}{\mu M} \approx 0.68663$ b. $t = -\frac{\ln 3}{\mu M} \approx -0.68663$ c. Just substitute and expand. Geometrically, this means that the graph is point symmetric with respect to $(0, M/2)$.

23.31 Substitute and expand.

23.32 $\ln\left(\frac{P}{P-M}\right) = \mu Mt + c$ for some $c$. Put $A = e^c$ then $\frac{P}{P-M} = Ae^{\mu Mt}$ with $P = P(t) = \frac{MA}{A - e^{-\mu mt}}$ as solution. $P_0 = 2M$ yields $A = 2$. $\lim_{t\to\infty} P(t) = M$.

23.33 Below $a, b, c$ and $A$ are arbitrary constants.
a. $y^2 - x^2 = c$ b. $ax = by$ c. $y = Ae^{\frac{1}{2}x^2}$ d. $(x+c)y = -1$ and the line $y = 0$
e. $y = Ae^{\frac{1}{3}x^3}$

# Formula overview

## Algebra

$$a(b+c) = ab + ac$$

$$(a+b)(c+d) = ac + ad + bc + bd$$

$$(a+b)^2 = a^2 + 2ab + b^2$$

$$(a-b)^2 = a^2 - 2ab + b^2$$

$$a^2 - b^2 = (a+b)(a-b)$$

$$\frac{a}{c} + \frac{b}{c} = \frac{a+b}{c}$$

$$\frac{a}{b} + \frac{c}{d} = \frac{ad+bc}{bd}$$

$$\frac{a}{c} \times \frac{b}{d} = \frac{ab}{cd}$$

$$\frac{a}{b} : \frac{c}{d} = \frac{a}{b} \times \frac{d}{c} = \frac{ad}{bc}$$

$$\sqrt[n]{a} = a^{1/n}$$

$$a^r \times a^s = a^{r+s}$$

$$a^r : a^s = a^{r-s}$$

$$(a^r)^s = a^{rs}$$

$$(a \times b)^r = a^r \times b^r$$

$$\left(\frac{a}{b}\right)^r = \frac{a^r}{b^r}$$

## Sequences and equations

Pascal's triangle:

$$
\begin{array}{ccccccccccccccc}
 & & & & & & & 1 & & & & & & & \leftarrow n=0 \\
 & & & & & & 1 & & 1 & & & & & & \leftarrow n=1 \\
 & & & & & 1 & & 2 & & 1 & & & & & \leftarrow n=2 \\
 & & & & 1 & & 3 & & 3 & & 1 & & & & \leftarrow n=3 \\
 & & & 1 & & 4 & & 6 & & 4 & & 1 & & & \leftarrow n=4 \\
 & & 1 & & 5 & & 10 & & 10 & & 5 & & 1 & & \leftarrow n=5 \\
 & 1 & & 6 & & 15 & & 20 & & 15 & & 6 & & 1 & \leftarrow n=6 \\
1 & & 7 & & 21 & & 35 & & 35 & & 21 & & 7 & & 1 \quad \leftarrow n=7 \\
 & & & & & & & \cdots & & & & & & & \cdots
\end{array}
$$

Factorials:

$$
\begin{aligned}
0! &= 1 \\
k! &= 1 \times \cdots \times k \quad \text{for each positive integer } k
\end{aligned}
$$

Binomial coefficients: $\displaystyle\binom{n}{k} = \frac{n!}{k!(n-k)!}$

Newton's binomial formula:

$$
\begin{aligned}
(a+b)^n &= \binom{n}{0}a^n + \binom{n}{1}a^{n-1}b + \cdots + \binom{n}{n-1}ab^{n-1} + \binom{n}{n}b^n \\
&= \sum_{k=0}^{n} \binom{n}{k} a^{n-k}b^k
\end{aligned}
$$

Summation arithmetic sequence: $\displaystyle\sum_{k=1}^{n} a_k = \frac{1}{2}n(a_1 + a_n)$

Summation finite geometric sequence: $\displaystyle\sum_{k=0}^{n-1} ar^k = \frac{a(1-r^n)}{1-r}$ if $r \neq 1$

Summation infinite geometric sequence: $\displaystyle\sum_{k=0}^{\infty} ar^k = \frac{a}{1-r}$ if $-1 < r < 1$

Limits: $\displaystyle\lim_{n\to\infty} \frac{n^p}{a^n} = 0$ if $a > 1$, $\quad\displaystyle\lim_{n\to\infty} \frac{a^n}{n!} = 0$, $\quad\displaystyle\lim_{n\to\infty} \frac{n!}{n^n} = 0$

The *abc*-formula: If $ax^2 + bx + c = 0$, $a \neq 0$ and $b^2 - 4ac \geq 0$, then

$$x = \frac{-b \pm \sqrt{b^2 - 4ac}}{2a}$$

## Geometry in the plane

Equation of a line with normal vector $\begin{pmatrix} a \\ b \end{pmatrix}$: $\quad ax + by = c$

Equation of the line through the points $A = (a_1, a_2)$ and $B = (b_1, b_2)$:

$$(a_1 - b_1)(y - b_2) = (a_2 - b_2)(x - b_1)$$

Distance $d(A, B)$ of the points $A = (a_1, a_2)$ and $B = (b_1, b_2)$:

$$d(A, B) = \sqrt{(a_1 - b_1)^2 + (a_2 - b_2)^2}$$

Inner product of $\mathbf{a} = \begin{pmatrix} a_1 \\ a_2 \end{pmatrix}$ and $\mathbf{b} = \begin{pmatrix} b_1 \\ b_2 \end{pmatrix}$: $\quad \langle \mathbf{a}, \mathbf{b} \rangle = a_1 b_1 + a_2 b_2$

Length of $\mathbf{a}$: $\quad |\mathbf{a}| = \sqrt{\langle \mathbf{a}, \mathbf{a} \rangle} = \sqrt{a_1^2 + a_2^2}$

If $\varphi$ is the angle between $\mathbf{a}$ and $\mathbf{b}$, then: $\quad \langle \mathbf{a}, \mathbf{b} \rangle = |\mathbf{a}||\mathbf{b}| \cos \varphi$

Equation of the circle with centre $(m, n)$ and radius $r$:

$$(x - m)^2 + (y - n)^2 = r^2$$

Triangles:

$\sin \alpha = \frac{a}{c}, \qquad \cos \alpha = \frac{b}{c}, \qquad \tan \alpha = \frac{a}{b}$

$a^2 + b^2 = c^2$ (Pythagoras' theorem)

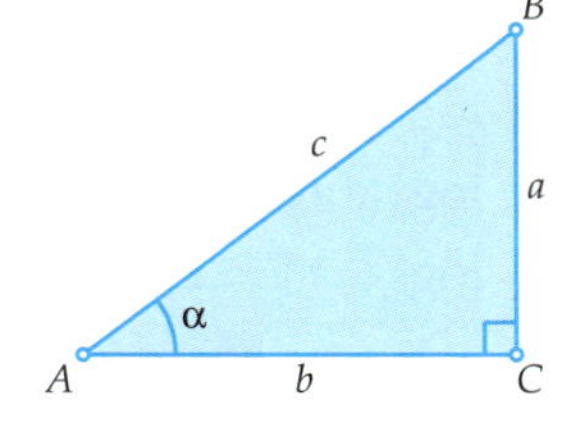


$\frac{a}{\sin \alpha} = \frac{b}{\sin \beta} = \frac{c}{\sin \gamma}$ (rule of sines)

$a^2 = b^2 + c^2 - 2bc \cos \alpha$ (cosine rule)

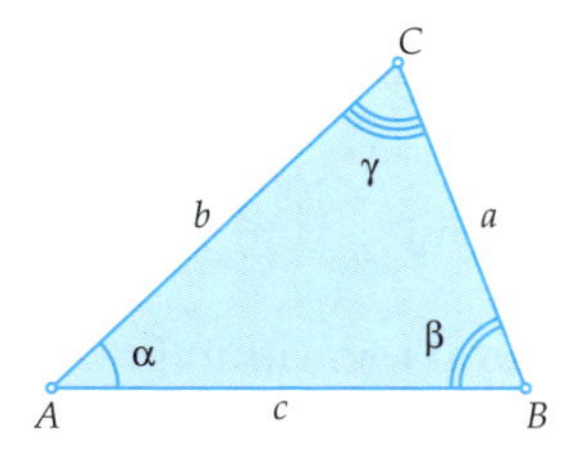


$O = \frac{1}{2} bc \sin \alpha = \frac{1}{2} ca \sin \beta = \frac{1}{2} ab \sin \gamma$ (area formula)

## Geometry in space

Equation of a plane with normal vector $\begin{pmatrix} a \\ b \\ c \end{pmatrix}$: $\quad ax + by + cz = d$

Distance $d(A, B)$ of the points $A = (a_1, a_2, a_3)$ and $B = (b_1, b_2, b_3)$:

$$d(A, B) = \sqrt{(a_1 - b_1)^2 + (a_2 - b_2)^2 + (a_3 - b_3)^2}$$

Inner product of $\mathbf{a} = \begin{pmatrix} a_1 \\ a_2 \\ a_3 \end{pmatrix}$ and $\mathbf{b} = \begin{pmatrix} b_1 \\ b_2 \\ b_3 \end{pmatrix}$: $\quad \langle \mathbf{a}, \mathbf{b} \rangle = a_1 b_1 + a_2 b_2 + a_3 b_3$

Length of $\mathbf{a}$: $\quad |\mathbf{a}| = \sqrt{\langle \mathbf{a}, \mathbf{a} \rangle} = \sqrt{a_1^2 + a_2^2 + a_3^2}$

If $\varphi$ is the angle between $\mathbf{a}$ and $\mathbf{b}$, then: $\quad \langle \mathbf{a}, \mathbf{b} \rangle = |\mathbf{a}||\mathbf{b}| \cos \varphi$

Equation of a sphere with centre $(m_1, m_2, m_3)$ and radius $r$:

$$(x - m_1)^2 + (y - m_2)^2 + (z - m_3)^2 = r^2$$

## Exponential and logarithmic functions

For each $a > 0$ with $a \neq 1$, we have $\quad \log_a x = y \iff a^y = x$

$$\log_a(xy) = \log_a x + \log_a y$$

$$\log_a(x/y) = \log_a x - \log_a y$$

$$\log_a (x^y) = y \log_a x$$

$$\log_a x = \frac{\log_g x}{\log_g a}$$

$$\lim_{x \to +\infty} \frac{x^p}{a^x} = 0 \quad \text{if} \quad a > 1$$

$$\lim_{x \to +\infty} \frac{\log_a x}{x^q} = 0 \quad \text{if} \quad q > 0$$

Natural logarithm (base e): $\quad \ln x = y \iff \mathrm{e}^y = x$

$$\lim_{x \to 0} \frac{\mathrm{e}^x - 1}{x} = 1$$

$$\lim_{x \to 0} \frac{\ln(1 + x)}{x} = 1$$

## Trigonometric formulas

$$\tan x = \frac{\sin x}{\cos x}$$

$$\cos^2 x + \sin^2 x = 1$$

$$1 + \tan^2 x = \frac{1}{\cos^2 x}$$

$$\begin{aligned}\sin 2x &= 2\sin x\cos x\\ \cos 2x &= \cos^2 x - \sin^2 x\\ &= 2\cos^2 x - 1\\ &= 1 - 2\sin^2 x\\ \tan 2x &= \frac{2\tan x}{1-\tan^2 x}\end{aligned}$$

$$\begin{aligned}\sin(x+y) &= \sin x\cos y + \cos x\sin y\\ \sin(x-y) &= \sin x\cos y - \cos x\sin y\\ \cos(x+y) &= \cos x\cos y - \sin x\sin y\\ \cos(x-y) &= \cos x\cos y + \sin x\sin y\end{aligned}$$

$$\tan(x+y) = \frac{\tan x + \tan y}{1 - \tan x\tan y}$$

$$\tan(x-y) = \frac{\tan x - \tan y}{1 + \tan x\tan y}$$

$$\sin x + \sin y = 2\sin\frac{x+y}{2}\cos\frac{x-y}{2}$$

$$\sin x - \sin y = 2\sin\frac{x-y}{2}\cos\frac{x+y}{2}$$

$$\cos x + \cos y = 2\cos\frac{x+y}{2}\cos\frac{x-y}{2}$$

$$\cos x - \cos y = -2\sin\frac{x+y}{2}\sin\frac{x-y}{2}$$

$$\sin(\tfrac{\pi}{2} - x) = \cos x$$

$$\cos(\tfrac{\pi}{2} - x) = \sin x$$

$$\lim_{x\to 0}\frac{\sin x}{x} = 1$$

$$\begin{aligned}x = \arcsin y &\iff y = \sin x \text{ and } -\tfrac{1}{2}\pi \le x \le \tfrac{1}{2}\pi\\ x = \arccos y &\iff y = \cos x \text{ and } 0 \le x \le \pi\\ x = \arctan y &\iff y = \tan x \text{ and } -\tfrac{1}{2}\pi < x < \tfrac{1}{2}\pi\end{aligned}$$

## Derivatives

Rules for differentiable functions:

$$\begin{aligned}(c\,f(x))' &= c\,f'(x) \quad \text{for any constant } c\\ (f(x)+g(x))' &= f'(x)+g'(x)\\ (f(g(x)))' &= f'(g(x))g'(x) \quad \text{(chain rule)}\\ (f(x)g(x))' &= f'(x)g(x)+f(x)g'(x) \quad \text{(product rule)}\\ \left(\frac{f(x)}{g(x)}\right)' &= \frac{f'(x)g(x)-f(x)g'(x)}{(g(x))^2} \quad \text{(quotient rule)}\end{aligned}$$

Standard functions and their derivatives:

| $f(x)$ | $f'(x)$ | |
|---|---|---|
| $x^p$ | $p\,x^{p-1}$ | for each $p$ |
| $a^x$ | $a^x \ln a$ | for each $a > 0$ |
| $\mathrm{e}^x$ | $\mathrm{e}^x$ | |
| $\log_a x$ | $\dfrac{1}{x \ln a}$ | for each $a > 0, a \neq 1$ |
| $\ln x$ | $\dfrac{1}{x}$ | |
| $\sin x$ | $\cos x$ | |
| $\cos x$ | $-\sin x$ | |
| $\tan x$ | $\dfrac{1}{\cos^2 x}$ | |
| $\arcsin x$ | $\dfrac{1}{\sqrt{1-x^2}}$ | |
| $\arccos x$ | $-\dfrac{1}{\sqrt{1-x^2}}$ | |
| $\arctan x$ | $\dfrac{1}{1+x^2}$ | |

## Differentials, integrals and applications

If $y = f(x)$ then $dy = d(f(x)) = f'(x)\,dx$

Substitution rule: if $y = g(x)$ then

$$\int f(g(x))g'(x)\,dx = \int f(g(x))\,d(g(x)) = \int f(y)\,dy$$

As definite integral:

$$\int_{x=a}^{x=b} f(g(x))g'(x)\,dx = \int_{x=a}^{x=b} f(g(x))\,d(g(x)) = \int_{y=g(a)}^{y=g(b)} f(y)\,dy$$

Partial integration: $\displaystyle\int f\,dg = f\,g - \int g\,df$

As definite integral:

$$\int_a^b f(x)\,dg(x) = (f(b)g(b) - f(a)g(a)) - \int_a^b g(x)\,df(x)$$

Primitive functions of the standard functions:

| $f(x)$ | $F(x)$ | |
|---|---|---|
| $x^p$ | $\frac{1}{p+1}\,x^{p+1}$ | if $p \neq -1$ |
| $a^x$ | $\frac{1}{\ln a}\,a^x$ | for any $a > 0,\ a \neq 1$ |
| $e^x$ | $e^x$ | |
| $\frac{1}{x}$ | $\ln\lvert x\rvert$ | |
| $\sin x$ | $-\cos x$ | |
| $\cos x$ | $\sin x$ | |
| $\frac{1}{\cos^2 x}$ | $\tan x$ | |
| $\frac{1}{\sqrt{1-x^2}}$ | $\arcsin x$ | |
| $\frac{1}{1+x^2}$ | $\arctan x$ | |

If $F(x)$ is a primitive function of $f(x)$, then also $F(x) + c$ is a primitive function of $f(x)$ for each constant $c$.

Radius vector of a plane or solid parametric curve:

$$\mathbf{r}(t) = \begin{pmatrix} x(t) \\ y(t) \end{pmatrix} \quad \text{or} \quad \mathbf{r}(t) = \begin{pmatrix} x(t) \\ y(t) \\ z(t) \end{pmatrix}$$

Velocity vector (tangent vector):

$$\mathbf{v}(t) = \mathbf{r}'(t) = \begin{pmatrix} x'(t) \\ y'(t) \end{pmatrix}$$

or, in space

$$\mathbf{v}(t) = \mathbf{r}'(t) = \begin{pmatrix} x'(t) \\ y'(t) \\ z'(t) \end{pmatrix}$$

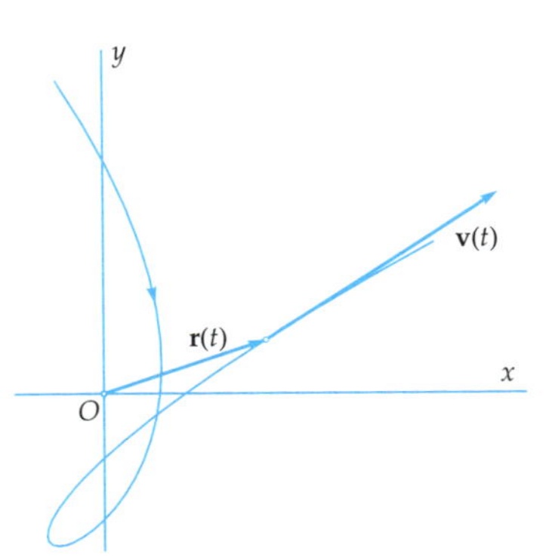


Scalar velocity:

$$v(t) = |\mathbf{v}(t)| = \sqrt{(x'(t))^2 + (y'(t))^2} \quad \text{or} \quad \sqrt{(x'(t))^2 + (y'(t))^2 + (z'(t))^2}$$

Length of a parametric curve in the plane or in space between $\mathbf{r}(a)$ and $\mathbf{r}(b)$:

$$L = \int_a^b v(t)\,dt = \int_a^b \sqrt{(x'(t))^2 + (y'(t))^2}\,dt \quad \text{or}$$
$$L = \int_a^b v(t)\,dt = \int_a^b \sqrt{(x'(t))^2 + (y'(t))^2 + (z'(t))^2}\,dt$$

Volume of the solid of revolution that results from rotating the graph of the function $z = f(y)$ around the $y$-axis:

$$V = \int_a^b \pi\, f(y)^2\,dy$$

Surface area of the same solid of revolution:

$$O = \int_a^b 2\pi f(y)\sqrt{1 + (f'(y))^2}\,dy$$

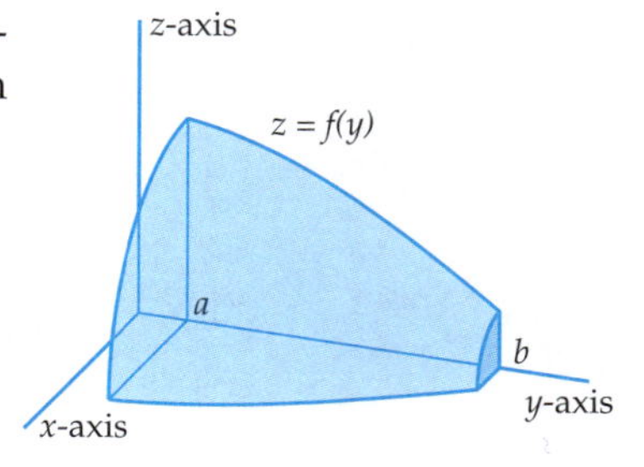


Exponential growth, differential equation: $dP = \lambda P\,dt$
Solution functions: $P(t) = P_0 e^{\lambda t}$

Logistic growth, differential equation: $dP = \mu(M - P)\,dt$

Solution functions for $0 < P < M$: $P(t) = \dfrac{M}{1 + e^{-\mu M(t - t_0)}}$

# Index